Rolf Isermann

Fault-Diagnosis Systems

Rolf Isermann

Fault-Diagnosis Systems

An Introduction from Fault Detection to Fault Tolerance

With 227 Figures

Springer

Professor Dr. Rolf Isermann
TU Darmstadt
Institut für Automatisierungstechnik
Fachgebiet Regelungstechnik und Prozessautomatisierung
Landgraf-Georg-Str. 4
64283 Darmstadt
Germany
risermann@iat.tu-darmstadt.de

Library of Congress Control Number: 2005932861

ISBN-10   3-540-24112-4 Springer Berlin Heidelberg New York
ISBN-13   978-3-540-24112-6 Springer Berlin Heidelberg New York

Springer is a part of Springer Science+Business Media
springeronline.com

© Springer-Verlag Berlin Heidelberg 2006
Printed in Germany

Typesetting: Digital data supplied by author
Final processing by PTP-Berlin Protago-T$_E$X-Production GmbH, Germany
Cover-Design:
Printed on acid-free paper   62/3141/Yu – 5 4 3 2 1 0

Preface

With increasing demands for efficiency and product quality and progressing integration of automatic control systems in high-cost and safety-critical processes, the field of supervision (or monitoring), fault detection and fault diagnosis plays an important role. The classical way of supervision is to check the limits of single variables and alarming of operators. However, this can be improved significantly by taking into account the information hidden in all measurements and automatic actions to keep the systems in operation.

During the last decades theoretical and experimental research has shown new ways to detect and diagnose faults. One distinguishes *fault detection* to recognize that a fault happened, and *fault diagnosis* to find the cause and location of the fault. Advanced methods of fault detection are based on mathematical signal and process models and on methods of system theory and process modelling to generate fault symptoms. Fault diagnosis methods use causal fault-symptom-relationships by applying methods from statistical decision, artificial intelligence and soft computing. Therefore, efficient supervision, fault detection and diagnosis is a challenging field by encompassing physical oriented system theory, experiments and computations. The considered subjects are also known as fault detection and isolation (FDI) or fault detection and diagnosis (FDD).

A further important field is *fault management*. This means to avoid shut-downs by early fault detection and actions like *process condition-based maintenance* or *repair*. If sudden faults, failures or malfunctions cannot be avoided, *fault-tolerant systems* are required. Through methods of fault detection and reconfiguration of redundant components, break-down and in the case of safety-critical processes accidents may be avoided.

The book is intended to give an introduction to advanced supervision, fault detection and diagnosis and fault-tolerant systems for processes with mainly continuous, sampled signals. Of special interest is an application-oriented approach with methods which have proven their performance in practical applications.

The material is the result of many own research projects during the last 25 years on fault detection and diagnosis, but also of publications by many other research groups. The development of the field can especially be followed by the IFAC-

Symposia series "SAFEPROCESS", which was initiated 1991 in Baden-Baden and then repeated all three years in Helsinki, Hull, Budapest, Washington, Beijing, and the IFAC-Workshop "On-line fault detection and supervision in the chemical industries" started in Kyoto (1986), and then held in Newark, Newcastle, Folaize and Cheju, but also by other conferences.

The book is dedicated as an introduction in teaching the field of fault detection and diagnosis, and fault-tolerant systems for graduate students or students of higher semesters of electrical and electronic engineering, mechanical and chemical engineering and computer science. As the treated field is in a phase of increasing importance for technical and also non-technical systems, it has been tried to present the material in an easy to understand and transparent way and with realistic perspectives for the application of the treated and discussed methods. Therefore the book is also oriented towards practising engineers in research and development, design and manufacturing. Preconditions are basic undergraduate courses of system theory, automatic control, mechanical and/or electrical engineering.

The author is greatful to his research associates, who performed many theoretical and practical research projects on the subject of this book since 1975, among them H. Siebert, L. Billmann, G. Geiger, W. Goedecke, S. Nold, U. Raab, B. Freyermuth, S. Leonhardt, R. Deibert, T. Höfling, T. Pfeufer, M. Ayoubi, P. Ballé, D. Füssel, O. Moseler, A. Wolfram, F. Kimmich, A. Schwarte, M. Vogt, M. Münchhof, D. Fischer, F. Haus and I. Unger. Following chapters or sections were worked out by: 8.4.4: F. Kimmich, 9.2.3: M. Vogt, 10.4.2: P. Ballé, 11.4.2: I. Unger, 13: F. Haus, 15.2, 16, 17.3 and 23.2: D. Füssel. I appreciate these contributions highly as valuable inputs to this book.

Finally, I especially would like to thank Brigitte Hoppe for the laborious and precise text setting, including the figures and tables in camera-ready form.

Darmstadt, February 2005

Rolf Isermann

Contents

Part II Fault-Detection Methods

List of symbols

Only frequently used symbols and abbreviations are given.

Letter symbols

a	parameters of differential of difference equations
b	parameters of differential or difference equations
c	spring constant, constant, concentration, stiffness
d	damping coefficient
e	equation error, control deviation $e = w - y$, number $e = 2.71828\ldots$
f	fault, frequency ($f = 1/T_p$, T_p period), function $f(\ldots)$
g	gravitational acceleration, function $g(\ldots)$, impulse response
i	integer, gear ratio, index, $\sqrt{-1}$ (imaginary unit)
j	integer, index
k	discrete number, discrete time $k = t/T_0 = 0, 1, 2, \ldots$ (T_0: sampling time)
l	index, length
m	mass, order number
n	rotational speed, order number, disturbance signal
p	pressure, index, controller parameter, probability density function, process parameter
q	controller parameter, failure density
r	index, radius, reference variable, residual
s	Laplace variable $s = \delta + i\omega$, symptom
t	continuous time, principal component coordinate
u	input signal change ΔU
v	speed, specific volume, disturbance signal
w	reference value, setpoint
x	space coordinate, state variable
y	output signal change ΔY, space coordinate, control variable change ΔY, signal

z	space coordinate, disturbance variable change ΔZ, z-transform variable $z = \exp T_0 s$
$\hat{x}$	estimated or observed variable
$\tilde{x}$	estimation error
$\bar{x}$	average, steady-state value
x_0	amplitude
x_{00}	value in steady state
A	area
B	magnetic flux density
C	capacitance
D	damping ratio, diameter
E	module of elasticity, energy, potential, bulk modulus
F	filter transfer function, force
G	weight, transfer function
H	magnetic field strength, height
I	electrical current, mechanical momentum, torsion, second moment of area
J	moment of inertia
K	constant, gain
L	inductance
N	discrete number, windings number
P	power, probability
Q	generalized force, heat, unreliability function
R	electrical resistance, covariance or correlation function, reliability function, risk number
S	spectral density, sum, performance criterion
T	absolute temperature, torque, time constant
T_0	sampling time
U	input variable, manipulated variable (control input), voltage
V	volume
X	space coordinate
Y	output variable, space coordinate, control variable
Z	space coordinate, disturbance variable
$\mathbf{a}$	vector
$\mathbf{A}$	matrix
$\mathbf{A}^T$	transposed matrix
$\mathbf{I}$	identity matrix
$\boldsymbol{\theta}$	parameter vector
$\mathbf{P}$	covariance matrix
$\boldsymbol{\psi}$	data vector
α	coefficient, angle
β	coefficient, angle
γ	specific weight, correcting factor

δ	decay factor, impulse function
ϕ	correlation function
η	efficiency
ϑ	temperature
λ	thermal conductivity, forgetting factor, failure rate
μ	friction coefficient, permeability, membership function
ν	kinematic viscosity, index
π	number $\pi = 3.14159\ldots$
ρ	density
σ	standard deviation, σ^2 variance
τ	time
φ	angle
ω	angular frequency, $\omega = 2\pi/T_p; T_p$ period
Δ	change, deviation
θ	parameter
Π	product
Σ	sum
Θ	magnetic flux
Ψ	magnetic flux linkage

Mathematical abbreviations

$\exp(x) = e^x$	
$E\{\}$	expectation of a statistical variable
dim	dimension
adj	adjoint
det	determinant
Re	real part
Im	imaginary part
$\dot{Y}$	dY/dt (first derivative)
var [ ]	variance
cov [ ]	covariance
$\mathcal{F}$	Fourier transform

Abbreviations

ACF	Auto Correlation Function
ARMA	Auto Regressive Moving Average process
CCF	Cross Correlation Function
DFT	Discrete Fourier Transform
ETA	Event Tree Analysis
FDD	Fault Detection and Diagnosis
FDI	Fault Detection and Isolation
FMEA	Failure Mode and Effects Analysis

FFT   Fast Fourier Transform
FTA   Fault Tree Analysis
HA   Hazard Analysis
LS   Least Squares
MLP   Multilayer Perceptron
MTBF   Mean Time Between Failures
MTTF   Mean Time To Failure $= 1/\lambda$
MTTR   Mean Time To Repair
NN   Neural Net
PCA   Principal Component Analysis
PRBS   Pseudo Random Binary Signal
RBF   Radial Basis Function
RLS   Recursive Least Squares
rms (...)   root of mean squared of (...)

1

Introduction

Since about 1960 the influence of automation on the operation and the design of technical processes increased progressively. This development of expanding process automation was caused by an increasing demand on the process performance or the product quality, the independence of process operation from the presence of human operators, the relieve of operators from monotonic tasks and because of rising wages. The degree of automation was pushed forward drastically since around 1975 when relatively cheap and reliable microcomputers were available and could solve many automation problems in one device. This was paralleled by further progress in the areas of sensors, actuators, bus-communication systems, and human-machine interfaces. The improvement in the theoretical understanding of processes and automation functions also played a large role.

1.1 Process automation and process supervision

Figure 1.1 shows a simplified scheme for process automation of two coupled process parts. The lower level contains the sequential control, feedforward and feedback control. Supervision can be assigned to a medium level. The higher levels comprise more global acting tasks like coordination, optimization and management. Important information about the process is displayed at the operator's console.

Great progress could be observed for *digital sequence control* and *digital (continuous) control*. An enormous activity has shown up in the theory, development and implementation of feedback control systems. Especially process model-based control systems, which comprise state observers and parameter estimation and compensate for nonlinearities have shown large improvements. Herewith, the modern theory of dynamic systems and signals has had a great influence and many processes with difficult behavior can now much better be controlled than earlier.

However, the better the control functions are performed in the lower levels, the more important become the *supervision functions*, because operators are removed from the process. An acting human operator does not only control the process with regard to setpoints or time schedules. He or she also supervises a process, especially

operating console   automation units

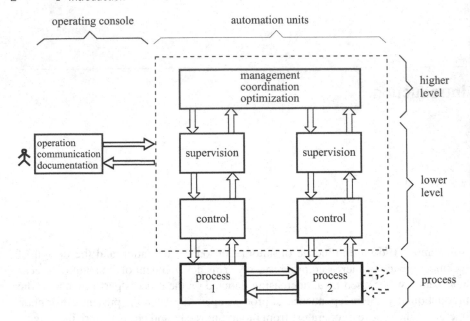

Fig. 1.1. Simplified scheme of process automation

if there is a direct contact with the process. Therefore, with the improvement of lower level control functions the supervising functions must be improved, too

Another reason is the *integration* of control systems into the process, where process and its control become autonomous units. This is, for instance, the case in *mechatronic systems* and shows, for example, up in *drive-by-wire* aircraft and vehicles. Here not only the control tasks itself but also the reliability and safety depends on the correct functioning of all process parts, actuators, sensors and control computers. Faults in the system must then immediately be displayed to the operator (pilot, driver) and redundant or reconfigurable components must be activated by a fault-management system for critical processes.

The *automatic supervision* in the past was mostly realized by *limit checking* (or threshold checking) of some important process variables, like, e.g. force, speed, pressure, liquid level, temperatures. Usually *alarms* are raised if limit values are exceeded and operators have to act or *protection systems* act automatically. This is in many cases sufficient to prevent larger failures or damages. However, faults are detected rather lately and a detailed fault diagnosis is mostly not possible with this simple method. Methods of modern systems theory show the systematic use of mathematical process and signal models, identification and estimation methods and methods of computational intelligence. With these methods it is possible to *develop advanced methods of fault detection and diagnosis*. The goals of these methods are, for example:

- early detection of small faults with abrupt or incipient behavior;
- diagnosis of faults in actuators, processes, components and sensors; fault detection in closed loops;
- supervision of processes in transient states;
- process condition-based maintenance and repair;
- deep quality control of assembled products in manufacturing;
- teleservices like remote fault detection and diagnosis;
- basis for fault management;
- basis for fault-tolerant and reconfigurable systems.

1.2 Contents

To treat these advanced supervision, fault detection and diagnosis methods the book is divided in five parts:

I   Fundamentals
II Fault-detection methods
III Fault-diagnosis methods
IV Fault-tolerant systems
V   Application examples

The first chapter of *Part I*, *Chapter 2*, describes the **basic tasks** of supervision, monitoring, automatic protection, fault detection and fault diagnosis up to fault management. As the treated subject is distributed over many different technological areas, the used **terminology** is not unique. Therefore an attempt is made to give definitions to frequently used terms like faults, failures, malfunctions, reliability, availability, safety, dependability and integrity, with reference to international standards. One goal of advanced supervision is the improvement of **reliability, availability** and **maintainability**. Therefore some basics are summarized in *Chapter 3*. Measures for the reliability like failure rate and MTTF, and for maintainability and availability are given together with numerical examples. For safety related systems, special analysis and synthesis methods are required, which are covered by the terms **safety, system integrity** and **dependability**, *Chapter 4*. A brief summary is given of event tree analysis, fault tree analysis, failure mode and effects analysis (FMEA), hazard analysis and risk classification.

Part II treats the basic fault-detection methods. As advanced methods of fault detection are using mathematical **process** and **signal models**, see Figures 1.2 and 1.3, *Chapters 5* and *6* describe some basic continuous-time and discrete-time models. An important issue in this connection is the mathematical modelling of faults. Different kinds of frequently used fault models, different time behavior and their influence on process models is discussed. Examples are given how faults can be modelled for actuators, processes and sensors.

Static and dynamic process models are considered and it is shown how additive (offset) and multiplicative (parametric) faults influence the measurable signals. Then some models for periodic and stochastic signals are given which are suitable for fault

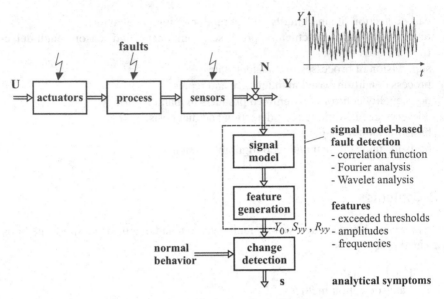

Fig. 1.2. Scheme for the fault detection with signal models

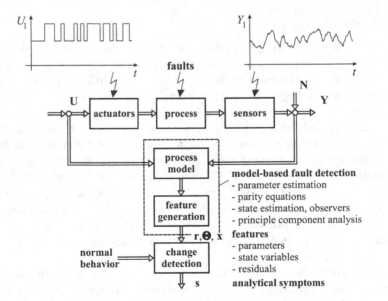

Fig. 1.3. Scheme for the fault detection with process models

detection with signal-analysis methods. *Chapters 7* and *8* treat fault-detection methods based on the measurement of single signals, see Figure 1.4 left. *Chapter 7* gives a survey on the most frequently used way for fault detection, the **limit checking**. This is usually applied for measurable absolute values and their trends. Then more sophisticated **change-detection methods** are considered. A basic method is, of course, the real-time estimation of the mean and variance of observed stochastic variables. Also statistical tests like hypothesis testing, t-test, run-sum test, F-test, likelihood ratio test are discussed. Furtheron fuzzy thresholds, adaptive thresholds and plausibility checks are described.

The **fault detection with signal models** in *Chapter 8* considers first periodic signals. Classical methods like Fourier and correlation analysis including FFT and spectral estimation are summarized, followed by the identification of non-stationary periodic signals with short-time Fourier transform and wavelet transform and the identification of stochastic signals. The goal is to detect changes of the signal behavior caused by process faults, see Figure 1.2.

The following chapters treat the **fault detection with process models**, see Figure 1.4 right. As faults may change the behavior of processes between input and output signals, changes in the behavior of the processes can be used to indicate inherent faults which are not directly measurable. Therefore attempts are made to extract changes in the process behavior by using several measurements, see the scheme in Figure 1.3. This also means that "analytical redundancy" between measured signals is used, expressed by process models.

Chapter 9 considers **fault detection with process-identification methods**. Here the process models adapt to the individual process behavior by using cross-correlation or parameter estimation. Especially the recursive least squares parameter estimation method including their modifications is described for linear time-invariant and time-variant processes with discrete-time and continuous-time signals. A great advantage is that powerful methods exist for the identification of nonlinear processes because most real processes are nonlinear. Parameter estimation for static and dynamic non-linear processes is considered and an extract for applicable neural networks for static and dynamic systems and implementation as look-up tables is given. Fault symptoms then reflect as parameters or output signal deviations.

The methods of **parity equations** are using fixed process models, *Chapter 10*. They can be designed with transfer functions leading to output or equation errors which are called primary residuals, or with state-space models. In order to make the residuals more sensitive and robust to certain faults, enhanced residuals can be generated, giving the residuals special structures or directions. Depending of the process model structure and the kind of faults, strongly isolating or weakly isolating residuals can be distinguished.

A further alternative for model-based fault detection are **state observers** and **state estimation**, *Chapter 11*. Changes in the input/output behavior of a process lead to changes of the output error and state variables. Therefore they can be used as residuals. Enhanced residuals are obtained with fault-detection filters or bank of observers. Similar approaches for noisy processes are possible by state estimation with Kalman filters. Output or unknown input observers result from a transformation to new state variables and outputs such that unknown inputs have no influence on the residuals. A comparison of the computational form of the residual equations and simulations show similarities between the parity equation and observer-based methods.

A special chapter is devoted to the **fault detection of control loops**, *Chapter 12*. As detuned controller parameters, actuator or sensor faults and some disturbances have a similar effect on the control performance, it is rather difficult to detect and diagnose faults in closed loop. A **comparison** of model-based fault-detection is shown in *Chapter 14* by considering the assumptions made and simulations. The suitability of the individual methods for special types of faults is discussed and certain combinations are proposed.

Chapter 13 gives an introduction to the **principal component analysis** which under the assumption of linearity, analyzes the fluctuations of input and output variables of large scale processes and reduces the number of variables to those being uncorrelated while preserving most of the information. Changes of the new variables are then used to form residuals. Figure 1.4 summarizes the treated fault-detection methods.

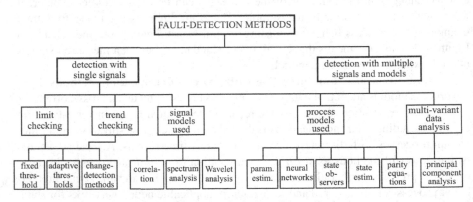

Fig. 1.4. Survey on fault-detection methods

Part III provides an overview of most important **fault-diagnosis methods**. Based on the symptoms of the fault-detection methods, with binary or fuzzy thresholds, and different kind of residuals or features, the task is to find the cause of the faults, see Figure 1.5. The symptoms may be analytical or heuristic, which are observed by humans and expressed as linguistic terms, and exist in form of numbers which are calculated. After an **introduction** into the basic problems in *Chapter 15*, fault diagnosis with **classification methods** is described, from classical pattern recognition, geometric classifiers to neural networks, *Chapter 16*. With more information, fault trees can be established, allowing **inference methods** for approximate reasoning, forward and backward chaining, *Chapter 17*. Hybrid neuro-fuzzy systems then are used to identify fault trees with if-then rules. Figure 1.6 gives an overview of the treated fault-diagnosis methods.

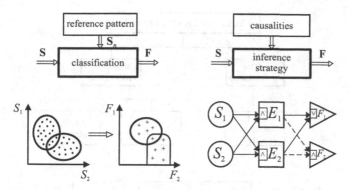

Fig. 1.5. Methods of fault diagnosis. **S**: symptoms, **F**: faults, **E**: events

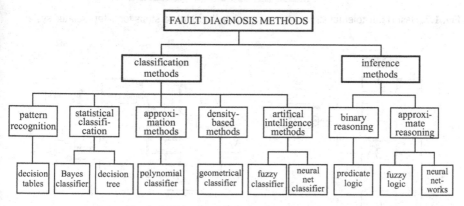

Fig. 1.6. Survey on fault-diagnosis methods

Fault-tolerant systems which tolerate appearing faults are treated in *Part IV*. *Chapter 18* presents the fault-tolerant design with basic static and dynamic redundant structures, e.g. with voters, hot and cold standby, see Figure 1.7. The various degradation steps with states of fail safe, fail operational and fail silent are considered. Then it is shown how fault-tolerant sensors with hardware or analytical redundancy and fault-tolerant actuators can be built, like the example in Figure 1.8. Finally, **fault-tolerant components** and **fault-tolerant control systems** are briefly discussed in *Chapter 19*. The last two chapters consider parts of a general **fault management**.

The last *Part V* shows some **application examples** for model-based fault detection in detail. Experimental results are shown for a DC motor, *Chapter 20*, an AC motor driven centrifugal pump, *Chapter 21* and an automotive suspension, *Chapter 22*. A more comprehensive presentation of many other technical applications is provided by another book, [1.25].

Summing up, after a summary of basics in the area of reliability and safety the chapters describe in the form of an introduction basic methods for fault detection, fault diagnosis and an extract of a general fault management. These methods can be

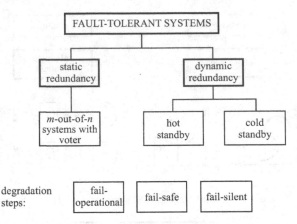

Fig. 1.7. Basic fault-tolerant structures and possible degradation steps for safety-related systems

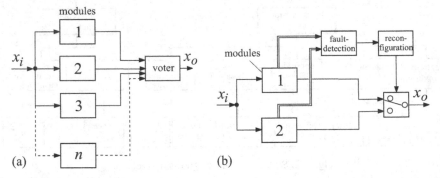

Fig. 1.8. Redundancy schemes for electronic hardware (extract): (a) static redundancy; (b) dynamic redundancy (hot standby)

applied to many different types of actuators, processes and sensors in open or closed loop, as indicated in Figure 1.9.

1.3 Historical notes

The *historical development* of the various supervision, fault-detection and diagnosis methods is difficult to describe because original contributions are very much distributed in the technical literature. Limit checking is probably as old as the instrumentation of machines dating back to the end of the 19th century. For the supervision of plants the use of ink and later point printing recorders was standard equipment since about 1935. Later, around 1960, analog controllers with transistor-based amplifiers (operation amplifiers) and sequential controllers with hardwired devices became available and then still used limit checking. Signal model-based methods like spectral analysis could be realized with analog bandpass filters and oscilloscopes.

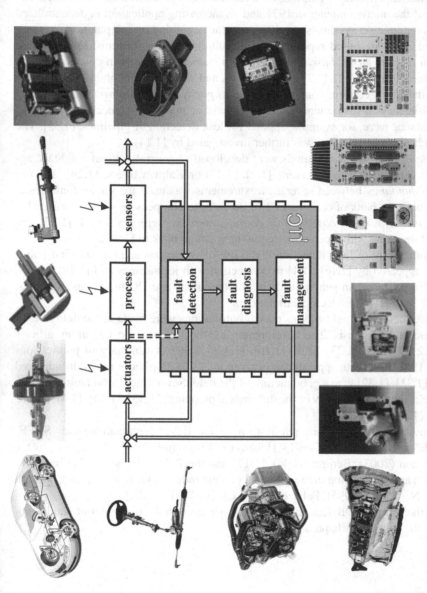

Fig. 1.9. Signal flow for fault detection, fault diagnosis and fault management and pictures of some actuators, processes and sensors (Sources of pictures: VDA, Bosch, Bosch-Rexroth, Continental-Teves)

The implementation of on-line operating process computers in 1960 then opened the way for improved supervision methods, like trend analysis. In 1968 first programmable logic controllers were introduced to replace hardwired controllers for electromechanical relays. This made the realization of protection systems easier. The advent of the microcomputer in 1971 and its increasing application in decentralized process automation systems since 1975 was the beginning of computationally more involved, software-based supervision and fault-detection algorithms. First publications on *process model-based fault-detection methods* appeared in connection with aerospace systems, [1.2], [1.27], [1.37], [1.6], and with chemical plants, [1.15]. Several of these first concepts can be classified as *parity relation approaches*, checking the consistency of instrument readings or mass or material balances. Residuals of mass balance were, for example, applied for leak detection of pipelines, [1.35]. The approach of parity relations was further investigated by [1.14].

State observer-based methods were developed to generate output residuals, applying Luenberger state observers, [1.2], [1.27] or Kalman filters, [1.28]. The *analytic redundancy* between several measurements was used for sensor fault detection, applying a bank of observers, [1.6]. In order to compensate for not-measurable inputs, *unknown input observers* or *output observers* were developed, [1.36] and [1.11]. Observers with eigen-structure assignment go back to [1.29].

Another way of fault detection is the use of *parameter estimation*. First publications appeared by [1.16], [1.1] in connection with jet turbines and [1.19], [1.20], [1.21] for processes in general and circulation pumps and DC motors as examples, and [1.8] and [1.7] for electrical motors.

Since these early publications many contributions were made in the field of fault detection and diagnosis. The development can be followed up by survey articles like [1.21], [1.22], [1.23], [1.9], [1.10], [1.12], [1.30]. A summary of publications during 1991-1996 with applications is given in [1.26]. Furtheron the multi-authored books [1.31], [1.32] give a good picture of the field. Several books on fault detection provide a valuable summary of the different approaches: [1.33], [1.15], [1.4], [1.34], [1.24], [1.3], [1.13], [1.5].

Another source for many publications is the IFAC Symposium series SAFEPROCESS, Baden-Baden (1991), Helsinki (1994), Hull (1997), Budapest (2000), Washington (2003), Beijing (2006), [1.17] and the IFAC Workshop "On-line fault detection and supervision in the chemical process industries", Kyoto (1986), Newark (1992), Newcastle (1995), Folaize (1998), Cheju (2001), [1.18].

Further original publications are cited together with the treatment of single subjects in the following chapters.

Part I

Fundamentals

2

Supervision and fault management of processes – tasks and terminology

The *supervision* of technical processes is aimed at showing the present state, indicating undesired or unpermitted states, and taking appropriate actions to avoid damage or accidents. The deviations from normal process behavior result from *faults* and *errors*, which can be attributed to many causes. They may result in some shorter or longer time periods with malfunctions or failures if no counteractions are taken. One reason for supervision is to avoid these malfunctions or failures. In the following sections the basic tasks of supervision are shortly described. This is then followed by a closer look on some terms and the terminology used in this field.

2.1 Basic tasks of supervision

A process P or a product is considered which operates in open loop, Figure 2.1 a. $U(t)$ and $Y(t)$ are each a measurable input and output signal, respectively. A fault can now appear due to external or internal causes. Examples for external causes are environmental influences like humidity, dust, chemicals, electromagnetic radiation, high temperature leading, e.g. to corrosion, pollution. Examples for internal causes are missing lubrication and therefore higher friction or wear, overheating, leaks, shortcuts. These faults $\mathbf{F}(t)$ then affect first internal process parameters Θ by $\Delta\Theta(t)$ like changes of resistances, capacitances or stiffness and/or internal state variables $\mathbf{x}(t)$ by $\Delta\mathbf{x}(t)$ like changes of mass flows, currents or temperatures, which are frequently not measurable. According to the dynamic process transfer behavior, the faults influence the measurable output $Y(t)$ by a change $\Delta Y(t)$. However, it has to be taken into account that also natural process disturbances and noise $N(t)$ and also changes of the manipulated variable $U(t)$ influence $Y(t)$.

A remaining fault $f(t)$ generally results for a process operating in open loop in a permanent offset of $\Delta Y(t)$, as shown in Figure 2.2. In case of a *closed loop* the behavior is different. Depending on the time history of parameter changes $\Delta\Theta(t)$ or state variable changes $\Delta\mathbf{x}(t)$ the output shows only a more or less shorter and vanishing small deviation $\Delta Y(t)$ if a control with integral behavior (e.g. a PI-controller)

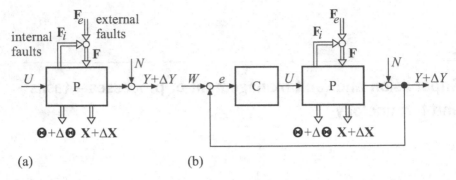

(a)   (b)

Fig. 2.1. Scheme of a process influenced by faults **F**: (a) process in open loop; (b) process in closed loop

is used. But then the manipulated variable shows a permanent offset $\Delta U(t)$ for proportionally acting processes. If only the output $Y(t)$ is supervised, the fault may not be detected because of the small and short deviation, furthermore corrupted by noise. The reason is that a closed loop is not only able to compensate for disturbances $N(t)$ but also to compensate for parameter changes $\Delta\Theta(t)$ and state changes $\Delta x(t)$ with regard to the control variable $Y(t)$. This means that faults $F(t)$ may be compensated by the closed loop. Only if the fault grows in size and causes the manipulated variable to reach a restriction value (saturation) a permanent deviation ΔY may arise. For processes in closed loop, therefore $U(t)$ should be monitored, as well as $Y(t)$ what is frequently not realized. Mostly only $Y(t)$ and the control deviation $e(t)$ are supervised.

The supervision of technical processes in normal operation or the quality control of products in manufacturing is usually performed by *limit-checking* or *threshold-checking* of some few measurable output variables $Y(t)$, like pressures, forces, liquid levels, temperatures, speeds, oscillations. This means one checks if the quantities are within a tolerance zone $Y_{min} < Y(t) < Y_{max}$. An alarm is then raised if the tolerance zone is exceeded. Hence, a first task in supervision is, compare Figure 2.3:

1. *Monitoring*: Measurable variables are checked with regard to tolerances, and alarms are generated for the operator. After an alarm is triggered the operator then has to take appropriate counteractions.

However, if exceeding of a threshold means a dangerous process state, the counteraction should be generated automatically. This is a second task of supervision, Figure 2.3:

2. *Automatic protection*: In the case of a dangerous process state, the monitoring function automatically initiates an appropriate counteraction. Usually, the process is then commanded to a fail-safe state, which is frequently an emergency shut down. Table 2.1 shows some examples.

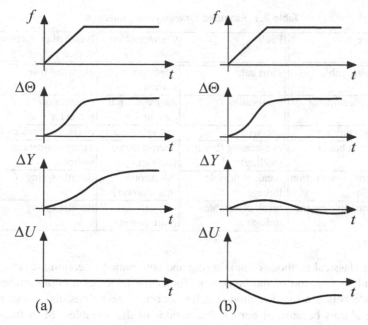

Fig. 2.2. Time behavior of a parameter change $\Delta\Theta$ and measurable signals $Y(t)$ and $U(t)$ after appearance of fault f: (a) open loop; (b) closed loop

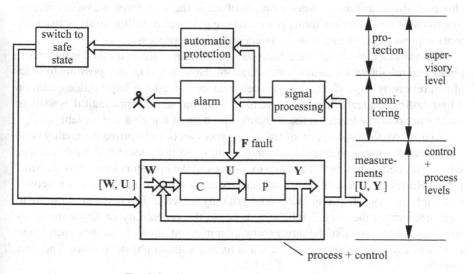

Fig. 2.3. Monitoring and automatic protection

Table 2.1. Examples for automatic protection

Process	Fault	Counteraction (safe state)	Name of protection device
electric cable	short cut	interruption of current	electrical fuse
electrical motor	overheating	interruption of current	temperature protector
steam turbine	overspeed	fast valve closes	overspeed protector
heating boiler	overheating (boiling) (boiling)	interruption of fuel supply	safety temperature switch
aircraft combustion engine	break of flexible linkage	full throttle (max. power)	throttle spring
automotive engine	break of flexible linkage	idle gas (min. power)	throttle spring

These classical methods of monitoring and automatic protection are suitable for the overall supervision of the processes. To set the tolerances, compromises have to be made between the detection size for abnormal deviations and unmeasurable or wrong alarms because of normal fluctuations of the variables. Most frequently, limit checking with fixed thresholds is applied which works well if the process stays in steady-state or the monitored variable does not depend on the operating point. However, the situation becomes more involved if the monitored variable changes dynamically with other operating points, like, e.g. forces in rolling mills or machine tools or pressures and temperatures in chemical batch processes.

The advantage of the classical limit-value based supervision method is their simplicity and reliability for steady-state situations. However, it is only possible to react after a relatively large change of a process feature, i.e. after a large sudden fault or a long-lasting gradually increasing fault. In addition, an in-depth fault diagnosis is usually not possible based on the threshold violation of one or a few variables.

To improve the supervision of technical processes or to improve the quality control of manufactured products a first step could be to implement additional sensors with are related to expected faults and to implement the operators know-how in computers. However, the use of additional sensors, cables, transmitters, plugs for getting better information of special faults does not only increase the costs but at the same time deteriorates the overall reliability because the probability of faults increases with more elements. Also the direct software implementation of operator knowledge is not an easy task and does not lead much further without physically-based process models.

For large-scale processes with many monitored and limit-checked values, there is another problem: after a severe process fault or failure several alarms may be triggered in short time sequence, known as "alarm-shower". The operators then are overloaded with regard to their immediate reactions and to finding the causes of the faulty behavior.

Therefore *advanced methods of supervision, fault detection and fault diagnosis* are required which satisfy the following requirements:

(i) early detection of small faults with abrupt or incipient time behavior;
(ii) diagnosis of faults in the processes or process parts and their manipulating devices (actuators) and measurement equipment (sensors);
(iii) detection of faults in closed loops;
(iv) supervision of processes in transient states.

The goal for the early fault detection and diagnosis is to have enough time for counteractions such as other operations, reconfiguration, planned maintenance or repair.

Figure 2.4 shows a general scheme how, in addition to the classical monitoring and automatic protection, these goals can be reached by automatic means. The intention is to generate more information about the process by using all available measurements and to relate them together in form of mathematical process models. If not only output signals $\mathbf{Y}(t)$ are measured but also the corresponding input signals $\mathbf{U}(t)$, some accessible state variables $\mathbf{x}(t)$ and maybe disturbance signals, then changes of the static and dynamic behavior of the processes by the faults can be used as important information source. Then also changes of output signals $\Delta\mathbf{Y}(t)$ which are not caused by faults but by input signals $\Delta\mathbf{U}(t)$ or measurable disturbances are automatically taken into account and therefore make the observed comparison variables more sensible to faults. This means that the effects on the outputs $\mathbf{Y}(t)$ by either normal disturbances or faults are automatically separated.

The general scheme in Figure 2.4 shows in the third level the following tasks:

3. *Supervision with fault diagnosis*

(a) *feature generation* by, e.g. special signal processing, state estimation, identification and parameter estimation or parity relations;
(b) *fault detection and generation of symptoms*;
(c) *fault diagnosis* by using analytical and also heuristic symptoms and their relations to faults, e.g. by classification methods or reasoning methods via fault-symptom trees. The goal is to determine the kind, size and location of the fault;
(d) *fault evaluation* with regard to classify the faults into different hazard classes;
(e) *decision on actions* dependent on the hazard class and possible degree of danger. This may be done either automatically or by the operator. Some examples for hazard classes are shown in Table 2.2.

Based on the gained in-depth information about the condition of the process, further tasks are necessary in order to improve the reliability or safety:

4. *Supervision actions and fault management*: Depending on the hazard classes of the diagnosed fault(s) the following actions can be taken:

(a) *safe operation*, e.g. shut down if there is an immanent danger for the process or the environment;

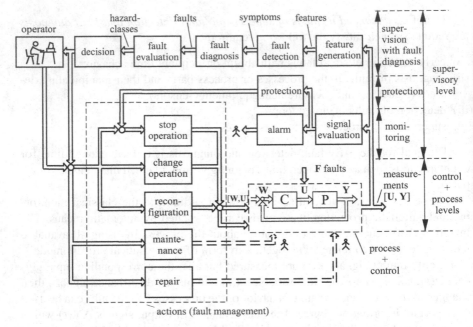

Fig. 2.4. General scheme of different supervision methods with fault management (supervisory loop)

(b) *reliable operation*, e.g. by hindering a further fault expansion through changes of operation state, e.g. operation with lower load, speed, pressure, temperature;

(c) *reconfiguration*, e.g. by using other sensors, actuators or redundant (standby) components to keep the process in operation and under control with a "reconfigured" structure;

(d) *inspection* to perform a detailed diagnosis by additional measures;

(e) *maintenance*, e.g. instantaneously or by next possibility to tune process parameters or exchange worn parts;

(f) *repair*, e.g. instantaneously to remove a fault or at next possibility (overhaul or revision).

These actions are also called *fault management* and may incorporate several intermediate actions in the case of redundant systems if the process is in a dangerous state, as, e.g. for aircraft, power plants, chemical plants or automatic guided vehicles.

Hence, the advanced methods of supervision and following actions are means to improve both the reliability and the safety of technical systems. Of course, these improvements by better information processing and computational intelligence have to be accompanied on the process side by further improving the reliability of all hardware components, by, e.g. proper materials, stress and overall design. Some further interesting developments are

Table 2.2. Examples for the evaluation of faults and related actions (fault management)

Hazard class	Affection of		Actions		Example
	safety	reliability	during operation	after operation	
1	high	high	emergency fail-safe (e.g. stop)	repair	bearing without lubrication
2	medium	medium	other operation state	maintenance	bearing with too low oil pressure
3	medium	high	reconfiguration	repair	one pump of a duplex pump system overheated
4	no	high	transfer to reliable state	repair	unbalanced wheels
5	no	small	no	maintenance	leaking seals

- *maintenance on request* (process condition);
- *tele-diagnosis* with modern communication;
- 100 % *quality control* of products.

As especially maintenance costs resemble in most cases a high percentage (e.g. ≤ 20 %) of overall operating costs, the advanced supervision and diagnosis may help to reduce maintenance effort and costs and improve the life time of the processes.

The general scheme in Figure 2.4 shows that there exists a feedback system from faults, signals, features, symptoms, decisions over various actions to compensate for faults. Therefore, this can be called *supervisory loop* or *fault management loop*. However, different to feedback control the signals or states are not all in continuous action. Some parts of information processing like signal evaluation, feature generation and symptom generation may operate continuously, but fault diagnosis, decision making and actions act as discrete events in the case of fault appearance. Hence, the supervisory loop is a hybrid continuous and discrete event system.

The known literature on the state-of-the-art of supervision and fault management is mostly related to special processes: Examples are

- machines: [2.39], [2.10], [2.2], [2.9], [2.21];
- electrical motors: [2.6], [2.7];
- pumps: [2.7], [2.25], [2.42], [2.4], [2.11];
- steam turbines: [2.34];
- manufacturing: [2.33];
- bearings and machinery: [2.5], [2.43], [2.40], [2.21];
- aircraft: [2.27], [2.24], [2.23];
- automotive systems: [2.31], [2.20], [2.18];
- chemical processes: [2.12], [2.32].

Books on model-based methods for fault detection are: [2.28], [2.29], [2.16], [2.8], [2.3], [2.35], [2.1].

The subject of fault-tolerant systems is treated in [2.22], [2.37].

2.2 Faults, failures, malfunctions

As the treated field from faults and failures through reliability, safety and fault-tolerant systems is distributed over many different technological areas, the used terminology is not unique. Various efforts have been made to come to a standardization, for example, the RAM (reliability, availability and maintainability) dictionary, [2.26], in contributions [2.14] and several German standards as DIN and VDI/VDE-Richtlinien (guidelines). The IFAC-Technical Committee SAFEPROCESS has made an effort to come to accepted definitions, [2.19], see also Chapter 23.1. A survey on related standardization literature is given in the bibliography of Chapter 23.1. The next sections describe the terminology used in this book, taking into account the mentioned literature.

Fault:

"A fault is an unpermitted deviation of at least one characteristic property (feature) of the system from the acceptable, usual, standard condition."

Remarks:

- a fault is a state within the system;
- the unpermitted deviation is the difference between the fault value and the violated threshold of a tolerance zone for its usual value;
- a fault is an abnormal condition that may cause a reduction in, or loss of, the capability of a functional unit to perform a required function [2.13];
- there exist many different types of faults, e.g. design fault, manufacturing fault, assembling fault, normal operation fault (e.g. wear), wrong operation fault (e.g. overload), maintenance-fault, hardware-fault, software-fault, operator's fault. (Some of these faults are also called errors, especially if directly caused by humans);
- a fault in the system is independent of whether the system is in operation or not;
- a fault may not effect the correct functioning of a system (like a small rent in an axle);
- a fault may initiate a failure or a malfunction;
- frequently, faults are difficult to detect, especially, if they are small or hidden;
- faults may develop abruptly (stepwise) or incipiently (driftwise).

Failure:

"A failure is a permanent interruption of a system's ability to perform a required function under specified operating conditions."

Remarks:

- a failure is the termination of the ability of a functional unit to perform a required function, [2.13];
- a failure is an event;

- a failure results from one or more faults;
- different types of failures can be distinguished:
    - number of failures: single, multiple;
    - predictability:
        - random failure (unpredictable, as, e.g. statistically independent from operation time or other failures);
        - deterministic failure (predictable for certain conditions);
        - systematic failure or causal failure (dependent on known conditions);
- usually a failure arises after begin of operation or by increasingly stressing the system.

Malfunction:

"A malfunction is an intermittent irregularity in the fulfillment of a system's desired function."

Remarks:

- a malfunction is a temporary interruption of a system's function;
- a malfunction is an event;
- a malfunction results from one or more faults;
- usually a malfunction arises after begin of the operation or by increasingly stressing the system.

Figure 2.5 shows the relation of faults, failures and malfunctions. The fault may develop abruptly, like a step-function, or incipiently, like a driftlike function. The corresponding feature of the system related to the fault is assumed to be proportional to the fault development. After exceeding the tolerance of normal values, the feature indicates a fault. Dependent on its size, a failure or a malfunction of the system follows at time t_e. Table 2.3 shows some example.

2.3 Reliability, availability, safety

With regard to the overall functioning of elements, components, processes and systems the terms reliability, availability and safety play an important role. These terms are considered in more detail in Chapters 3 and 4.

Reliability:

"Ability of a system to perform a required function under stated conditions, within a given scope, during a given period of time."

Remarks:

- short version: ability to perform a required function for a certain period of time;

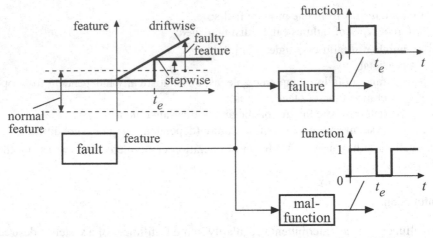

Fig. 2.5. Development of the events "failure" or "malfunction" from a fault which causes a stepwise or driftwise change of a feature

Table 2.3. Examples for faults, failures and malfunctions

Process	Fault	Feature	Failure	Malfunction
electrical illumination	switch with corroded contacts	occasionally no electrical conductivity	–	interrupted light
	broken wire in cable	no electrical conductivity	no light	–
electrical DC motor	worn brushes	armature resistance high	–	occasionally interrupted torque and changing speed
	broken wire in excitation coil	no electrical flux	no torque, no speed	–
machine tool belt drive	belt with too low pretension	no continuous torque transfer	–	sluggish dynamics piecewise motion
	broken belt threads	no torque transfer	standstill of feed drive	–
pneumatic valve	leak in supply air pressure	slow motion, limited position range	–	closed loop does not follow setpoint for some time
	corroded shaft	mechanical friction too high	no motion permanent control deviation	–

- reliability is quality for some time;
- the reliability can be affected by malfunctions and failures;
- a measure for reliability is the Mean Time To Failure MTTF $= 1/\lambda$, where λ is the rate of failures per time unit (see Chapter 5)

Safety:

"Ability of a system not to cause danger to persons or equipment or the environment."

Remarks:

- short version: ability not to cause danger;
- the safety is concerned with the dangerous effects of faults, failures and malfunctions;
- the safety can usually be seen as a status, where the risk is not larger than a specified risk limit (risk threshold).

The measures to improve the reliability are oriented towards avoiding faults, failures and malfunctions. Measures for improving safety are subject to avoid a dangerous effect of failures and malfunctions. An improvement of the reliability generally improves also safety. However, an improvement of safety can result in a deterioration of the reliability if, e.g. the number of components increase. Note that safety and security have similar meanings. *Safety* usually deals with life, equipment or environment, whereas *security* deals with privacy, property, community or state.

Availability:

"Probability that a system or equipment will operate satisfactorily and effectively at any period of time."

Remarks:

- availability is of major importance for the user of a system;
- availability takes into account that failures and malfunctions happen and need some time for repair;
- a measure for availability is $A = \frac{MTTF}{MTTF+MTTR}$ where MTTF is the Mean Time To Repair, see Chapter 5;
- to reach a high availability MTTF must be large in comparison to MTTR. This can be reached by
  - large operation time MTTF
    - $\rightarrow$ perfection: high reliable components;
    - $\rightarrow$ tolerance: tolerable faults through redundant structure
  - small repair time MTTR
    - $\rightarrow$ fast and reliable fault diagnosis;
    - $\rightarrow$ fast and reliable remove of faults (maintenance repair);

- fault detection and fault diagnosis can improve the availability by early fault detection in combination with maintenance on demand (larger MTTF) and by fast and reliable diagnosis (smaller MTTR).

Dependability:

The term dependability seems not to be clearly defined. Therefore different meanings are cited:

(i) "A form of availability that has the property of always being available when required (and not at any time). It is the degree to which a system is operable and capable of performing its required function at any randomly chosen time during its specific operating time, provided that the system is available at the start of the period"
This definition excludes non-operation related influences, [2.26];

(ii) "Dependability is a property of a system that justifies placing one's reliance on it. It covers reliability, availability, safety, maintainability and other issues of importance in critical systems", [2.37].

The [2.13] standard on safety-related systems does not define dependability, only safety integrity.

Integrity:

According to [2.37], the term integrity was earlier defined as:

"The integrity of a system is the ability to detect faults in its own operation and to inform a human operator".

Over the years the meaning was broadened and associated with critical systems. Integrity is frequently used as a synonym for dependability. According to [2.13] it is defined as:

"Safety integrity is the probability of a safety-related system satisfactorily performing the required safety functions under all the stated conditions within a period of time".

Some other expressions like accident, hazard, risk are defined in Chapter 4.

2.4 Fault tolerance and redundancy

After applying reliability and safety analysis for the improvement of the design, testing of the product and also corresponding quality control methods during manufacturing, the appearance of certain faults and failures cannot be avoided totally. Therefore, these unavoidable faults should be tolerated by additional design efforts. Hence, high-integrity systems must have the ability of *fault tolerance*. This means that faults are compensated in such a way that they do not lead to system failures. After the application of principles to improve the perfection of the components the remaining obvious way to reach this goal is to implement *redundancy*. This means that in addition to the considered module one or more modules exist as back-up modules usually in a parallel configuration, see Figure 2.6.

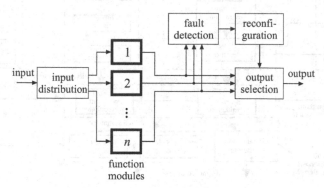

Fig. 2.6. Basic scheme of a fault-tolerant system with parallel function modules as redundance

The function modules can be hardware components or software parts, either identical or diverse. Different arrangements of fault-tolerant systems exist with static or dynamic redundancy, cold or hot standby. In general, the function modules are supervised with fault-detection capability followed by a reconfiguration mechanism to switch off failed modules and to switch on spare modules (dynamic redundancy). The modules are, e.g. actuators, sensors, computers, motors or pumps. For electronic hardware simpler schemes exist with $n \geq 3$ modules and majority voters to build up, e.g. 2-out-of-3 systems (static redundancy). These redundant systems are treated in Part IV.

2.5 Knowledge-based fault detection and diagnosis

As fault detection and fault diagnosis are fundamental for advanced methods of supervision and fault management, these tasks will be considered briefly. Fault detection and diagnosis, in general, are based on measured variables by instrumental and observed variables and states by human operators. The automatic processing of measured variables for fault detection requires analytical process knowledge and the evaluation of observed variables requires human expert knowledge which is called heuristic knowledge. Therefore fault detection and diagnosis can be considered within a knowledge-based approach, [2.30], [2.38]. Figure 2.7 shows an overall scheme, [2.15], [2.17].

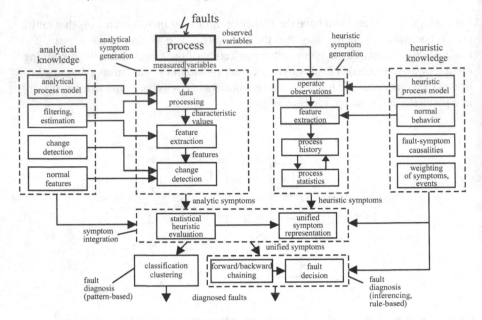

Fig. 2.7. Overall scheme of knowledge-based fault detection and diagnosis

2.5.1 Analytic symptom generation

The analytical knowledge about the process is used to produce quantifiable, analytical information. To do this, data processing based on measured process variables has to be performed to generate first the characteristic values by

- limit value checking of direct, measurable signals. The characteristic values are the violated signal tolerances;
- signal analysis of directly measurable signals by the use of signal models like correlation functions, frequency spectra, autoregressive moving average (ARMA) or the characteristic values as, e.g. variances, amplitudes, frequencies or model parameters;
- process analysis by using mathematical process models together with parameter estimation, state estimation and parity equation methods. The characteristic values are parameters, state variables or residuals.

In some cases, special features can then be extracted from these characteristic values, e.g. physically defined process coefficients, or special filtered or transformed residuals. These features are then compared with the normal features of the non-faulty process. For this, methods of change detection and classification are applied. The resulting changes (discrepancies) in the mentioned directly measured signals, signal models or process models are considered as analytic symptoms.

2.5.2 Heuristic symptom generation

In addition to the symptom generation using quantifiable information, heuristic symptoms can be produced by using qualitative information from human operators. Through human observation and inspection, heuristic characteristic values in the form of special noises, colors, smells, vibration, wear and tear, etc., are obtained. The process history in the form of maintenance performed, repairs, former faults, life-time and load measures, constitutes a further source of heuristic information. Statistical data (e.g. MTBF, fault probabilities) achieved from experience with the same or similar processes can be added. In this way heuristic symptoms are generated, which can be represented as linguistic variables (e.g. small, medium, large) or as vague numbers (e.g. around a certain value).

2.5.3 Fault diagnosis

The task of fault diagnosis consists in determining the type, size and location of the most possible fault, as well as its time of detection.

Fault-diagnosis procedures use the analytic and heuristic symptoms. Therefore they should be presented in an unified form like confidence-numbers, membership functions of fuzzy sets or probability density functions after a statistical evaluation over some time. Then either classification methods can be applied, if a learned pattern-based procedure is preferred, to determine the faults from symptom patterns or clusters. If, however, more information of fault-symptom-relations, e.g. in form of logic fault-symptom-trees or if-then rules are known, reasoning methods with forward and backward chaining can be applied.

The terminology used in this field is described in [2.19], based on definitions of the IFAC Technical Committee SAFEPROCESS, see Chapter 23.1.

The related methods for knowledge-based fault detection and fault diagnosis are considered in detail in Part II and III of this book.

2.6 Implementation issues

The development of fault detection and diagnosis systems (FDD-systems) and fault-tolerant systems can be represented as a "V"-diagram, which originate probably from the [2.36] and is especially used for the design of mechatronic systems, [2.41], see Figure 2.8. This diagram indicates important steps of the development process in a sequential manner. However, though the steps are given in logical order, some steps are performed in parallel or in iterative ways. Each phase has usually an outcome, called "deliverable".

The development starts with stating the *requirements*. The expected functions are summarized, such as the faults to be detected and diagnosed (a fault list), the smallest replaceable units which can be replaced if they contain a fault, the allowable cost for the development and the final product. A reliability and safety analysis at this stage may be very useful to find the weak points and risks of the considered product or

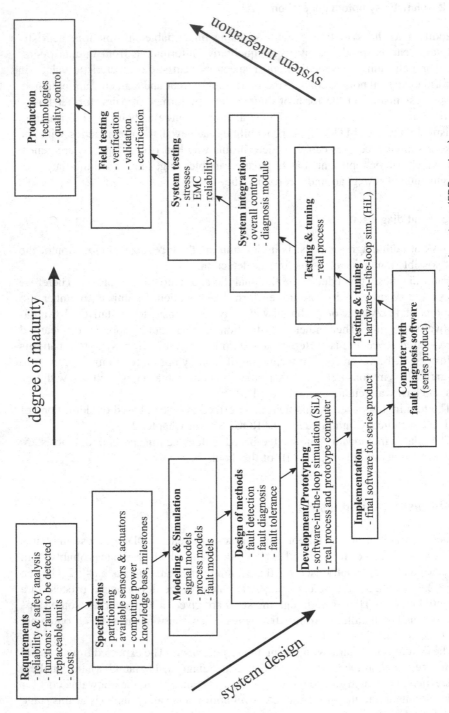

Fig. 2.8. A "V" development scheme for fault detection and diagnosis systems (FDD-system)

process. Because the requirements are the basis for the top-level design procedure, great care should be taken for their preparation and it should continuously be updated during the design process if basic changes happen.

Based on the requirements the *specifications* are formulated. Items are stated how the requirements are fulfilled, by partitioning of the functions or parts, the available sensors and actuators, the available computing power, use of further knowledge and definition of milestones.

The fault-diagnosis system design begins frequently with mathematical *modelling* of the process, its signals and expected faults. This includes also *simulation* of the behavior without and with faults. On the basis of these considerations the *design of the methods* for fault detection and diagnosis (FDD) and, if required, of fault-tolerance are performed. This is followed by the *development* of the FDD-methods with software-in-the-loop simulations (SiL). Herewith common software systems are used for the simulation of the process, the faults and the FDD-functions. Next, a powerful real-time prototype computer together with the real process can be applied (also called prototyping). The FDD-system is then mature for the *implementation* of the final software for the series product microcomputer.

Then the various *testing and tuning procedures* for the FDD-microcomputer hardware and software begins. First tests can be made by *hardware-in-the-loop simulation* (HiL). Here the microcomputer works together with other real parts, like actuators and real-time simulation of the process and another powerful computer. This requires a special sensor-simulation-interface. HiL is performed if expensive tests with the real system can be saved or experiments with faults are made which are not allowed with the real process. Otherwise tests are made with the *real process* directly.

If the FDD-system is implemented together with other functions, e.g. of automatic control, *system integration* takes place considering the functional dependencies of all control levels, from lower level control to top level process management, including documentation of FDD-results. The next tests are *system tests*, including verification and validation. *Verification* examines if the system meets its specifications, i.e. fulfills the functions of the specifications correctly. *Validation* considers the system as a whole with regard to satisfy the requirements, i.e. examines if the system is appropriate for its intended purpose. Therefore, it includes consideration of the correctness of the specification. For critical systems external regulating authorities have to be convinced to achieve *certification*. Here stated standards and guidelines are checked and tests have to show, e.g. the fault coverage for given operating conditions. *Field tests* are usually undertaken to test the system under many different operating conditions, production tolerances of the processes, and hard environmental conditions, before the system is given to series production.

2.7 Problems

1) What are the basic tasks for the supervision of technical processes?
2) State the differences between supervision and monitoring.

3) Find other examples for automatic protection as in Table 2.1.
4) Which tasks are included in advanced supervision compared to classical supervision?
5) State the following-up of tasks for advanced supervision.
6) What are the differences between faults, failures and malfunctions? Give one example for each case.
7) State the definitions of reliability, safety and availability in a table.
8) Find an example for a system with fault tolerance.
9) What are the differences between analytical and heuristic symptoms? Give some examples.
10) How can the tasks of fault detection and fault diagnosis be described and differentiated?

3

Reliability, Availability and Maintainability (RAM)

Reliability, availability and maintainability are very important properties of modern products. Therefore, this chapter describes these properties by the help of characteristic quantities. Historically, reliability studies were first applied to electrical and electronic components and quality control in manufacturing around the 1940s and then extended to other areas like military equipment, aerospace vehicles, nuclear power systems, automobiles, see, e.g. [3.4], [3.22]. In the following use is made of standards like [3.10] and [3.6]. It is tried to use generally accepted terms and definitions.

3.1 Reliability

Definition
"Reliability is the ability of a component, process or a system to perform a required function correctly under stated condition within a given scope, during a given period of time."

The reliability is affected by faults and failures. Therefore the reliability analysis depends on the kind of faults and failures.

3.1.1 Type of faults

Faults usually show a characteristic behavior for the various components. They may be distinguished by their form, time behavior and extent, compare Table 3.1. The *form* can be either systematic or random. The *time behavior* may be described by permanent, transient, intermittent, noise or drift, see Figure 5.3. The extent of faults is either local or global and includes the size.

Table 3.1 gives an overview of a variety of fault types in dependence on the system components. *Electronic hardware* shows systematic faults if they originate in specification or design mistakes. Once in operation faults in hardware components are mostly random with all kind of time behavior. The faults or mistakes in *software* (bugs) are usually systematic, e.g. by wrong specification, coding, logics, calculation overflows, etc. They are in general not random like faults in hardware.

Table 3.1. Main characteristics ($\sqrt{}$) of primary faults for different components

type of faults		components			
		mechanical components	electrical components	electronic hardware	software
form	systematic	$\sqrt{}$		$\sqrt{}$	$\sqrt{}$
	random		$\sqrt{}$	$\sqrt{}$	
time behavior	permanent			$\sqrt{}$	$\sqrt{}$
	transient	$\sqrt{}$	$\sqrt{}$	$\sqrt{}$	
	intermittent		$\sqrt{}$	$\sqrt{}$	$\sqrt{}$
	noise		$\sqrt{}$	$\sqrt{}$	
	drift	$\sqrt{}$	$\sqrt{}$	$\sqrt{}$	
extent	local	$\sqrt{}$	$\sqrt{}$	$\sqrt{}$	$\sqrt{}$
	global			$\sqrt{}$	$\sqrt{}$

Failures of *mechanical systems* can be classified into following failure mechanisms: distortion (buckling, deformation), fatigue and fracture (cycle fatigue, thermal fatigue), wear (abrasive, adhesive, cavitation), or corrosion (galvanic, chemical, biological), see, e.g. [3.19]. They may appear as driftlike changes (wear, corrosion) or abruptly (distortion fracture) at any time or after stress. *Electrical systems* usually consist of a large number of components with various failure modes, like short cuts, loose or broken connection, parameter changes, contact problems, contamination, EMC problems, etc. Generally, electrical faults appear more randomly than mechanical faults. Table 3.1 shows mainly the effect of primary faults for the different components and their typical behavior. The extent depends very much on the importance of the considered components and can of course be global for all cases even if the faults primarily appear locally.

Reliability analysis is usually based on the assumption of random faults. This holds then especially for electronic and electrical components and for large systems with many components and systematic faults which seem to appear randomly because of their large number, as for large mechanical systems and software systems, compare the discussion, e.g. in [3.22].

3.1.2 Reliability estimation

The reliability of a large number N of identical elements at the begin of operation is defined by the *reliability function*

$$R(t) = \frac{n(t)}{N} = \frac{\text{failure free elements}}{\text{number of all elements at the begin of operation}} \qquad (3.1)$$

relating to number $n(t)$ of correct functioning elements to N. It describes the probability that the elements function correctly until time t. The *unreliability function* is then

$$Q(t) = n_f(t)/N = 1 - R(t) \tag{3.2}$$

with number of failed elements

$$n_f = N - n$$

The *failure rate* is defined as the instantaneous rate of failing elements dn/dt related to still functioning elements $n(t)$

$$\lambda(t) = \frac{1}{n(t)} \frac{dn_f(t)}{dt} = \frac{1}{R(t)} \left(\frac{-dR(t)}{dt} \right) = \frac{1}{1 - Q(t)} \frac{dQ(t)}{dt} \tag{3.3}$$

or in words

$$\lambda(t) = \frac{1}{\text{number of functioning elements}} \frac{\text{number of failures}}{\text{time interval}} \tag{3.4}$$

Experience shows that electronic components and also electro-mechanical systems show a large decreasing failure rate after commissioning (infant mortality), then a constant value in the normal operation life (useful life) and finally an increasing value for the ageing period (wear out). This leads to the well known "bathtub-curve", Figure 3.1. During normal operation-life a constant failure rate λ can be assumed, leading to the *exponential failure law*

$$R(t) = e^{-\lambda t} \tag{3.5}$$

see Figure 3.2.

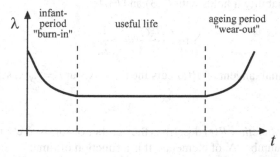

Fig. 3.1. Typical failure rate $\lambda(t)$ for randomly appearing faults in dependence of lifetime ("bathtub-curve")

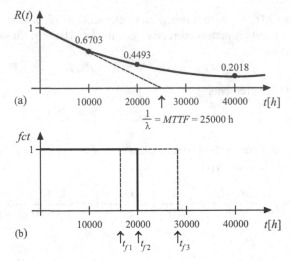

Fig. 3.2. (a) Exponential reliability function $R(t) = e^{-\lambda t}$ for constant failure rate $\lambda (\lambda = 4\cdot 10^{-5}$ [1/h]; (b) Time history of elements which function correctly ($fct = 1$) for $t < t_f$ and fails at $t = t_f (fct = 0)$

Example 3.1: Reliability function

For 1000 units and $\lambda = 4 \cdot 10^{-5}$ [1/h] the probability of correct functioning units at operating time $t = 20000$ h is $R = 0.449$. For the failed units it holds: $Q = 0.551$. This means that $n = 1000 \cdot 0.449 = 449$ units are still functioning (have survived) and $n_f = 1000 \cdot 0.551 = 551$ units have failed. The reliability function can also be called survival probability.

$\square$

For initial reliability it holds with (3.3) and (3.4)

$$\lim_{t \to 0} \frac{dR(t)}{dt} = -\frac{1}{\lambda} \qquad (3.6)$$

Therefore, the initial tangent of $R(t)$ cuts the time axis at $t = 1/\lambda$, see Figure 3.2a.

Remarks

a) the *reliability function* $R(t)$ describes the number of survived elements n relative to the initial number N of elements. It is a function of time;
b) the *failure rate* λ relates the number of failures per time interval to the number of survived elements. If $\lambda = $ const. then the failures per time interval become smaller with time. Therefore, the reliability function decays also with time.

Another measure for reliability is the *Mean Time To Failure* (MTTF). It is the average failure free (correct) operation time t_f until a failure

$$\text{MTTF} = E\{t_F\} = \lim_{V \to \infty} \frac{1}{N} \sum_{i=1}^{N} t_{fi} \tag{3.7}$$

It can also be calculated by dividing the number of function elements $n(t)$ through the number of failures per time interval

$$\text{MTTF} = \frac{n(t)}{\frac{dn_f(t)}{dt}} = \frac{\text{number of functioning elements}}{\frac{\text{number of failures}}{\text{time interval}}} \tag{3.8}$$

Comparison with (3.3) shows that MTTF is the reciprocal to the failure rate

$$\text{MTTF} = \frac{1}{\lambda} \tag{3.9}$$

For constant failure rate it follows from the exponential reliability function

$$\text{MTTF} = \int_0^\infty R(t)dt = \int_0^\infty e^{-\lambda t} dt = \frac{1}{\lambda} \tag{3.10}$$

The MTTF represents the average operating time of a large number of non-repairable elements before a failure happens, for the case that the failure rate is constant. However, this means that the probability of a system to function at $t = \text{MTTF} = 1/\lambda$ is

$$R(t) = e^{-1} = 0.37$$

Hence, after an operating time of MTTF only 37% probability of correct operation results. This underlines that MTTF represents an average life time of a large number of components. Table 3.2 and Table 3.3 show examples for failure rates of various electronic and mechanical components.

Table 3.2. Typical failure rates of mechanical and electromechanical elements, [3.19]

Mechanical elements	$\lambda[h^{-1}]$	Electromechanical elements	$\lambda[h^{-1}]$
ball bearing	$1.64 \cdot 10^{-6}$	actuator, general	$26 \cdot 10^{-6}$
sleeve bearing	$2.38 \cdot 10^{-6}$	brush, general	$9 \cdot 10^{-6}$
belt	$19.72 \cdot 10^{-6}$	cable, general	$1 \cdot 10^{-6}$
coupling	$5.54 \cdot 10^{-6}$	electric motor, general	$9 \cdot 10^{-6}$
gear	$4.69 \cdot 10^{-6}$	generator, general	$73 \cdot 10^{-6}$
pump	$43.65 \cdot 10^{-6}$	regulator, general	$13 \cdot 10^{-6}$
seal	$5.47 \cdot 10^{-6}$		
valve, hydraulic	$8.83 \cdot 10^{-6}$		

Another similar measure like MTTF is the *Mean Time Between Failure* (MTBF). According to [3.19] and [3.17] it is defined for repairable systems where all failed units are repaired periodically, see Section 3.3.

Table 3.3. Failure rates of electronic/electrical elements [3.14], stationary, 40°C, 60% power. π_E : multiplication factor for mobile application

electronic/electrical elements	$\lambda[\mathrm{h}^{-1}]$	mobile application π_E
discrete elements		
transistor	$1 - 70 \cdot 10^{-9}$	
diodes	$1 - 6 \cdot 10^{-9}$	18
thyristors	$36 - 360 \cdot 10^{-9}$	
resistances	$1 - 100 \cdot 10^{-9}$	
condensators	$5 - 150 \cdot 10^{-9}$	
analog integrated circuits		
operation amplifier	$0.3 - 0.9 \cdot 10^{-6}$	4.2
analog switch	$20 \cdot 10^{-6}$	
digital integrated circuits		
logic elements	$0.03 \cdot 10^{-6}$	
multiplexer	$0.05 - 0.2 \cdot 10^{-6}$	4.2
8-bit CPU (8080, 6800)	$2 - 5 \cdot 10^{-6}$	
flight computer (80286)	$10^{-4} \ldots 10^{-3}$	[3.18]

A measure for the unreliability is the *failure density*

$$q(t) = \frac{dQ(t)}{dt} = \frac{1}{N}\frac{dn_f(t)}{dt} = -\frac{dR(t)}{dt} \qquad (3.11)$$

which becomes for λ =const.

$$q(t) = \lambda e^{-\lambda t} \qquad (3.12)$$

It has the same time history as $R(t)$. Therefore, the failure rate is large for small t and small for large t. The probability of failed units during time period Δt becomes

$$\Delta n_f \approx N\lambda e^{-\lambda t}\Delta t \qquad (3.13)$$

Example 3.2: Failed units

If $\lambda = 10^{-5}$ [1/h] and $N = 10^6$ units the initial probability of failed units in the first year is

$$\Delta n_f = 10^6 \cdot 10^{-5} \cdot 0.916 \cdot 8760 = 80253 \triangleq 8.02\%$$

In the fifth year (after $t = 35040$ h) the failed units are

$$\Delta n_f = 10^6 \cdot 10^{-5} \cdot 0.705 \cdot 8760 \approx 61731 \triangleq 6.17\%$$

□

The considered exponential distribution function of reliability over time is the simplest one and it has at least been shown to approximate the behavior of electronic components well. Another frequently used reliability function is the *Weibull distribution*

$$R(t) = e^{\left(\frac{t-t_0}{t_L-t_0}\right)^\beta}$$ (3.14)

where

t_0 failure free period (location parameter);
t_L characteristic life time (63 % of distribution, scale parameter);
β shape parameter.

The Weibull function can be adjusted to a wide variety of different distributions, compare, e.g. [3.19]. For $\beta = 1$ and $t = 0$ it reduces to the exponential distribution and for $\beta \approx 3.55$ it approximates the normal distribution.

The reliability of *electronic components* depends much on the operating conditions, like environmental conditions and internal load. The failure rate therefore depends on the the external and internal temperature, voltage, current, power, vibration, dust, humidity. (In [3.27] it is shown, that the MTTF of IC's is inverse proportional the square of current density and depends exponentially on the temperature.)

According to [3.14], these influences are considered by factors

$$\lambda = \lambda_b \pi_Q \pi_L \pi_T \pi_P \pi_E$$ (3.15)

according to quality, learning, temperature, power, environment. Therefore, the failure rate may vary by $2 \ldots 3$ decades. For standstill 10 % of λ_b (basic λ) is taken. Temperature increase from 40°C to 70°C enlarges the failure rates by factors 2 to 15.

The reliability of *mechanical elements* is influenced much by distortion, fatigue, fraction, wear and corrosion. For more details see, e.g. [3.4], [3.14], [3.21], [3.20], [3.13].

The required MTTF of components depends of course very much on the product or process. For automotive electronics it is, for example, assumed that for reliability investigations following numbers are used, [3.9]: time in service: 10 years, operation time: 3000 h; average speed: 50 km/h; drive distance: 150000 km; number of rides: 50000. The required life time of some components follows from [number of rides in 10 years × demands per ride] and results in: airbag: 50 h; ABS: $5 \cdot 10^5$ h; window heating: 10^4 h; door locking: $5 \cdot 10^4$ h; wipers, lighting: $2.5 \cdot 10^5$ h.

3.1.3 Connected elements

Reliability analysis generally requires the evaluation of connected elements. This is based on reliability networks, representing the kind of connection also called combinational modelling. For *series connection* of elements, see Figure 3.3a, which may fail statistical independently from each other, it holds for the overall reliability that all elements operate correctly for constant failure rates λ_i

$$R_{tot}(t) = \prod_{i=1}^{m} R_i(t) = e^{-\sum_{i=1}^{m} \lambda_i t} = e^{-\lambda_{tot} t}$$ (3.16)

This is called the *product law of reliability*. Hence, the total failure rate and total MTTF become

$$\lambda_{tot} = \sum_{i=1}^{m} \lambda_i \ \text{ and } \text{MTTF}_{tot} = \left[\sum_{i=1}^{m} (\text{MTTF}_i)^{-1} \right]^{-1} \qquad (3.17)$$

In order to reach a small overall failure rate, all elements should be of similar small failure rate λ_i. Otherwise the largest λ_i will dominate and determine λ_{tot}.

(a)
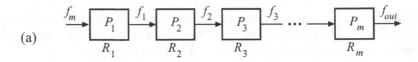

(b)
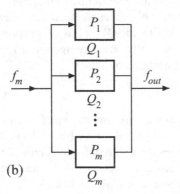

Fig. 3.3. Reliability network for: (a) series connection; (b) parallel connection The function f_i is either correct (1) or fail (0)

If the elements are arranged in *parallel connection*, see Figure 3.3b, i.e. they are redundant and the unreliability $Q_i(t) = 1 - R_i(t)$ describes the probability that one element fails, all m_i parallel elements have failed with probability

$$Q_{tot}(t) = \prod_{i=1}^{m} Q_i(t) = \prod_{i=1}^{m} (1 - R_i(t)) \qquad (3.18)$$

This is the *product law of unreliability*.

The reliability of the parallel connection is then

$$R_{tot}(t) = 1 - \prod_{i=1}^{m} (1 - R_i(t)) = 1 - \prod_{i=1}^{m} \left(1 - e^{-\lambda_i t} \right) \qquad (3.19)$$

The failure rate λ and the MTTF of parallel connected elements can be determined as follows. Using (3.9) for constant failure rate and assuming the same $\lambda_i = \lambda$ for all elements leads to

$$\text{MTTF} = \int_{0}^{\infty} R(t)dt = \int_{0}^{\infty} \left[1 - \left(1 - e^{-\lambda t} \right)^{m} \right] dt \qquad (3.20)$$

For $m = 3$ it holds

$$R(t) = 1 - \left(1 - e^{-\lambda t}\right)^3$$
$$= 1 - 3e^{-\lambda t} + 3e^{-2\lambda t} + e^{-3\lambda t}$$

and therefore with (3.20)

$$\text{MTTF} = 3\frac{1}{\lambda} + 3\frac{1}{2\lambda} + \frac{1}{3\lambda} = \frac{11}{6\lambda}$$

This can also be expresses as, [3.2]

$$\text{MTTF} = \frac{1}{\lambda} + \frac{1}{2\lambda} + \frac{1}{3\lambda} = \frac{1}{\lambda}\left[1 + \frac{1}{2} + \frac{1}{3}\right] = \frac{11}{6\lambda}$$

Generalizing leads to

$$\text{MTTF} = \frac{1}{\lambda}\sum_{i=1}^{m}\frac{1}{i} \tag{3.21}$$

or

$$\lambda_{tot} = \frac{\lambda}{\sum_{i=1}^{m}\frac{1}{i}} \tag{3.22}$$

Hence, MTTF increases for two parallel elements by factor 1.5, for three by 1.833, for 4 by 2.083.

Table 3.4 and Figure 3.4 show reliability numbers for identical connected elements in either series or parallel connection. The overall reliability decreases considerably with increasing number of elements. The MTTF, for example, is only half for two serially connected elements, one third for three elements, etc. However, redundancy by parallel elements leads to an improvement of MTTF by 50 % for two elements, 83 % for three elements, etc., The improvement is initially considerable, but the incremental advantage becomes smaller with each parallel component.

For series-parallel connections the system has to be divided in series and parallel connected arrangements, such that the product laws of reliability and unreliability can be applied consecutively, see, e.g. [3.22], [3.2].

3.2 Maintainability

Definition:
"*Maintenance* is understood as an action taken to retain a system in, or return a system to its designed operating condition. It extends the useful life of systems, ensures the optimum availability of installed equipments or equipment for emergency use."

Maintenance is quite often very expensive and may extend the investment cost with time.

Table 3.4. Overall reliability of m connected identical elements for equal failure rate $\lambda_i = 10^{-5}[\text{h}^{-1}]$, $i = 1, 2, ..., m$ and an operating time of $t = 10^4$ [h]

elements m	series connection			parallel connection		
	R_{tot} $(t = 10^4 \text{ h})$	λ_{tot} $[\text{h}^{-1}]$	MTTF$_{tot}$ [h]	R_{tot} $(t = 10^4 \text{ h})$	λ_{tot} $[\text{h}^{-1}]$	MTTF$_{tot}$ [h]
1	0.905	10^{-5}	10^5	0.905	10^{-5}	10^5
2	0.818	$2 \cdot 10^{-5}$	$0.5 \cdot 10^5$	0.9910	$0.667 \cdot 10^{-5}$	$1.499 \cdot 10^5$
3	0.741	$3 \cdot 10^{-5}$	$0.33 \cdot 10^5$	0.99914	$0.545 \cdot 10^{-5}$	$1.83 \cdot 10^5$
4	0.670	$4 \cdot 10^{-5}$	$0.25 \cdot 10^5$	0.99992	$0.480 \cdot 10^{-5}$	$2.08 \cdot 10^5$
5	0.606	$5 \cdot 10^{-5}$	$0.2 \cdot 10^5$	0.99999	$0.438 \cdot 10^{-5}$	$2.283 \cdot 10^5$
⋮	⋮	⋮	⋮	⋮	⋮	⋮
10	0.368	10^{-4}	10^4	0.99999	$0.341 \cdot 10^{-5}$	$2.929 \cdot 10^5$

Figure 3.5 gives a survey of the different kinds of maintenance activities. Usually maintenance is a planned action. *Preventive* or *scheduled maintenance* begins with inspection and includes, e.g. cleaning, adjustment, lubrication at predetermined intervals. Is also includes the replacement of minor components, subject to wear, before they fail, depending on inspection. *Corrective maintenance* includes minor repairs and planned regular overhauls, i.e. complete replacement of wearing components. Unplanned maintenance is performed in an emergency situation to avoid immediately drastic failures, shut-downs or losses, or for safety reasons.

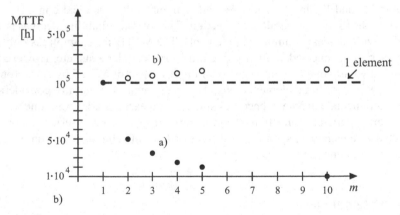

Fig. 3.4. Mean Time To Failure (MTTF) for m connected elements with equal failure rate $\lambda = 10^{-5}[\text{h}^{-1}]$: (a) series connection; (b) parallel connection

Furtheron, maintenance can be performed while the system is in operation or by shutting it down temporarily. The first one holds, for example, for ship engines or electric transmission lines. However, usually the system has to be taken out of service leading to down times.

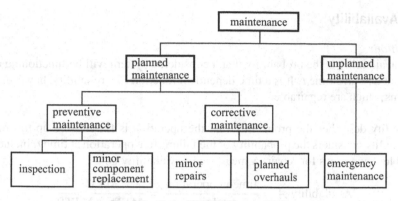

Fig. 3.5. Kinds of maintenance, [3.2]

Maintainability is a quality of several features which enable a system to be maintained. This includes proper design, required tools, standardized components, group subsystems (modularity), aids for trouble-shooting, space parts and logistics. The down time is the interval where the system is not in acceptable operation and can be separated in diagnostic time, active repair time, logistic time and administrative time.

A measure for maintainability is the probability that a failed system will be repaired within a time period T_R. Then, the expected repair time can be stated, called MTTR: *Mean Time To Repair*

$$\text{MTTR} = E\{T_R\} = \lim_{N \to \infty} \frac{1}{N} \sum_{i=1}^{N} T_{Ri} \qquad (3.23)$$

compare Figure 3.6

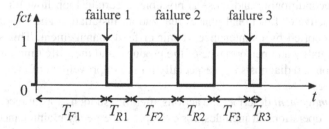

Fig. 3.6. Time history of the function *fct* for repairable elements. T_{Fi} : time to failure; T_{Ri} : repair time

3.3 Availability

Definition:
"The availability is the probability that a considered system will be functioning correctly at any give time t. It is a time dependent function like reliability, however, for systems which are repairable."

Reliability describes the probability that the operation is failure-free up to time t. Availability considers the probability of the failure-free operation at time t, including possible down times for repair. A measure for availability is

$$\text{Availability } A = \frac{\text{time in operation}}{\text{total time}} = \frac{\text{MTTF}}{\text{MTTF} + \text{MTTR}}$$
$$= \frac{1}{1 + \dfrac{\text{MTTR}}{\text{MTTF}}} \tag{3.24}$$

A large availability is obtained by large MTTF and small MTTR. For a repairable system the *Mean Time Between Failure* is

$$\text{MTBF} = \text{MTTF} + \text{MTTR}$$
$$\text{If MTTR} \ll \text{MTTF, then}$$
$$\text{MTBF} \approx \text{MTTF} \tag{3.25}$$

3.4 Fault management for total life cycles

The discussed properties of products, like reliability, availability and maintainability have to be seen in the context of the *overall life cycle*. Figure 3.7 shows the various steps, beginning with planning and design, through realization/manufacturing, installation/commissioning, operation, decommissioning up to waste treatment. If recycling, reconditioning and reuse is possible, a certain loop flow for some parts arises. The active or functional phases realization, installation and especially operation are supported by maintenance, repair or fault management. This support increases the quality and performance of the products and their life time. Supervision, fault detection and diagnosis help especially for the improvement of these functional product phases.

Fault management describes the actions after the (sudden) appearance of faults to maintain the operations. It includes, for example, emergency maintenance or repair, use of spare parts, or reconfiguration in the case of static or dynamic redundancy, see Chapter 19.

3.5 Some failure statistics

In order to obtain some quantitative information on the reliability, and understanding of the appearance and kind of faults in technical processes, some published statistics

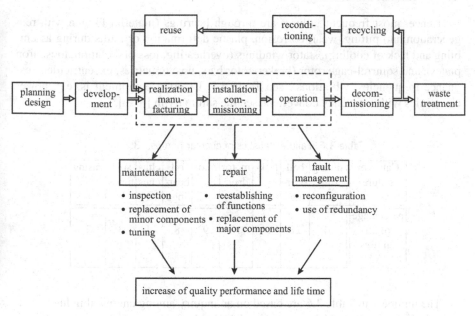

Fig. 3.7. Life cycle of technical products and measures to increase quality, performance and life time

are gathered in the following tables. The basis of the statistical numbers given is, of course, rather different. The numbers stem from publications of technical organizations, service associations, insurances, etc. They are divided in *components* and *systems*.

3.5.1 Statistics of components

Table 3.5. Failure statistics of AC motors (11 . . . 50 kW . . . 1 MW), [3.24]

Cause of failure	bearings	stator windings	external equipment	broken cage or rings	axle clutch	not defined
percentage of all failures [%]	51.1	15.8	15.6	4.7	2.4	10.4

Failure rates: low power (< 100 kW):   5% p.a. or $\lambda = 5.8 \cdot 10^{-6} [\text{h}^{-1}]$

                     high power (> 100 kW):   10% p.a. or $\lambda = 1.16 \cdot 10^{-5} [\text{h}^{-1}]$

References: [3.24], [3.8], [3.15].

Hence, most frequent failures are through bearings (material fatigue with rent generation and pitting, wear, corrosion, plastic deformation or faults during assembling and lack of cooling), stator windings (overheating, loss of isolation, loose iron plates), and squirrel-cage rotor (broken rotor bars, unsymmetries, excentricities, resonance frequencies, vibrations), see Table 3.5. External equipment is, e.g. power electronics with frequency converters, see the summary in [3.26].

Table 3.6. Failure statistics of circular pumps, [3.25]

Cause of failure	sliding ring seal	ball bearings	leak-age	motor drive	rotor	oil bearings	clut-ches	split tubes	casing
percentage of all failures [%]	31	22	10	10	9	8	4	3	3

The numbers in Table 3.6 are based on an inquiry among chemical industry, water and wastewater treatment companies. The kind of operation ist: permanent operation 59 %, daily operation 19 %, short operation 22 %. Inspection happens all three months, unplanned defects with repair all nine months. Most frequent failures are with sliding ring sealing and ball bearings. Causes for operational malfunctions are: cavitation, gases in liquids, blockage through closed valves, dry operation through missing fluid, wear through erosion (particles), corrosion, ball bearings, split flow, deposits and oscillations, see the summary in [3.26]. According to another inquiry with the customers, the priorities are: 1. reliability; 2. energy consumptions, 3. no leakages; 4. price; 5. noise; 6. control range.

References: [3.25], [3.16], [3.11].

Table 3.7. Failure statistics of hydraulic actuators (aircraft components)

Cause of failures	spool	valve	cylinder	mechanics	power components (pumps)	others
percentage of all failures [%]	32	19	16	16	3	14

The largest percentage of failures, Table 3.7, arises with 51 % in the spool valve which manipulates the oil flow to the cylinder. Failures of the spool valve are erosion at the edges (65 %), dirt (20 %) and external leakage (10 %). The failures of the valve are internal leakage (10 %), insufficient function (35 %) and external leakage (15 %). The cylinder shows failures through external leakage at the rod seals (58 %), internal leakage (14 %) or broken or cracked rod.

Table 3.8. Lifetime costs of sensors and actuators over 15 years of operation. 388 plants. 41 EUR/h labor cost, [3.7]

Statement of cost	pressure sensor		temperature sensor		flow sensor		valve	
	EUR	%	EUR	%	EUR	%	EUR	%
acquisition	816	31.5	230	27.1	1990	28.7	832	30.5
planning, assembling	490	18.9	388	45.8	1173	16.9	735	27.0
maintenance	460	17.7	153	18.1	918	13.2	153	5.6
repair	745	28.7	0	0	2704	39.0	534	33.9
others	82	3.2	77	9.0	153	2.2	82	3.0
total costs	2593	100	848	100	6938	100	2336	100
maintenance + repair / acquisitions	1.5		0.7		1.8		1.3	
overall cost / acquisitions	3.2		3.7		3.5		3.3	

The longtime consideration of *sensors and actuators* for plants of the chemical industry 3.8, shows several interesting effects. The total costs of flow sensors are highest, followed by pressure sensors and valves. Temperature sensors are cheapest. Maintenance and repair costs over 15 years are especially high for the flow and pressure sensors and valves. The overall costs are about 3.5 larger than the acquisition costs. These numbers underline the great significance of early fault detection to reduce at least part of costs for maintenance and repair.

3.5.2 Statistics of systems

The statistics of the German automobile club ADAC, Table 3.9, are based on approximately 20000 - 100000 breakdowns and service helps per year. Most frequent are (and increasing within the last 5 years) failures of the electrics, like battery, generator, V-belt, starter blockage, loosened cables or burned fuses. The ignition system showed defects because of the immobilizer (theft protection), ECU's, sparks

Table 3.9. Breakdown of passenger cars, ADAC Pannenstatistik 1999, [3.1]

Cause of failure	general electr.	ignition	motor	cooling system	fuel system	fuel injection	wheels tires
percentage of all failures [%]	31.3 (35.1)	14.1 (14.1)	12.2 (10.5)	8.4 (7.2)	6.6 (6.0)	6.2 (5.7)	6.8 (6.9)
Cause of failure	clutch gear	chassis	exhaust system	brake system	suspension		steering system
percentage of all failures [%]	5.8 (5.0)	4.2 (4.8)	2.2 —	1.4 —	0.6 —		0.3 —

or marten bites. Problems of the engine were broken toothed belts or chains of the camshaft, oil pump defects, too less oil and overheating.

Hydraulic brakes of passenger cars in Germany (1999) are responsible for about 900 accidents with injuries from about 36000, i.e. 2.5 % of all accidents and 18 % of all accidents due to technical faults (accidents during rain, snow and ice are about 48.5 %), [3.23], [3.5]. The reasons for brake failures are leakage or gas enclosures. 60 % are due to lacking maintenance (porous flexible brake tubes, corroded brake lines, cut seals at cylinder pistons), and 10 % due to wrong assembling and repair.

Table 3.10. Failure of the water cooling system of passenger cars, ADAC Pannenstatistik 1999, [3.1]

Cause of failure	V-belt	flexible pipes (heating cooling)	cylinder head seals	water pump	coolant liquid	thermo-stat	cooler expansion vessel
percentage of all failures [%]	29	20	16	10	6	6	6

The cooling system shares about 8 % of all vehicle break downs, see Table 3.10. Most of the failures are caused by V-belt, the flexible pipes for heating and cooling, cylinder heads and water pumps. By leak detection almost 48 % of all failures can be detected, e.g. by measuring the coolant liquid level.

Table 3.11. Damage statistics of components in the chemical industry during assembling and commissioning, [3.3], [3.12]

Failed components	pipes	columns, tanks, cooling towers	containers reactors	vessels filters mufflers	heat exchangers	ovens
percentage of all failures [%]	10	9	18	9	14	14

Table 3.12. Damage statistics of components in the process industries (refineries, petrochemical gas production, terminals, general process industries). Based on 2023 failures, [3.12]

Failed components	pipes	tanks	containers	reactors	heat exchangers	ovens, vessels heaters
percentage of all failures [%]	5-22	5-21	5-20	3-20	4-5	4-7

A summary of damages and accidents in the *process industries* is given in [3.12]. An extract is shown in Table 3.11 and 3.12. The relative frequency of accidents is especially high in following industries: paper, cellulose, alkali-chlorus, fertilizer, chemical (US industries, 1994-1999), with up to 200 accidents per year, e.g. in the paper industry. All these statistics are summarized in [3.12]

Hence, pipes and pipelines are the most frequent cause for damages (go even up to 33 % or 46 % due to other sources), followed by tanks, containers and reactors. The reasons for damages by the pipes are material weakening through erosion / corrosion or chemical corrosion, especially at elbows and because of unsuitable material. Other reasons are overpressure (20 %), corrosion (16 %) and human operating errors (31 %). All human errors are 41 % with a share of wrong maintenance (39 %), design (27 %) and operation (14 %). Also damages by valves are included within pipes, with about 4 % to 11 % of pipe damages.

Tanks and containers cause the next frequent damages, especially in the petrochemical industry, for example by overfilling or failing of welding seams. These damages of pipes, tanks and containers are the reason of about 30-50 % of all events. This means that especially the more simple components and not the more complicated ones increase the overall damages.

The costs per event are in an average for material and property/break of operations: petrochemical industry 20/28 Mill US $; refineries 15/17 Mill US $; chemical processes 11/15 Mill US; and machines, electronics 5/7 Mill US $. This is based on 2700 damages since 1984.

3.6 Problems

1) For 10000 devices 10 failures are observed each year. Determine the failure rate and MTTF.
2) The mortality rate of a 30 year old man is $\lambda = 10^{-3}$ [1/year]. Determine the estimated life time as MTTF if the man would stay in same health condition.
3) Estimate the infant phase, normal phase and ageing phase of a human and an automobile. Determine the reasons for failures for all three life phases.
4) The failure rate of a manufacturing unit be $\lambda_1 = 1$ failure per year. Determine the overall failure rate and MTTF if 2, 3, 5, 10 units with the same failure rate are connected in series. How are these numbers for parallel connection?
5) Passenger cars have about 300 hrs/year operation time. Given a failure rate of $\lambda = 10^{-3}$ [1/h], how many can fail for a population of 10^6 cars in the 1st and the 5th year?
6) The hydraulic brake system of passenger cars consists of redundant paths in a parallel configuration. The failure rate of one path is $\lambda = 10^{-5}$ 1/h. Determine the MTTF of the whole brake system.
7) Determine the availability of a machine tool for MTTF = 1 year and MTTR = 1 day.
8) Determine the failure rate λ and MTTF for an electric motor with gear for a series connection of a switch, cable, motor, gear according to the failure rates of Table 3.2 and 3.3.

9) One type of trucks has under same driving condition an MTTF = 8 months and 2 days repair, another type MTTF = 5 months and 1 day repair. Compare the availability.

10) Switching solenoid relays in automobiles show about 5 ppm faulty ones. How many cars are effected if they contain 50 relays each? What is the failure rate if these faulty relays show up in 1 year or 5 years?

Safety, Dependability and System Integrity

For safety-related systems all aspects of reliability, availability, maintainability and safety (RAMS) have to be considered because they are relevant for the responsibility of the manufacturers and the acceptability of customers. To meet safety requirements special procedures were developed in different technical disciplines like railway, aircraft, space, military and nuclear systems. These procedures are covered by the terms *system integrity or system dependability.*

The various kinds of safety requirements lead to different levels of integrity of safety-related systems, from lowest to highest requirements. In this context "integrity" means more precisely "safety integrity" with following definition:

Safety integrity is the probability of a safety-related system satisfactorily performing the required safety functions under all the stated conditions within a stated period of time" [4.2].

Safety and reliability are generally achieved by a combination of

- fault avoidance;
- fault removal;
- fault tolerance;
- fault detection and diagnosis;
- automatic supervision and protection.

Fault avoidance and removal has to be mainly accomplished during the design and testing phase. For investigating the effect of faults on the reliability and safety during the design and also for type certification a range of analysis methods were developed, such as:

- reliability analysis;
- event tree analysis (ETA);
- fault tree analysis (FTA);
- failure mode and effect analysis (FMEA);
- hazard analysis (HA);
- risk classification.

Some of the methods developed in special fields are more or less accepted in other fields. These methods are briefly reviewed in this chapter to find an appropriate procedure for developing safety-related systems, see also [4.13] and [4.2].

4.1 Reliability analysis

Reliability is usually described as the probability of a component or system to function correctly over a certain period of time and for certain operating conditions. Unreliability, therefore, arises if faults develop or appear which lead to a failure of the component or the system. Faults and failures may be random or systematic, [4.2], as discussed in Section 3.1, see Table 3.1. *Random faults* occur at random time subject to degradation mechanisms, especially in electronic hardware components. Degradation mechanisms depend on components' quality, manufacturing tolerances and operational stress. *Systematic faults* or failures are related in a deterministic way to a certain cause, like human errors, or in manufacturing processes or operational procedures. Random hardware failures can be predicted with reasonable accuracy, based on the experience with large numbers of pieces. However, systematic failures cannot be accurately predicted based on statistics.

The reliability of elements with random faults is described by the reliability function $R(t)$, or the failure rate $\lambda[h^{-1}]$ (e.g. failure per 10^6 hours) or Mean Time To Failures MTTF [h], see Chapter 3. Based on these measures, the reliability of connected elements can be estimated. Series connections deteriorate and parallel (redundant) connections improve the reliability.

Failures of *mechanical systems* arise through distortion, fatigue and fracture, wear and corrosion. The reliability of mechanical systems can be improved a lot by oversizing, protection (corrosion) and wear reduction and therefore generally need no redundancy. *Hydraulic systems* are more subject to wear and possible faults through the fluid and sealing. Therefore, redundancy plays, e.g. a significant role for aircraft. The reliability of *electrical systems* can be improved by almost all measures. The reliability of *electronic hardware components* depends greatly on the manufacturing process, environmental conditions and internal load and failures appear suddenly and more randomly. Therefore, the reliability can especially be improved by redundancy and protection. To cope with systematic *software faults* usually only redundancy with diversity helps and maintenance to reduce bugs. Hence, the ways to improve reliability are different for the different types of components, see the summary in Table 4.1, compare [4.9]. This influences the design of safety-related systems considerably.

4.2 Event tree analysis (ETA) and fault tree analysis (FTA)

The *event tree analysis* (ETA) begins with the event of a component (a basic fault) and progresses this through all the components in normal or fail operational mode

Table 4.1. Improvement of reliability for different components, compare [4.9]. ++ very large potential, + large potential, 0 small potential, not usable

improvement of reliability	components				
	mechanical	hydraulic	electrical	electronic hardware	software
oversizing	++	+	+	+	0
maintenance	++	++	+	0	+
protection	++	++	+	++	0
wear reduction	++	+	+	0	0
redundancy	0	+	+	++	+
• static	0	+	+	++	0
• dynamic	0	0	+	++	+
• diversity	0	0	0	++	++

and determines the consequence of the system's function, i.e. either active or inactive, Figure 4.1. Therefore, the possible expected (normal) and failed conditions of all connected components have to be considered, including logic AND and OR operations. As each event causes a branch in the diagram, a tree with N events will have 2^N branches. Therefore, the ETA results in very large trees because the normal as well as the failed functions are considered. Note that only binary states (yes or no) are taken into account.

The *fault tree analysis* (FTA) proceeds in the reverse direction of ETA. It begins with the failure of the system (as top event) and determines the possible causes in the components respective basic components failures including logic operations, Figure 4.2. As only the failed events are taken into account, the tree becomes smaller than for ETA. Also here only binary states are considered.

4.3 Failure mode and effects analysis (FMEA)

The FMEA is a formalized method to consider all components, their functions, failure modes and causal system failures. The method was developed in 1960s and used by NASA for the Apollo project, later for aerospace and nuclear power stations, [4.1]. Now it is a standard method also in the automotive industry [4.14], [4.3].

FMEA starts with listing all the components, their operating modes, and their failure modes. It then considers possible causes for each failure mode and describes their effects for the unit under consideration and for the complete system. Furtheron, counter actions are listed. Usually only single failures are considered. Because FMEA worksheets are used, it is a formalized method, Table 4.2. The method is used to detect weak spots of the design in early and later stages of development.

The procedure results in a tree-like network structure with binary states, similar as an event tree, however, without normal operating modes and without logic interconnection. Therefore, it does not blow up like ETA. The strength of FMEA is its completeness. However, it may result in a very time-consuming procedure.

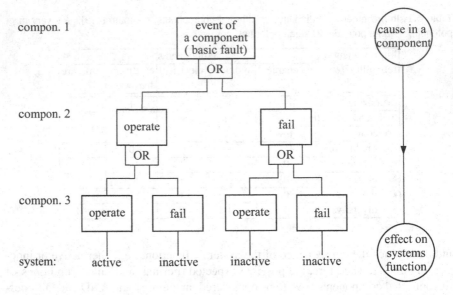

Fig. 4.1. Event tree analysis (ETA) for three components in series connection (inductive analysis from special cause to general system effects)

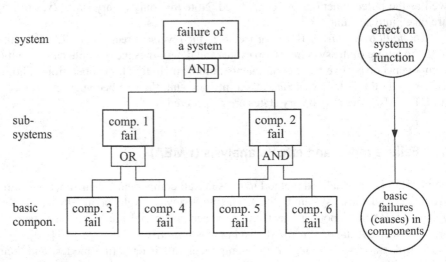

Fig. 4.2. Fault tree analysis (FTA) for six components with parallel connections for 1, 2 and 5, 6 and serial connection for 3, 5 (deductive analysis from special effects to causes)

Table 4.2. FMEA worksheet

Component	Failure mode	Failure causes	Failure effect on unit	Failure effect on system	Counter action

FMEA can favorably be combined with FTA, because it yields the possible system failures, which are the inputs of FTA. Therefore, FMEA and FTA complement each other.

Failure modes, effects and criticality analysis (FMECA) is an extension of FMEA. The importance of each component failure is taken into account by considering its probability of occurrence and effects on the system, i.e. by expressing the risk of operation.

A Failure Risk Priority Number (FRPN) can be stated to show the criticality of different failures

$$FRPN = A \times B \times E \qquad (4.1)$$

where

- A: probability of failure occurrence,
- B: effects on the system,
- E: detection rate.

For A, B, E ranking numbers between 1 and 10 are used, as shown in Table 4.3 and 4.4 , [4.7], [4.11], [4.14].

Table 4.3. Ranking numbers: A for the evaluation of the possibility of failure occurrence for passenger cars, [4.14]. The failure rate is calculated with the assumption of 300 operating hours per year

Ranking number A	Evaluation	Failures/year ν [ppm]	Failure rate $\lambda[\frac{1}{h}]$
10	very high	500000	$1.67 \cdot 10^{-1}$
9		100000	$0.33 \cdot 10^{-1}$
8	high	50000	$1.67 \cdot 10^{-2}$
7		10000	$0.33 \cdot 10^{-2}$
6		5000	$1.67 \cdot 10^{-3}$
5	medium	2000	$0.67 \cdot 10^{-3}$
4		1000	$0.33 \cdot 10^{-3}$
3	low	100	$0.33 \cdot 10^{-4}$
2		50	$1.67 \cdot 10^{-5}$
1	very low	1	$0.33 \cdot 10^{-6}$

Table 4.4. Ranking number B for the evaluation of the effects of failures for vehicles

Ranking number B	Fundamental effects	Effects for systems	Effects on occupants	Effects for vehicles
10, 9	catastrophic	system damage	death	destruction, fire
8, 7	major	system shut down	serious injury	serious damage
6, 5	severe	subsystem down	injury	failure
4, 3	minor	partly down	slight injury	minor failure
2, 1	no effect	no effect	no injury	no effect

4.4 Hazard-analysis (HA)

Hazards are undesirable system conditions with the potential to cause or to contribute to a damage or accident. Hazard-analysis is therefore a basic procedure to provide awareness and information of the systems safety-critical components and states, [4.6].

Based on the FMEA all safety-critical failures are extracted with the goal to identify hazards (potential sources of harm, i.e. physical injury or damage), their effects on the system and ways to minimize or avoid them. This can be represented in a similar worksheet like the one for FMEA. The results are given in binary state, i.e. yes or no. An example is given for an electromechanical brake in [4.4].

Once hazards are identified, their causes can be analyzed by proceeding with a fault tree analysis (FTA) starting with the hazard as top event.

A more detailed analysis is the *Hazard and Operatability Studies* (HAZOP) developed in the 1960s in the chemical industry, [4.13]. Here, the effects of deviations from normal operating conditions are investigated in a systematic way. For example, parametric changes and out-of-range values are considered if they could result in a hazard. Safety features to compensate hazards are also taken into account. HAZOP starts with identifying the interconnection of components, looks at the flow of materials or signals and defines attributes by using a limited number of guide words like "no, more, less, reverse, early, late". This means that not only binary values are considered but ranges of values in the sense of a stepwise classification. By analyzing causes and consequences for the system, possible hazards are identified. The results can then be stated in a fault tree.

It is typical for these reliability and safety analysis procedures that they are compiled by a team of engineers in consecutive sessions with expertise covering the whole system. These analysis procedures generally have to be repeated at several times during the project development.

4.5 Risk classification

As hazards represent situations of potential danger they may lead to an *accident*, which is an unintended event causing injury, death, environmental or material damage. In order to judge the relative importance of hazards and their possible accept-

Table 4.5. Accident probability ranges for military systems, [4.13]

Accident frequency	Occurrences during operational life considering all instances of the system
frequent	likely to be continually experienced
probable	likely to occur often
occasional	likely to occur several times
remote	likely to occur some time
improbable	unlikely, but may exceptionally occur
incredible	extremely unlikely that the event will occur at all

ability the associated risk has to be considered. *Risk* is herewith determined by the combination of the *probability* (frequency) and its *severity* (consequence) of a hazard. The risk can be classified by using either qualitative or quantitative methods, [4.13].

The probability of hazards is, e.g. classified in six levels, Table 4.5, [4.2]. Table 4.6 shows probability numbers for aircraft systems. The severity is subdivided into four classes from catastrophic to negligible. The risk of hazards is also classified into four classes, from intolerable to negligible risk, Table 4.7. The *qualitative risk classification* then follows by considering both, the probability and severity, Table 4.8.

Table 4.6. Aircraft systems hazard probabilities, [4.10], [4.13]

System criticality	Catagory of effect	Effect on aircraft	Qualitative probability term		Occurence per flight hour	Flight hours per occurence	Flight years per occurence
critical	loss of aircraft	catastropic	extremely improbable	extremely improbable	10^{-10}	10^{10}	
					10^{-9}	10^{9}	10^{5}
essential	large safety reduction	hazardous	improbable	extremely remote	10^{-8}	10^{8}	
					10^{-7}	10^{7}	
	significant safety reduction	major		remote	10^{-6}	10^{6}	10^{2}
non-essential	operating limitations emergency procedure			resonably probable	10^{-5}	10^{5}	
			probable		10^{-4}	10^{4}	
		minor			10^{-3}	10^{3}	10^{-1}
	normal or nuisance			frequent	10^{-2}	10^{2}	
					10^{-1}	10^{1}	
					1	1	

Quantitative risk measures can, e.g. be obtained by calculating a hazard risk number

Table 4.7. Interpretation of risk classes, [4.2]

Risk class	Interpretation
class I	intolerable risk
class II	undesirable risk, and tolerable only if risk reduction is impractical or if the costs are greatly disproportional to the improvement gained
class III	tolerable risk if the cost of the reduction would exceed the improvement gained
class IV	negligible risk

Table 4.8. Risk classification of hazards or accidents, [4.2]

Frequency	Consequence			
	catastrophic	critical	marginal	negligible
frequent	I	I	I	II
probable	I	I	II	III
occasional	I	II	III	III
remote	II	III	III	IV
improbable	III	III	IV	IV
incredible	IV	IV	IV	IV

$$R = F_H \times C \tag{4.2}$$

where F_H is the frequency (probability) of the hazard and C the consequence (severity), [4.2]. If the risk has to be reduced this can be accomplished by reducing both risk parameters, the probability and severity of hazards. The assignment of risk measures depends a lot on the technical area, like, e.g. nuclear, aircraft or heating and ventilating systems. Therefore, there exist industry specific standards.

Based on the probability of dangerous failures (high, constant value of C) the IEC has proposed four *safety integrity levels* (SIL) for electronic programmable systems, Table 4.9 Safety integrity is to be seen as a measure of the likelihood of the safety system correctly performing its tasks.

Table 4.9. Safety integrity levels (SIL) for safety related electronic programmable systems and dangerous failures

Safety integrity level (SIL)	Failure probability per hour	Operating hours per failure	Operating years per failure
4	$10^{-9}...10^{-8}$	$10^8...10^9$	$10^4...10^5$
3	$10^{-8}...10^{-7}$	$10^7...10^8$	$10^3...10^4$
2	$10^{-7}...10^{-6}$	$10^6...10^7$	$10^2...10^3$
1	$10^{-6}...10^{-5}$	$10^5...10^6$	$10...100$

If the effect of a safety-critical failure depends on the operational state, the risk number can be modified by the frequency of the operational state F_{OP}

$$R_{OP} = F_H \times F_{OP} \times C \tag{4.3}$$

see, e.g. [4.8], [4.9]. This applies, e.g. for vehicles with operational states like acceleration, cruising with high or low speed, braking with engine or brake, cornering with or without acceleration or deceleration.

4.6 Integrated reliability and safety design

The existing methods for analyzing the reliability and safety can now be combined appropriately. Figure 4.3 shows an overall scheme. The FMEA identifies all components, failures, causes and effects. The single failures proceed to a FTA to determine the causes and their logic interconnection on a component level. The failure causes are then used to design the overall reliability. Remaining failures which cannot be avoided are then classified and determine the maintenance procedure.

Based on the FMEA the hazard analysis extracts safety critical failures. Their presentation in a (reduced) fault tree determines the causes with logic interconnections, [4.12], i.e. dangerous faults leading to hazards. Based on this, the safety system at lower levels can be designed. Remaining dangerous failures then undergo a risk classification and determine the supervision and safety methods at higher levels to reduce the risk to an acceptable measure.

In addition, ways of fault tolerance can be implemented at component and unit level to improve both, reliability and safety. Figure 4.3 indicates the integrated reliability and safety procedure *during the design and testing phases*.

The unavoidable failures have to be covered by maintenance and on-line supervision and safety methods *during operation*, including fault tolerance, protection and supervision with fault detection and diagnosis and appropriate safety actions. These methods are discussed in the following chapters.

4.7 Problems

1) What are the steps for achieving a system with high safety integrity? Which steps have to be performed during design and which ones during operation?
2) Which methods are known to investigate the effect of faults?
3) Give some examples for systematic faults and random faults.
4) What kind of faults are typical for mechanical and electronic components and for software?
5) Draw an event tree and a fault tree for an electromagnetic valve and a DC motor.
6) What are the differences between an event tree and a FMEA worksheet?
7) What are the differences between reliability analysis methods and hazard analysis?
8) Hazard risk numbers R_{op} according to (4.3) have to be calculated for an aircraft with an automatic control system and a passenger car with a steer-by-wire control system and the assumption that a total breakdown of the control systems causes a major effect. The assumed parameters are:

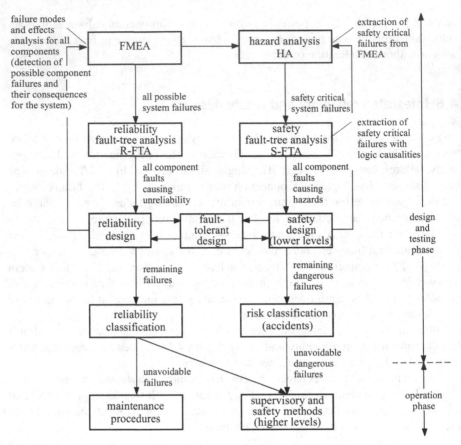

Fig. 4.3. Integrated design procedures for system reliability and safety to result in high system integrity, [4.5]

- one passenger aircraft with $F_H = 10^{-8}$ [1/h]; $C = 200$ people; $F_{op} = 5000$ h/year;
- one passenger car with $F_H = 10^{-6}$ [1/h]; $C = 4$ people; $F_{op} = 300$ h/year.

What safety integrity levels (SIL) are recommended for one passenger aircraft and one passenger car if the number of injured people is 1 in 100 years or 1000 years?

9) How does the hazard risk number R_{op} of Problem 8 change if total fleets of aircraft and cars are considered with fleet size $n_{aircraft} = 10^3$; $n_{cars} = 10^6$?

Part II

Fault-Detection Methods

5

Process Models and Fault Modelling

Model-based methods of fault detection use the relations between several measured variables to extract information on possible changes caused by faults. These relations are mostly analytical relations in form of process model equations but can also be causalities in form of, e.g. if-then rules. Figure 5.1 shows a general scheme for process model-based fault detection. The relations between the measured input signals $\mathbf{U}$ and output signals $\mathbf{Y}$ are represented by a mathematical process model. Fault-detection methods then extract special features, like parameters θ, state variables $\mathbf{x}$ or residuals $\mathbf{r}$. By comparing these observed features with their nominal values, applying methods of change detection, analytical symptoms $\mathbf{s}$ are generated.

These symptoms are the basis for fault diagnosis. For the application of model-based fault-detection methods the process configurations according to Figure 5.2 have to be distinguished. With regard to the inherent dependencies used for fault detection and the possibilities for distinguishing between different faults, the situation improves from a to b or c by the availability of more measurements, as will be shown later. The applied process models can be classified according to

- continuous models;
- discrete-event models;

both in

- continuous time;
- discrete time.

The continuous models are, in general, equation-based with further subclasses as linear, nonlinear, time-variant. Discrete event models are, e.g. finite state machines, functional diagrams or Petri-nets.

In the following some basic continuous models are considered. It is of special importance for the development of fault-detection methods how faults can be represented within these process models.

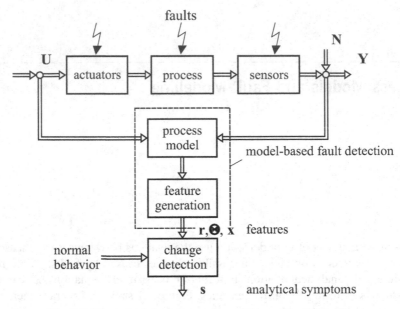

Fig. 5.1. General scheme of process model-based fault detection

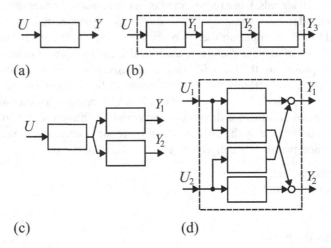

Fig. 5.2. Process configuration for model-based fault detection: (a) SISO (single-input single-output); (b) SISO with intermediate measurements; (c) SIMO (single-input multi-output); (d) MIMO (multi-input multi-output)

5.1 Fault models

A suitable modelling of faults is important for the right functioning of fault-detection methods. A realistic approach presupposes the understanding between the real physical faults and their effect on the mathematical process models. This can usually only be provided by the inspection of the considered real process, the understanding of the physics and a fault-symptom-tree analysis. There are many reasons for the appearance of faults. They stem for examples form

1) wrong design, wrong assembling;
2) wrong operation, missing maintenance;
3) ageing, corrosion, wear during normal operation.

With regard to the operation phase they may be already present or they may appear suddenly with a small or large size or in steps or gradually like a drift. They can be considered as *deterministic faults*. Especially disadvantageous are usually intermittent faults which appear as *stochastic faults*.

5.1.1 Basic fault models

A fault was defined in Chapter 2 as an unpermitted deviation of at least one characteristic property, called feature, from an usual condition. The feature can be any physical quantity. If the quantity is part of a physical law $Y(t) = g[U(t), \mathbf{x}(t), \boldsymbol{\theta}]$ in the form of an equation, and measurements of $U(t)$ and $Y(t)$ are available, the feature expresses itself either as input variable $U(t)$, output variable $Y(t)$, state variable $x_i(t)$ (time-dependent function) or parameter θ_i (usually constant value). Hence, faults may appear as changes of signals or parameters. The time dependency of faults may show up as, compare Figure 5.3,

- abrupt fault (stepwise);
- incipient fault (drift-like);
- intermittent fault (with interrupts).

With regard to the corresponding signal flow diagrams, see Figure 5.4, the changes of signals are *additive faults*, because a variable $Y_u(t)$ is changed by an addition of $f(t)$

$$Y(t) = Y_u(t) + f(t) \tag{5.1}$$

and the changes of parameters are *multiplicative faults*, because another variable $U(t)$ is multiplied by $f(t)$

$$Y(t) = (a + \Delta a(t))U(t) = aU(t) + \Delta a(t)U(t)$$
$$= Y_u(t) + f(t)U(t) \tag{5.2}$$

For the additive fault the detectable change $\Delta Y(t)$ of the variable is independent on any other signal

$$\Delta Y(t) = f(t) \tag{5.3}$$

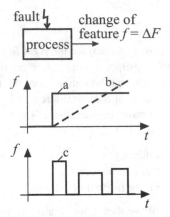

Fig. 5.3. Time dependency of faults: (a) abrupt; (b) incipient; (c) intermittent

(Instead of the output signal $Y(t)$, the input signal $U(t)$ or a state variable $x_i(t)$ can be influenced).

However, for the multiplicative fault, the detectable change of the output $\Delta Y(t)$ depends on the input signal $U(t)$

$$\Delta Y(t) = f(t)U(t) \qquad (5.4)$$

This means, if the signal $Y(t)$ can be measured, the additive fault is detectable for any $Y_u(t)$ but the multiplicative fault can only be detected if $U(t) \neq 0$. The size of the change $\Delta Y(t)$ then depends on the size of $U(t)$.

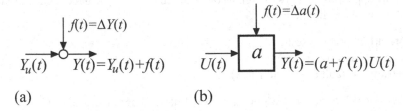

Fig. 5.4. Basic fault models: (a) additive fault for an output signal; (b) multiplicative fault

5.1.2 Examples for fault models

Because the kind of faults and their modelling depends primarily on the actual process, some typical examples are considered.

Example 5.1: Sensor faults

Sensors and measurement systems are dynamic transfer elements for which only the output $Y(t)$, the measurement variable is accessible. Without additional calibration equipment, the real physical input $Y_0(t)$ is unknown. The static behavior of a sensor may be linear

$$Y(t) = c_0 + c_1 Y_0(t)$$

or nonlinear

$$Y(t) = c_0 + c_1 Y_0(t) + c_2 Y_0^2(t) + \dots$$

The dynamic behavior of a sensor can for small changes frequently be approximated by a linear model with transfer function

$$G_s(s) = \frac{\Delta Y(s)}{\Delta Y_0(s)} = \frac{y(s)}{y_0(s)} = \frac{B_s(s)}{A_s(s)}$$

and in simple cases by a first order lag

$$G_s(s) = \frac{K_s}{1 + T_1 s}$$

with the gain

$$K_s = c_1 + 2c_2 + \dots$$

The sensor output is now usually influenced by several kind of disturbances. According to [5.19], one can distinguish between external and internal disturbances, see Figure 5.5.

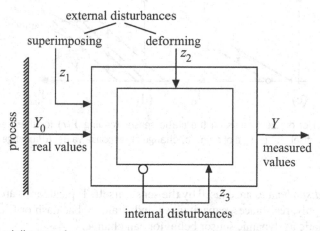

Fig. 5.5. Blockdiagram of a sensor or measurement equipment influenced by disturbances

External disturbances are generated by the sensor environment. Frequently, superimposed disturbances $z_1(t)$ arise, e.g. by induced electromagnetic influence

$$Y(t) = c_1[y_0(t) + z_1(t)] + c_0$$

The size of the sensor fault

$$\Delta Y(t) = c_1 z_1(t)$$

is independent on the true value Y_0, see Figure 5.6a. Environmental changes $z_2(t)$ like temperature, fluid flow velocity, contamination deform the transfer behavior by changing the gain and time constant

$$[T_1 + \Delta T_1(z_2)]\dot{y}(t) + y(t) = [K_s + \Delta K_s(z_2)]y_0(t)$$

This leads to a static and dynamic deviation

$$\Delta y(t) = -\Delta T_1(z_2)\dot{y}(t) + \Delta K_s(z_2)y_0(t)$$

and results in parametric faults.

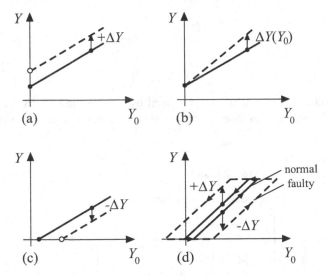

Fig. 5.6. Effect of different faults on the static sensor reading $Y(t)$ for the measured value $Y_0(t)$: (a) zero offset; (b) change of gain; (c) change of response value; (d) change of hysteresis

Internal disturbances are caused by the sensor itself. Typical faults are changes in the power supply, resistances, capacitances, inductances, backlash or friction. Then, as well the static as dynamic sensor behavior can change.

Figure 5.6 summarizes the impact of external and internal disturbances on the sensor output signals in the form of *static characteristics*. Mainly three different sensor faults can be distinguished:

a) constant offset ΔY
b) change of gain ΔK_s resulting in value-dependent offset $\Delta Y(Y_0)$
c) direction-dependent offset (hysteresis) $\Delta Y(\text{sign } \dot{Y}_0)$

Hence, the cases a and b can be modelled as additive faults, c as direction dependent additive fault.

Some faults result not only in a deterministic change but change also the stochastic character of the sensor output. Therefore, also the mean value and variance of the output are fault models

$$E\{Y(t)\} \text{ and } \sigma_y^2 = E\left\{\left[Y(t) - \overline{Y}\right]^2\right\}$$

A change of the dynamic behavior of the sensors, e.g. by a time constant change ΔT_1, has to be modelled together with the process and treated as a multiplicative fault.

□

Example 5.2: Actuator faults

An electromagnetic proportional acting flow valve is considered, see Figure 5.7. The static behavior depends on the characteristics of the current $I = f_1(U)$, the position $z = f_2(I)$, the valve area $A_F = f_3(z)$ and the mass flow $\dot{m} = f_4(A_F)$. The overall static behavior then results from $\dot{m} = f(U)$. Assuming a polynomial characteristic, the overall function can be approximated, e.g. by

$$\dot{m} = c_0 + c_1 U + c_2 U^2$$

assuming that the pressure p_1 does not change, $p_1 = $ const.

Additive faults are those who arise in a parallel shift of the characteristic, which is lumped in the constant c_0. The corresponding physical faults are, e.g. offset in U, change of spring counter position or spring pretension, change of flow area A_F in zero position $A_F(z = 0)$. Dry (Coulomb) friction can be modelled as direction dependent offset of the input

$$\Delta U = U_{co} \text{ sign } \dot{U}$$

and backlash as

$$\Delta U = U_{blo} \text{ sign } U$$

see [5.11].

Multiplicative faults are contained in the parameters c_1 and c_2. Examples are: change of flux in coils (e.g. resistance of coil wires), air gap in electromagnet, friction at shaft or magnet, gain of power electronics, change of valve piston geometry, change of pressure difference $\Delta p = p_1 - p_2$. Friction of the shaft or magnet or backlash can also be modelled as direction dependent parameter. (Corresponding equations for the components follow, e.g. from [5.11]). Hence, many faults belong to the class of multiplicative faults.

□

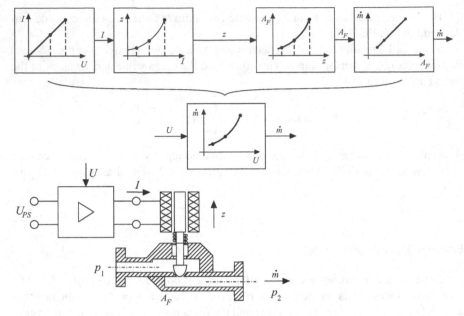

Fig. 5.7. Electromagnetic flow valve

U_{PS}	voltage power supply	z	position (height) of the valve
I	coil current	A_F	cross sectional area of valve opening
U	manipulated variable	$\dot{m}$	mass flow
p_1, p_2	fluid pressure before and after valve		

Example 5.3: Electrical cable connection

An electrical cable connection is considered with total resistance $(R_1 + R_2)$ which supplies a consumer resistance R_4, see Figure 5.8. A fault now happens in form of a shortcut with resistance R_3 between the resistances R_1 and R_2, resulting in a shortcut current $I_3(t) = f_1(t)$. This shortcut can now be modelled as changing resistance $R_3 \in (0, \infty)$.

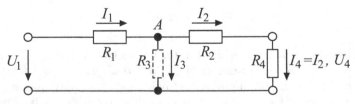

Fig. 5.8. Electrical cable connection with consumer R_4 and shortcut fault through R_3

a) Measurement of $I_1(t)$ and $I_4(t)$

If the fault is interpreted as a changing resistance $R_3 = f_R$, then from Kirchhoff's node and mesh law follows

$$I_4(t) = \frac{1}{1 + \frac{1}{f_R}(R_2 + R_4)} I_1(t) = g_1(f_R)I_1(t)$$

and the fault becomes a parametric fault because the "gain" $g_1(f_R)$ changes. If, however, only the current balance equation (mesh law)

$$I_4(t) = I_1(t) - f_I(t)$$

is applied, the shortcut-current sums to appear as additive fault. But then the offset $f_I(t)$ depends on the applied input current $I_1(t)$, as

$$f_I(t) = I_4(t) - I_1(t) = [g_1(f_R) - 1]I_1(t)$$

This means that the size of the "additive" fault depends on the input, which is not in agreement with its definition. Only for constant operation point $\bar{I}$, the shortcut can be considered as additive fault on $I_4(t)$.

b) Measurement of $U_1(t)$ and $I_4(t)$

By use of Kirchhoff's law, one obtains for the current without shortcut

$$I_4(t) = I_2(t) = \frac{1}{R_1 + R_2 + R_4} U_1(t) = \frac{1}{R_{tot1}} U_1(t)$$

If a shortcut with resistance $R_3 = f_R$ happens between R_1 and R_2, the consumer current becomes

$$I_4(t) = \frac{1}{(1 + \frac{R_1}{f_R})(R_2 + R_4) + R_1} U_1(t) = \frac{1}{R_{tot2}} U_1(t) = g_2(f_R)U_1(t)$$

Hence, also in this equation a shortcut expresses itself as a multiplicative fault. By parameter estimation R_{tot2} can be determined and the size of the fault f_R can be calculated if R_1, R_2 and R_4 are known. (For $f_R \to \infty$ there is no shortcut).

□

Example 5.4: Pipeline

A liquid flow through a pipeline, Figure 5.9, is considered. It is assumed that the flow at the input $\dot{m}_1$ and the output $\dot{m}_4$ and the pressure p_1 and p_4 relative to the absolute air pressure can be measured. A fault now happens in form of a leak with flow $\dot{m}_3(t) = f_m(t)$, compare [5.20], [5.9]. (According to the outflow equation of an orifice the leak flow $\dot{m}_3$ is proportional to the leak area A_F and $\sqrt{p_3}$).

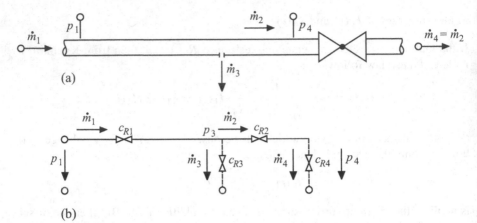

Fig. 5.9. Pipeline with a leak $\dot{m}_3$: (a) scheme with measurements; (b) scheme with analogy to the electrical circuit of Figure 5.8

a) Measurement of $\dot{m}_1(t)$ and $\dot{m}_4(t)$

The mass balance equation leads to

$$\dot{m}_4(t) = \dot{m}_1(t) - f_m(t)$$

Therefore, for a constant operation point $\dot{m}_1 = $ const., the leak flow shows up as an additive fault.

Now an interconnection of resistance can be made analog to the electrical circuit in Figure 5.9, assuming, simplifying, a laminar flow through the pipeline and its resistances and therefore a linear resistance law like

$$p_4 - p_1 = (c_{R1} + c_{R2})\dot{m}_2$$

in the case of no leak. If the leak hole in the pipeline has a resistance $c_{R3} = f_R$ one obtains analog to the electrical short cut

$$\dot{m}_4(t) = \frac{1}{1 + \frac{1}{f_R}(c_{R2} + c_{R4})}\dot{m}_1(t) = g_1(f_R)I_1(t)$$

The fault is therefore a multiplicative fault by considering arbitrary operating points $\bar{\dot{m}}_1$.

b) Measurement of $p_1(t)$ and $\dot{m}_4(t)$

Analog to the electrical shortcut-example, one obtains without leak

$$\dot{m}_4(t) = \frac{1}{c_{R1} + c_{R2} + c_{R4}}p_1(t) = \frac{1}{c_{RT1}}p_1(t)$$

and with a leak of resistance $c_{R3} = f_R$

$$\dot{m}_4(t) = \frac{1}{(1 + \frac{c_{R1}}{f_R})(c_{R2} + c_{R4}) + c_{R1}} p_1(t) = \frac{1}{c_{RT2}} p_1(t) = g_2(f_R) p_1(t)$$

Herewith, the leak appears as a multiplicative fault. As in the electrical example, the leak appears only with the balance equation and constant operation point as an additive fault and with the more explicit modelling in form of a resistance law for the leak hole and the whole operation range, as a multiplicative fault.

Another type of fault is *clogging*. Then the resistance, e.g. c_{R1} becomes larger and also c_{RT1}. Hence, also clogging is a multiplicative fault.

□

These examples show that the kind of fault, additive or multiplicative, depends on the used models and the physical nature of the fault. Several *sensor faults* can be modelled as additive faults. *Actuators* show more multiplicative than additive faults. *Processes* may show additive faults, if only balance equations are used. If, however, constitutive and phenomenological laws are applied to model processes in more detail, multiplicative faults appear frequently. In summary modelling of faults requires the consideration of the underlying physical effects.

5.2 Process models

5.2.1 Theoretical and experimental modelling

Mathematical models of dynamic processes are primarily obtained by either theoretical/physical modelling or experimentally by identification methods. For *theoretical modelling*, also called theoretical analysis or modelling by first principles, the model is set up on the basis of mathematically formulated laws of nature. For this, first the process elements are considered. By combining their models, one obtains models of subprocesses and overall processes. The theoretical modelling always begins with simplifying assumptions about the process, which simplifies the calculations or enables them at all with a tolerable expenditure. One can distinguish the following types of basic equations:

(1) balance equations for stored masses, energies and impulses;
(2) constitutive equations (physical-chemical state equations) of special elements;
(3) phenomenological equations, if irreversible processes (equalizing processes) take place (e.g. equations for thermal conduction, diffusion or chemical reaction);
(4) entropy balance equations, if several irreversible processes take place (if not already considered by (3);
(5) connection equations (describe the interconnection of the process elements).

For distributed parameter systems, the dependency on the space and time has to be considered. This usually leads to partial differential equations. If the space dependency is negligible, the systems can be considered with lumped parameters. These are described by ordinary differential equations as a function of time.

By summarizing the basic equations of all process elements, one receives a system of ordinary and/or partial differential equations of the process. This leads to a theoretical process model with a certain structure and certain parameters, if it can be solved explicitly. Frequently, this model is extensive and complicated, so it must be simplified for further applications.

The simplifications are made by linearization, reduction of the model order or approximation of systems with distributed parameters by lumped parameters when limiting on fixed locations. The first steps of these simplifications can be already made by simplifying assumptions while stating the basic equations.

But also if the set of equations cannot be solved explicitly, the individual equations supply important hints for the model structure. So, e.g. balance equations are always linear and some phenomenological equations are linear in wide areas. The constitutive equations often introduce nonlinear relations.

During experimental modelling, which is called identification, one obtains the mathematical model of a process from measurements. Here, one always proceeds from a priori knowledge, which was gained, e.g. from the theoretical analysis or from preceding measurements. Then, input and output signals are measured and evaluated by means of identification methods in such a way that the relation between the input and output signal is expressed in a mathematical model. The input signals can be naturally operating signals (occurring in the system) or artificially introduced test signals. Depending upon the application purpose, one can use identification methods for parametric or nonparametric models. The result of the identification then is an experimental model. A detailed description of the different techniques can be found, e.g. in [5.3], [5.13] and [5.16].

The theoretical and the experimental model can be compared, provided both types of modelling can be realized. If both models do not agree, then one can conclude from the type and size of the differences which particular steps of the theoretical or experimental modelling have to be corrected.

Theoretical and experimental modelling thus mutually complete themselves. The theoretical model contains the functional description between the physical data of the process and its parameters. Therefore, one will use this model, e.g. if the process is to be favorably designed with regard to dynamical behavior or if the process behavior has to be simulated before construction. The experimental model on the other hand, contains parameters as numeric values whose functional relation with the physical basic data of the process remains unknown. In many cases, the real dynamic behavior can be described more exactly or it can be determined at smaller expenditure by experimentally obtained models, which, e.g. is better suited to the adjustment of a feedback controller, the prediction of signals or for fault detection.

Theoretical models are also called "white-box models" and experimental models are called "black-box models". However, in practical cases frequently exist some types of models which are in between these two types, Figure 5.10. If, for example, the physical laws are known, but the parameters not and have to be determined experimentally, the resulting models can be called "light-grey models". If only physical if-then rules are known and the the parameters have to be determined by experiments "dark-grey models" result.

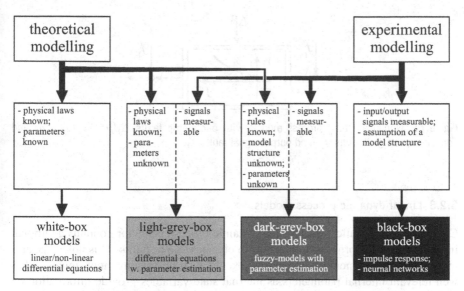

Fig. 5.10. Different kind of mathematical process models

The procedure of the theoretical modelling of technical processes is treated extensively in [5.11]. Despite the large variety of existing process elements, it is possible to reach a certain systematic. This is supported by many similarities and analogies between not only the mechanical and electrical, but also thermal, thermodynamic and chemical processes.

5.2.2 Static process models

The static behavior (steady-state) of processes is usually described by graphical represented characteristic curves, which are obtained either experimentally or by calculations from static analytical process models. In many cases they are expressed or approximated by polynomials like

$$Y = \beta_0 + \beta_1 U + \beta_2 U^2 + \ldots + \beta_q U^q$$
$$Y = \boldsymbol{\psi}_s^T \boldsymbol{\theta}_s$$
$$\boldsymbol{\theta}_s^T = [\beta_0 \beta_1 \beta_2 \ldots \beta_q]$$
$$\boldsymbol{\psi}_s^T = [1 \; U \; U^2 \ldots U^q] \tag{5.5}$$

Changes of β_0 are additive faults and changes of β_i $i = 1, \ldots, q$ are multiplicative faults. Input signal faults f_U and output signal faults f_Y are additive faults, compare Figure 5.11. (It holds here that $f_Y = \Delta\beta_0$). The parameters β_i frequently depend on different physical process parameters p_j, see Section 5.2.3 and examples, for, e.g. characteristics of flow valves or the circulation pump, Chapter 21.

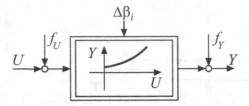

Fig. 5.11. Nonlinear static process models with parametric faults $\Delta\beta_i(i = 1,\ldots,q)$ and additive input signal faults f_U and output signal faults f_Y

5.2.3 Linear dynamic process models

Compared to the static behavior, the dynamic process behavior contains additional information on the process with regard to changes due to faults. This holds because mass, energy or momentum storages are excited by time-varying input signals and as well relevant internal parameters as internal state variables provide information on changes caused by faults, [5.8].

As introduction into this topic a spring-mass-damper-system example is considered.

Example 5.5: Spring-mass-damper-system

A mechanical oscillator with a serial connection of spring, mass and damper or parallel connection of spring and damper leads to the differential equation

$$m\ddot{z}(t) + d\dot{z}(t) + cz(t) = \Delta F_1(t)$$

where ΔF_1 is the excitation force at the mass and z is the deviation of the mass from a steady-state value.

The corresponding transfer function is obtained by Laplace-transformation

$$G_{zF}(s) = \frac{z(s)}{\Delta F_1(s)} = \frac{1}{ms^2 + ds + c} = \frac{\frac{1}{c}}{\frac{m}{c}s^2 + \frac{d}{c}s + 1}$$

$$= \frac{K}{\frac{1}{\omega_0^2}s^2 + \frac{2D}{\omega_0}s + 1} = \frac{K}{T_2^2s^2 + T_1s + 1} = \frac{K}{a_2s^2 + a_1s + 1}$$

with $K = 1/c; a_1 = d/c$ and $a_2 = m/c$. The static behavior reduces to

$$z = \frac{1}{c}\Delta F_1$$

and therefore only information on the spring constant c is obtained. However, for dynamic excitation the behavior is described by three parameters: m, c and d. Hence, the two parameters mass m and damping d (or natural frequency ω_0 and damping ratio D) give additional information on the process condition by considering the dynamic behavior.

□

a) Continuous-time dynamic process models

The general description of a dynamically excited process with lumped parameters follows by ordinary differential equations which generally are nonlinear. By considering small deviations around an operating point $(Y_{00}|U_{00})$ the input/output behavior can usually be simplified to a linear ordinary differential equation

$$y(t) + a_1 y^{(1)}(t) + \ldots + a_n y^{(n)}(t)$$
$$= b_0 u(t) + b_1 u^{(1)}(t) + \ldots + b_m u^{(m)}(t) \tag{5.6}$$

$$y(t) = Y(t) - Y_{00}; \quad u(t) = U(t) - U_{00} \tag{5.7}$$

where $y^{(n)}(t) = d^n y(t)/dt^n$ are derivatives.

This linear process model can also be written in vector form

$$y(t) = \boldsymbol{\psi}^T(t)\boldsymbol{\theta} \tag{5.8}$$
$$\boldsymbol{\theta}^T = [a_1 \ldots a_n \, b_0 \ldots b_m] \tag{5.9}$$
$$\boldsymbol{\psi}^T(t) = [-y^{(1)}(t) \ldots - y^{(n)} u(t) \ldots u^{(m)}(t)] \tag{5.10}$$

The corresponding transfer function becomes through Laplace transformation

$$G_p(s) = \frac{y(s)}{u(s)} = \frac{B(s)}{A(s)} = \frac{b_0 + b_1 s + \ldots + b_m s^m}{1 + a_1 s + \ldots + a_n s^n} \tag{5.11}$$

As shown in Figure 5.12 an input signal fault f_u and an output signal fault f_y are additive faults and the parameter faults Δa_i, Δb_j are multiplicative faults. Additive input faults f_u and output faults f_y lead to changes of the output signal

$$\Delta y(s) = \frac{B(s)}{A(s)} f_u(s) \tag{5.12}$$

$$\Delta y(s) = f_y(s) \tag{5.13}$$

compare the time histories in Figure 5.13. Herewith, in both cases deviations of the output result despite the fact that the input $U(t)$ is constant. To discuss the influence of parameter faults, first an example is considered.

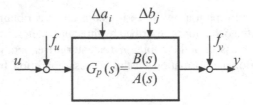

Fig. 5.12. Linearized dynamic process model with parametric faults $\Delta a_i, \Delta b_j$ and additive faults f_u and f_y

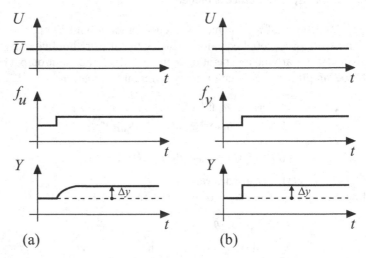

Fig. 5.13. Responses of a first order system with constant input $\bar{U}$ =const. to stepwise: (a) additive input change $f_u(t)$; (b) additive output change $f_y(t)$

Example 5.6: Time-varying parameters of a first order process

Given the linear first order process with constant parameters

$$\bar{a}_1 \dot{Y}(t) + Y(t) = \bar{b}_0 U(t) + \bar{b}_1 \dot{U}(t)$$

Setting the derivatives to zero yield the steady state

$$\bar{Y} = \bar{b}_0 \bar{U}$$

Now changes of the signals are introduced

$$Y(t) = \bar{Y} + \Delta Y(t); \quad U(t) = \bar{U} + \Delta U(t)$$

leading to

$$\bar{a}_1 \Delta \dot{Y}(t) + \Delta Y(t) = \bar{b}_0 \, \Delta U(t) + \bar{b}_1 \Delta \dot{U}(t)$$

or

$$\bar{a}_1 \dot{y}(t) + y(t) = \bar{b}_0 u(t) + \bar{b}_1 \dot{u}(t)$$

where $y(t) = \Delta Y(t)$ and $u(t) = \Delta U(t)$. The influence of parameter changes on the output

$$y(t) = -a_1\dot{y}(t) + b_0 u(t) + b_1\dot{u}(t)$$

is

$$\Delta y(t) = \frac{\partial y(t)}{\partial a_1}\Delta a_1 = -\dot{y}(t)\Delta a_1 - a_1\Delta\dot{y}(t)$$

$$\Delta y(t) = \frac{\partial y(t)}{\partial b_0}\Delta b_0 = -a_1\frac{\partial\dot{y}(t)}{\partial b_0}\Delta b_0 + u(t)\Delta b_0$$

$$\Delta y(t) = \frac{\partial y(t)}{\partial b_1}\Delta b_1 = -a_1\frac{\partial\dot{y}(t)}{\partial b_1}\Delta b_1 + \dot{u}(t)\Delta b_1$$

Hence, step changes of Δa_1 result in no change of the output if $u(t)$ and $y(t)$ are in steady state. Δb_0 changes the gain and therefore a lasting deviation $\Delta y(t)$ occurs. Δb_1 is a change of the lead time and does not result in an output change if $u = 0$, compare Figure 5.14. However, the situation changes if an input change $u(t) = \Delta U(t)$ occurs. Then passing deviations are observed for parameter changes Δa_1 and Δb_1 as shown in Figure 5.15.

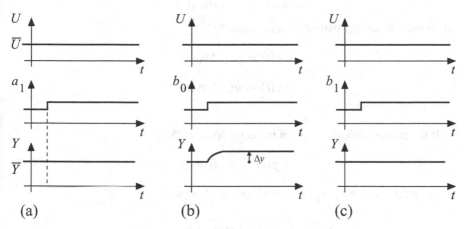

Fig. 5.14. Responses of a first order system to stepwise parameter changes for constant input $U = \bar{U} =$const.

This example shows that parameter changes of the gain can be detected from observing the output signal in steady state, but that the input signal must change to observe deviations of other parameters, like Δa_1 and Δb_1.

If the parameters change only slowly with regard to the process dynamics the above equations can be simplified. It holds, for example,

$$\Delta y(t) = -\dot{y}(t)\Delta a_1(t) + a_1\frac{\partial\Delta\dot{y}(t)}{\partial a_1}\Delta a_1(t)$$

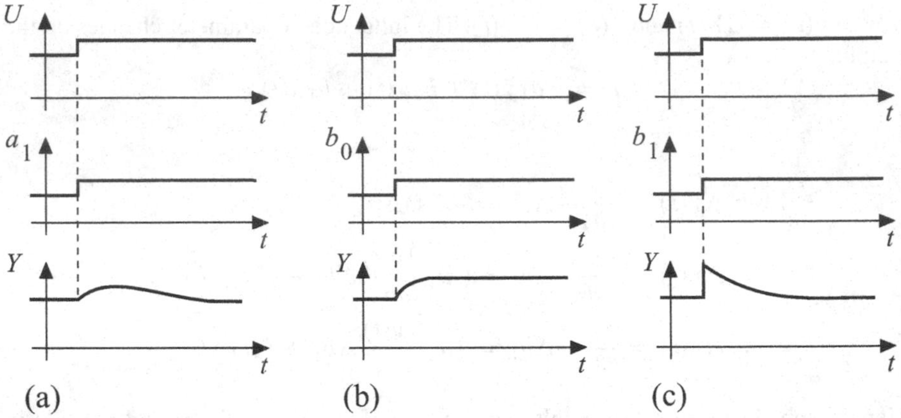

Fig. 5.15. Responses of a first order system to stepwise parameter changes and input changes $u(t) = \Delta U(t)$ at the same time instant

and for slowly changing $\Delta a_1(t)$ the term $\partial \Delta \dot{y}(t)/\partial a_1$ is small. Therefore the simpler equation

$$\Delta y(t) = -\dot{y}(t)\Delta a_1(t)$$

can be used as an approximation and similarly

$$\Delta y(t) = u(t)\Delta b_0(t)$$

$$\Delta y(t) = \dot{u}(t)\Delta b_1(t)$$

$\square$

If the process model is written in vector form as (5.8)

$$y(t) = \boldsymbol{\psi}^T(t)\boldsymbol{\theta} \tag{5.14}$$

then small parameter changes θ_i in the process lead to

$$\Delta y(t) = \frac{\partial}{\partial \theta_i}\left[\boldsymbol{\psi}^T(t)\boldsymbol{\theta}\right]\Delta\theta_i$$
$$= \frac{\partial \boldsymbol{\psi}^T(t)}{\partial \theta_i}\boldsymbol{\theta}\,\Delta\theta_i + \boldsymbol{\psi}^T(t)\Delta\theta_i \tag{5.15}$$

and for slow parameter changes, this simplifies to

$$\Delta y(t) = \boldsymbol{\psi}^T(t)\Delta\boldsymbol{\theta} \tag{5.16}$$

The differential equation (5.6) can be transformed into a state-space representation by defining a state vector $\mathbf{x}(t)$

$$\dot{\mathbf{x}}(t) = \mathbf{A}\mathbf{x}(t) + \mathbf{b}\,u(t) \tag{5.17}$$

$$y(t) = \mathbf{c}^T \mathbf{x}(t) \qquad (5.18)$$

see Figure 5.16. Additive faults of the state variable model are usually modelled as input or state faults f_l or output faults f_m. Parameter faults $\Delta b_j, \Delta a_i$ or Δc_j are multiplicative faults. Additive faults f_l and output faults f_m change the state space model according to

$$\dot{\mathbf{x}}(t) = \mathbf{A}\mathbf{x}(t) + \mathbf{b}u(t) + \mathbf{l}\, f_l(t)$$
$$y(t) = \mathbf{c}^T \mathbf{x}(t) + \mathbf{m}\, f_m(t) \qquad (5.19)$$

and parametric faults with slow changes result in

$$\dot{\mathbf{x}}(t) = (\mathbf{A} + \Delta\mathbf{A})x(t) + (\mathbf{b} + \Delta\mathbf{b})u(t)$$
$$y(t) = (\mathbf{c} + \Delta\mathbf{c})^T \mathbf{x}(t) \qquad (5.20)$$

Similar representations hold for nonlinear differential equations and nonlinear state

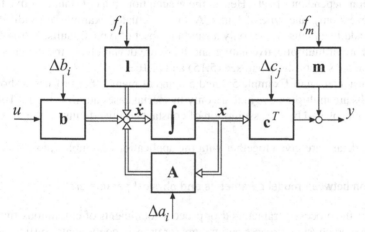

Fig. 5.16. Linearized dynamic process model with single-input single-output (SISO), additive state faults f_l and output faults f_m, and parametric faults $\Delta b_i, \Delta a_i$ and Δc_j

space models, and for multi-input and multi-output processes, also in discrete time.

On the basis of these basic static and dynamic process models and fault models, different methods of model-based fault detection can be developed. This is treated in the following chapters.

b) Discrete-time dynamic process models

If the process signals are sampled with sampling time T_0 and discrete time $k = t/T_0 = 0, 1, 2, \ldots$ the continuous-time differential equations can be discretized, if T_0 is small, or the sampled data can be approximated by amplitude modulated δ-functions for larger T_0. In both cases difference equations result

$$y(k) + a_1 y(k-1) + \ldots + a_m y(k-m)$$
$$= b_0 u(k) + b_1 u(k-1) + \ldots + b_m u(k-m) \qquad (5.21)$$

which, by applying the z-transform ($z = e^{T_0 s}$), leads to the z-transfer function

$$G_p(z) = \frac{y(z)}{u(z)} = \frac{B(z)}{A(z)} = \frac{b_0 + b_1 z^{-1} + \ldots + b_m z^{-m}}{1 + a_1 z^{-1} + \ldots + a_n z^{-n}} \qquad (5.22)$$

$G_p(z)$ may also include a holding element. For more details see, e.g. [5.7], [5.1].

The relation between the parameters of $G(z)$ and $G(s)$ are only simple for orders $n = 1$ or 2. For higher orders they become complicated.

Remarks on the modelling of faults
The additive and multiplicative faults in Figures 5.11, 5.12 and 5.16 are just ide-alistic assumptions in which way real faults enter into the model. Example 5.1 has shown that sensor faults can be modelled as constant offsets, value-dependent offsets or direction-dependent offsets. Hence, the assumption of a constant additive fault f_y or f_m covers only one type of faults. Also the actuator example 5.2 indicates that constant additive faults f_U are only a smaller subset of typical faults. Most of actua-tor faults are of multiplicative nature and therefore deviations in the actuator output depend on the size of the input, see (5.15) and (5.16).

A consideration of Example 5.3 and 5.4 and of many other processes shows that most faults are multiplicative and are only in special cases additive faults. The faults are then not covered by the assumption of constant additive faults on the measurable signals.

More details are given together with the applications examples in Part V.

c) Relation between model parameters and physical parameters

Generally, the process parameters θ or process coefficients of continuous-time mod-els, depend on physical process parameters (or process coefficients) $\mathbf{p}$ (like stiffness, damping factor, resistance, capacitance)

$$\theta = f(\mathbf{p}) \qquad (5.23)$$

via nonlinear algebraic relationships, see, e.g. Example 5.5. If the inversion of this relationship

$$\mathbf{p} = f^{-1}(\theta) \qquad (5.24)$$

exists, the changes Δp_i of the process parameters based on model parameter esti-mates $\hat{\theta}$ respectively $\Delta \hat{\theta}_j$ can be calculated. This enables then to localize the faults better, if they influence only specific physical parameters p_i, and therefore easies the fault diagnosis. However, certain conditions must be met for a unique inversion of (5.23), [5.18], [5.10], [5.12]. The corresponding *identifiability condition* for the physical process parameters is given in Chapter 23.4. It states that if the parameters estimates θ are related with products of physical parameters $\mathbf{z}$

$$\theta = \mathbf{C}\mathbf{z} \qquad\qquad (5.25)$$

where

$$z_\mu = \prod_{\mu=1}^{q} c_{i\mu} z_\mu$$

the implicit functional theorem for

$$\mathbf{q} = \theta - \mathbf{C}\mathbf{z} = 0$$

requires that the functional determinant must satisfy

$$det \; \mathbf{Q}_p = det \frac{\partial \mathbf{q}^T}{\partial \mathbf{p}} \neq 0. \qquad\qquad (5.26)$$

5.2.4 Nonlinear process models

Many processes show a nonlinear static and dynamic behavior, especially if wide areas of operation are considered. In general, the nonlinear models follow directly from theoretical modelling by applying basic physical laws as balance equations, constitutive equations and phenomenological laws, see, e.g. [5.15], [5.22], [5.14], [5.21], [5.2], [5.4] and [5.11]. Based on these so-called first principles, the nonlinear structure of the models arise and it can be seen if the models can be used directly for developing the fault-detection methods described for linear systems. As a multitude of nonlinear models exist, it is not possible to describe them here. However, some typical models are summarized in the following, especially those that have shown to be good approximators for general nonlinear behavior or well suitable for process identification. They will be given in the form of discrete-time systems.

a) Classical nonlinear dynamic models

Classical approaches for the treatment of nonlinear dynamic systems are frequently based on polynomial approximators. One distinguishes between general approaches, e.g. Volterra-series or Kolmogorov-Gabor polynomials, and approaches that involve special structure assumptions such as Hammerstein, Wiener or nonlinear difference equation (NDE) models, [5.3], [5.5], see also [5.13], [5.13] and Section 9.3.2

Another class are *bilinear systems*. They contain a multiplication of the input signal $U(k)$ with the state variables $\mathbf{x}(k)$ in the form

$$\mathbf{x}(k+1) = \mathbf{A}\,\mathbf{x}(k) = \mathbf{B}\,\mathbf{u}(k) + \sum_{i=1}^{P} \mathbf{A}_i \, U_i(k)\, \mathbf{x}(k) \qquad\qquad (5.27)$$

see, e.g. [5.17], [5.23]. An externally excited DC motor is an example for a bilinear system because the excitation current I_F is multiplied with the external state variables, speed ω and armature current I_A, [5.6].

b) Artificial neural networks

For a general applicable modelling of nonlinear systems are those methods of interest that do not require specific knowledge on the process structure. Artificial neural networks fulfil this requirement and are described in Chapter 9.

5.3 Problems

1) What type of faults can be distinguished with regard to the behavior depending on time?
2) What are the differences of additive and multiplicative faults? Take a sensor as an example.
3) What kind of faults are typical for temperature, pressure and flow sensors?
4) What type of faults can be modelled for contamination of a thermocouple implemented in the kernel of a high temperature fluid flow or in the wall of the pipe?
5) What type of faults are typical for pneumatic diaphragm actuators?
6) How can leaks in an oil pipeline be modelled by using mass balance equations or resistance laws?
7) Consider an electrical circuit with parameters R, L and C. Which faults in these parameters can be detected based on dynamic or static models?
8) Consider a second order mechanical system as in Example 5.5. Draw the signals for the input ΔF_1 and the output Δz for
   a) additive stepwise faults of the input and output variable and constant input signal;
   b) parametric faults of c and d with and without stepwise input excitation.
9) What are the advantages and disadvantages of continuous-time or discrete-time models for model-based fault-detection methods?
10) Consider an electrical RC-circuit. Which if of the parameters R and C can be determined by parameter estimation with two measured signals $U_{in}(t)$ and $U_{out}(t)$ or $I(t)$?

Signal models

Many processes are characterized by their oscillating or cyclic time behavior. This holds, for example, for rotating machines or alternating currents. The resulting signals are then *periodic signals* or contain periodic parts. Random processes like acoustic noise, driving over road surfaces, turbulence flow, on-off switching of many consumers in electrical or water networks result in *stochastic signals*. Both signal types can be used for fault detection if changes in their models are caused by faults in the processes. Therefore, the generation of these signals and some basic signal models are considered.

6.1 Harmonic oscillations

For undamped periodic signals with cycle duration T_p, the following expression is generally valid

$$y(t) = y\left(t + T_p\right) \tag{6.1}$$

6.1.1 Single oscillations

A harmonic steady state oscillation is described by a phase-shifted sine function

$$y(t) = y_0 \sin\left(2\pi f_0 t + \varphi\right) = y_0 \sin\left(\omega_0 t + \varphi\right) \tag{6.2}$$

with amplitude y_0, frequency $f_0 = 1/T_p$, angular frequency $\omega_0 = 2\pi f_0$ and phase angle φ. A damped harmonic oscillation is denoted by

$$y(t) = y_0 e^{-\delta t} \sin\left(\omega_0 t + \varphi\right) \tag{6.3}$$

with the damping constant δ.

Examples of the formation of such oscillations with mechanical systems have been shown in [6.6]. In the following, the combination of different harmonic oscillations and their models in the time domain are considered.

6.1.2 Superposition

The simplest form of combined oscillations results from the superposition (addition)

$$y(t) = \sum_{\nu=1}^{m} y_{0\nu}\, e^{-\delta_\nu t}\, \sin\,(\omega_\nu t + \varphi_\nu) \tag{6.4}$$

The superposition of two undamped oscillations with the angular frequencies 1 and 2 yields the amplitude spectrum shown in Figure 6.1

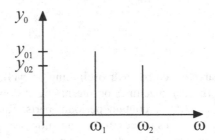

Fig. 6.1. Amplitude spectrum for the superposition of two oscillations

6.1.3 Amplitude modulation

An amplitude-modulated oscillation is obtained if the amplitude y_{01} of the carrier signal with angular frequency ω_1 is altered by a second oscillation, the modulation oscillation, with amplitude y_{02} and angular frequency ω_2. This results in the multiplicative operation

$$y(t) = y_1(t)y_2(t) = y_{01}\,[y_{02}\sin\,(\omega_2 t + \varphi_2)]\sin\,(\omega_1 t + \varphi_1) \tag{6.5}$$

Using the trigonometric relation

$$\sin\alpha\sin\beta = \frac{1}{2}\,[\cos\,(\alpha - \beta) - \cos\,(\alpha + \beta)] \tag{6.6}$$

one obtains

$$y(t) = \frac{1}{2}y_{01}y_{02}\,[\cos\,((\omega_1 - \omega_2)\,t + \varphi_1 - \varphi_2) - \cos\,((\omega_1 + \omega_2)\,t + \varphi_1 + \varphi_2)] \tag{6.7}$$

Thus, two oscillation components of the same amplitude with the difference and sum frequency appear as shown in Figure 6.2

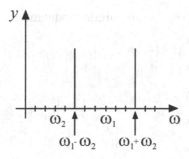

Fig. 6.2. Amplitude spectrum with amplitude modulation

6.1.4 Frequency and phase modulation

A frequency modulation of the carrier oscillation is obtained by

$$y(t) = y_{01} \sin \left[\omega_1 \left(y_{02} \sin \left(\omega_2 t + \varphi_2 \right) \right) t + \varphi_1 \right] \qquad (6.8)$$

and a modulation of the phase angle by

$$y(t) = y_{01} \sin \left(\omega_1 t + y_{02} \sin \left(\omega_2 t + \varphi_2 \right) + \varphi_1 \right) \qquad (6.9)$$

These modulations are particularly used in communication engineering, since the useful information is contained in the frequency and phase of the carrier signal. In this case, disturbances of the amplitude y_{01} have practically no influence on the reconstruction of the useful signal in a receiver (demodulation).

Example 6.1:

Figure 6.3 shows the results of the amplitude, phase and frequency modulation for a signal composed of two undamped partial oscillations.

$\square$

6.1.5 Beating (Libration)

Now, the superposition of two oscillations, with angular frequencies ω_1 and ω_2 with a minor difference of $\Delta \omega = \omega_2 - \omega_1$ but the same amplitudes, is considered

$$y_1(t) = y_0 \sin \left(\omega_1 t + \varphi_1 \right)$$
$$y_2(t) = y_0 \sin \left[\left(\omega_1 + \Delta \omega \right) t + \varphi_2 \right]$$

Using the trigonometric relation

$$\sin \alpha + \sin \beta = 2 \sin \left(\frac{\alpha + \beta}{2} \right) \cdot \cos \left(\frac{\alpha - \beta}{2} \right)$$

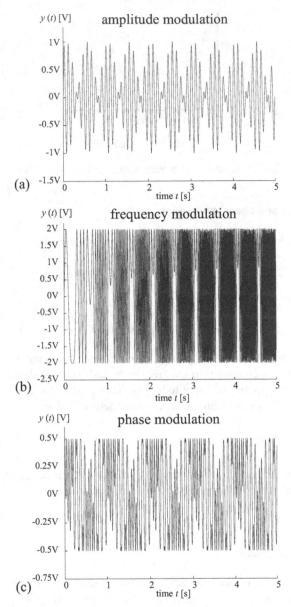

Fig. 6.3. Time course of an oscillation with: (a) amplitude modulation $y(t) = \sin(2\pi \cdot 10\,Hz \cdot t) \cdot \sin(2\pi \cdot 1\,Hz \cdot t + \pi/3)$ (b) frequency modulation $y(t) = 2\sin[(2\pi \cdot 10\,Hz \cdot t \cdot 0.5) \cdot \sin(2\pi \cdot 1\,Hz \cdot t + \pi/3)]$ (c) phase modulation $y(t) = 0.5\sin(2\pi \cdot 1\,Hz \cdot t + 2 \cdot \sin(2\pi \cdot 10\,Hz \cdot t) + \pi/3)$

one then obtains

$$y(t) = y_1(t) + y_2(t)$$
$$= y_0(t) \sin\left[\left(\omega_1 + \frac{\Delta\omega}{2}\right)t + \varphi\right]$$
$$= y_0(t) \sin\left[\frac{\omega_1 + \omega_2}{2}t + \varphi\right] \tag{6.10}$$

with

$$y_0(t) = 2\cos\left[\frac{\Delta\omega t - \varphi_1 + \varphi_2}{2}\right] = 2\cos\left[\frac{\omega_2 - \omega_1}{2} - \frac{\varphi_1 + \varphi_2}{2}\right]$$

$$\varphi = \frac{1}{2}(\varphi_1 + \varphi_2) \tag{6.11}$$

It yields a sinusoidal oscillation with an averaged frequency $(\omega_1 + \omega_2)/2$ and an amplitude $y_0(t)$ that changes co-sinusoidally with the half difference frequency $\Delta\omega/2$, a so-called *beating*. A superposition of oscillations with adjacent frequencies leads to an amplitude-modulated signal, whose carrier signal has the frequency $(\omega_1 + \omega_2)/2$ and whose modulation signal is the frequency $(\omega_2 - \omega_1)/2$.

Example 6.2:

The superposition of the oscillations

$$y_1(t) = \sin(2\pi \cdot 1\,\text{Hz})\,t$$
$$y_2(t) = \sin(2\pi \cdot 1.01\,\text{Hz})\,t$$

yields a beating with a frequency of $\Delta f/2 = 0.005\,Hz$, as shown in Figure 6.4

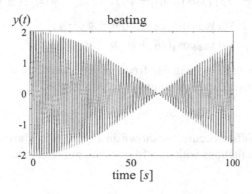

Fig. 6.4. Time course of a beating

6.1.6 Superposition and nonlinear characteristics

Now, the case of a signal $y(t)$ composed of two superimposed oscillations is considered

$$y(t) = y_1(t) + y_2(t)$$
$$y_1(t) = y_{01} \sin(\omega_1 t + \varphi_1)$$
$$y_2(t) = y_{02} \sin(\omega_2 t + \varphi_2) \tag{6.12}$$

This signal is fed into a succeeding nonlinear characteristic curve

$$z(y) = y^2 \tag{6.13}$$

Then,

$$\begin{aligned}
z(t) &= y_{01}^2 \sin^2(\omega_1 t + \varphi_1) + 2 y_{01} y_{02} \sin(\omega_1 t + \varphi_1) \sin(\omega_2 t + \varphi_2) \\
&\quad + y_{02}^2 \sin^2(\omega_2 t + \varphi_2) \\
&= y_{01}^2 \sin^2(\omega_1 t + \varphi_1) + y_{02}^2 \sin^2(\omega_2 t + \varphi_2) + 2 y_{01}[y_{02} \sin(\omega_2 t + \varphi_2)] \\
&\quad \sin(\omega_1 t + \varphi_1)
\end{aligned} \tag{6.14}$$

applies to the resulting signal. Thus, squared-sinusoidal oscillations for each fundamental frequency and an amplitude-modulated sinusoidal oscillation emerge. A further transformation by means of the trigonometric function

$$\sin^2 \alpha = \frac{1}{2}[1 - \cos 2\alpha]$$

yields

$$\begin{aligned}
z(t) &= \frac{1}{2}(y_0 1^2 + y_0 2^2) \\
&\quad - \frac{1}{2} y_{01}^2 \cos(2\omega_1 t + \varphi_1) - \frac{1}{2} y_{02}^2 \cos(2\omega_2 t + \varphi_2) \\
&\quad + y_{01} y_{02} \cos[(\omega_1 - \omega_2)t + \varphi_1 - \varphi_2] \\
&\quad - y_{01} y_{02} \cos[(\omega_1 + \omega_2)t + \varphi_1 + \varphi_2]
\end{aligned} \tag{6.15}$$

As two oscillations with the angular frequencies 1 and 2 pass through a squared nonlinear characteristic curve, the angular frequencies become

$$2\omega_1, 2\omega_2, \omega_1 - \omega_2, \omega_1 + \omega_2$$

and an additional offset occurs, as shown in Figure 6.5. Thus, nonlinear transfer elements lead to oscillations with new frequencies at the output.

6.2 Stochastic signals

The treatment of stochastic signals depends on the consideration in continuous time or discrete time. Therefore, both cases are briefly discussed.

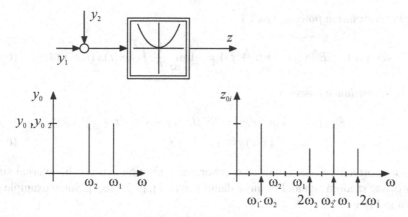

Fig. 6.5. Effect of a square characteristic curve on the amplitude spectrum of two superimposed oscillations

6.2.1 Continuous-time stochastic signals

The time history of stochastic signals is random and can therefore not be described precisely. With the aid of statistical methods, probability calculus and averaging some properties of these signals can be stated. Measurable stochastic signals are not completely random but show internal relations which can be expressed by mathematical signal models.

Because of the accidental character one signal source provides a family (ensemble) of random functions

$$\{x_1(t), x_2(t), \ldots, x_n(t)\}$$

This ensemble of signals represents a *stochastic process*. One random function is called a *sample function*.

This statistical treatment of stochastic signals uses probability density functions and results in simplified equations for *stationary signals* if the probability density functions become independent on time. Applying the ergodic hypothesis means that the same statistical information as for ensemble averaging can be obtained by averaging of one single sample function $x(t)$ for infinite long time intervals. Then the following averaged characteristic values of a stationary stochastic signal can be given

- *average value*

$$\bar{x} = E\{x(t)\} = \lim_{T \to \infty} \frac{1}{T} \int_0^T x(t) dt \qquad (6.16)$$

- *quadratic average value (variance)*

$$\sigma_x^2 = E\left\{(x(t) - \bar{x})^2\right\} = \lim_{T \to \infty} \frac{1}{T} \int_0^T (x(t) - \bar{x})^2 \, dt \qquad (6.17)$$

- *auto-correlation function* (ACF)

$$\phi_{xx}(\tau) = E\{x(t) - x(t+\tau)\} = \lim_{T\to\infty} \frac{1}{T} \int_0^T x(t)x(t+\tau)dt \qquad (6.18)$$

- *auto-covariance function*

$$R_{xx}(\tau) = \text{cov}\,[x,\tau] = E\{(x(t) - \bar{x})(x(t+\tau) - \bar{x})\}$$
$$= E\{x(t)x(t+\tau)\} - \bar{x}^2 \qquad (6.19)$$

The auto-correlation function or auto-covariance function express the internal similarity of the random signal. For more detail see [6.7], [6.2], [6.4]. Some example are shown in Figure 6.6.

The statistic relation between two different stochastic signals $x(t)$ and $y(t)$ is expressed by the

- *cross-correlation function* (CCF)

$$\phi_{xy}(\tau) = E\{x(t)y(t+\tau)\} = \lim_{T\to\infty} \frac{1}{T} \int_0^T x(t)y(t+\tau)dt$$
$$= \lim_{T\to\infty} \frac{1}{T} \int_0^T x(t-\tau)y(t)dt \qquad (6.20)$$

Through Fourier transformation of the auto-covariance function, one obtains in the frequency domain the

- *power density*

$$S_{xx}(i\omega) = \mathcal{F}\{R_{xx}(\tau)\} = \int_{-\infty}^{\infty} R_{xx}(\tau)e^{-i\omega\tau}d\tau$$
$$= 2\int_0^{\infty} R_{xx}(\tau)\,e^{-i\omega\tau}d\tau = 2\int_0^{\infty} R_{xx}(\tau)\cos\omega\tau d\tau \qquad (6.21)$$

A special stochastic signal is the *white noise*. It is totally statistically independent, resulting in the covariance function

$$R_{xx}(\tau) = S_0\delta(\tau) \qquad (6.22)$$

This signal has constant power density S_0 =const. for all frequencies and is not realizable. Its variance is infinite.

The auto-correlation function can also be applied for periodic signals and becomes then also periodic. Hence, it distinguishes significantly from stochastic signals. Therefore, correlation functions are very well suited for the separation of periodic and stochastic signals, [6.5].

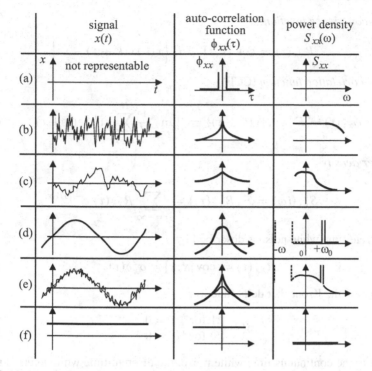

Fig. 6.6. Models of different stochastic and periodic signals: (a) white noise; (b) high-frequent noise; (c) low-frequent noise; (d) harmonic oscillation; (e) harmonic and stochastic noise; (f) DC value signal

6.2.2 Discrete-time stochastic signals

Discrete-time stochastic signals usually result from sampling of continuous time stochastic signals with sampling time T_0. Then the discrete time $k = t/T_0 = 0, 1, 2, ...$ is introduced. The signal model values follow directly from the previous once by replacing the integral through a sum.

- *average value*

$$\bar{x} = E\{x(k)\} = \lim_{N \to \infty} \frac{1}{N} \sum_{k=1}^{N} x(k) \qquad (6.23)$$

- *quadratic average value (variance)*

$$\sigma_x^2 = E\left\{x(k) - \bar{x}^2\right\} = \lim_{N \to \infty} \frac{1}{N} \sum_{k=1}^{N} [x(k) - \bar{x}]^2 \qquad (6.24)$$

- *auto-correlation function (ACF)*

$$\phi_{xx}(\tau) = E\{x(k)x(k+\tau)\} = \lim_{N \to \infty} \frac{1}{N} \sum_{k=0}^{N} x(k)x(k+\tau) \qquad (6.25)$$

- *auto-covariance function*

$$R_{xx}(\tau) = \text{cov}\,[x, \tau] = E\left\{x(k)x(k+\tau) - \bar{x}^2\right\} \qquad (6.26)$$

- *cross-correlation function* (CCF)

$$\phi_{xy}(\tau) = E\left\{x(k)y(k+\tau)\right\} = \lim_{N\to\infty} \frac{1}{N} \sum_{k=1}^{N} x(k)y(k+\tau) \qquad (6.27)$$

- *power density*

$$S_{xx}^*(i\omega) = \mathcal{F}\{R_{xx}(\tau)\} = \sum_{\tau=-\infty}^{\infty} R_{xx}(\tau)\,e^{-i\omega T_0 \tau} \qquad (6.28)$$

The discrete-time white noise is described by

$$R_{xx}(\tau) = \text{cov}\,[x, \tau] = \sigma_x^2 \delta(\tau) \qquad (6.29)$$

where $\delta(\tau)$ is the Kronecker delta function

$$\delta(\tau) \begin{cases} 1 \text{ for } \tau = 0 \\ 0 \text{ for } |\tau| = 0 \end{cases} \qquad (6.30)$$

Contrast to the continuous-time white noise, the discrete-time white noise is realizable and possesses a finite variance.

A further model representation of stochastic signals is possible by *stochastic difference equations*. They result by filtering of discrete-time white noise $v(k)$ by the difference equation

$$y(k) + c_1 y(k-1) + \ldots + c_n y(k-1)$$
$$= d_0 v(k) + d_1 v(k-1) + \ldots + d_m v(k-m) \qquad (6.31)$$

The signal $y(k)$ is herewith the output of a fictitious filter with the z-transfer function

$$G_F(z) = \frac{y(z)}{v(z)} = \frac{d_0 + d_1 z^{-1} + \ldots + d_m z^{-m}}{1 + c_1 z^{-1} + \ldots + c_n z^{-n}} = \frac{D(z)}{C(z)} \qquad (6.32)$$

and as input signal the white noise $v(z)$ with zero mean and variance $\sigma_v^2 = 1$. Specializations are the autoregressive process (AR) with

$$G_F(z) = \frac{d_0}{C(z)} \qquad (6.33)$$

and the moving average process (MA) with

$$G_F(z) = D(z) \qquad (6.34)$$

Therefore, (6.32) represents an ARMA process. For more details, see [6.3], [6.1], [6.5].

6.3 Problems

1) An acoustic harmonic signal with $f_1 = 1000$ Hz is amplitude-modulated with f_2 $= 50$ Hz. Determine the frequencies of the resulting oscillations.
2) The amplitude $y_{01} = 1$ of a harmonic signal with frequency $f_1 = 100$ Hz is modulated with $f_2 = 20$ Hz and amplitude $y_{02} = 0.5$. Determine the resulting frequencies and show them in a diagram for the amplitude spectrum.
3) The two engines of an aircraft run at 2500 and 2510 rpm. It is assumed that the six-cylinder four-stroke engines generate a noise with the ignition frequency. Which frequencies will be heard?
4) The electromagnetic force on the armature in a solenoid is proportional to the square of the magnetomotance and the current respectively. Which frequencies arise in the magnetic force for an alternative current of 50 Hz or 60 Hz?
5) State the differences between auto-correlation functions and auto-covariance functions.
6) Draw the autocorrelation for white noise in the case of continuous-time and discrete-time signals.
7) Which sensors can be used at the crankshaft-casing walls of a gasoline engine to detect knocking? The resulting oscillations are within 20 ... 30 kHz. Which methods can be used for knock detection? Write corresponding equations.

7

Fault detection with limit checking

The most simple and frequently used method for fault detection is the limit checking of a directly measured variable $Y(t)$. Herewith, the measured variables of a process are monitored and checked if their absolute values or trends exceed certain thresholds. A further possibility is to check their plausibility.

7.1 Limit checking of absolute values

Generally, two limit values, called thresholds, are preset, a maximal value Y_{max} and a minimal value Y_{min}. A normal state is when

$$Y_{min} < Y(t) < Y_{max} \tag{7.1}$$

which means that the process is in normal situation if the monitored variable stays within a certain tolerance zone. Exceeding of one of the thresholds then indicates a fault somewhere in the process, compare Figure 7.1. This simple method is applied in almost all process automation systems. Examples are the oil pressure (lower limit) or the coolant water (higher limit) of combustion engines, the pressure of the circulation fluid in refrigerators (lower limit) or the control error of a control loop. the thresholds are mostly selected based on experience and represent a compromise.

On one side false alarms through normal fluctuations of the variable should be avoided, on the other side faulty deviations should be detected early. Therefore a trade-off between too narrow and too wide thresholds exists.

7.2 Trend checking

A further simple possibility is to calculate the first derivative $\dot{Y} = dY(t)/dt$, the trend of the monitored variable and to check if

$$\dot{Y}_{min} < \dot{Y}(t) < \dot{Y}_{max} \tag{7.2}$$

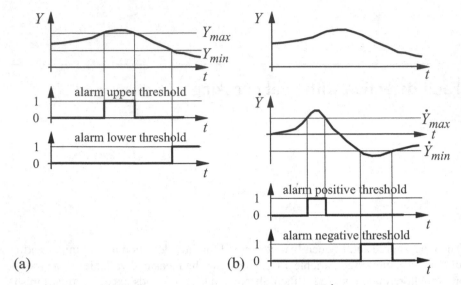

Fig. 7.1. Limit checking: (a) absolut value $Y(t)$; (b) trend $\dot{Y}(t) = dY(t)/dt$

If relatively small thresholds are selected, an alarm can be obtained earlier than for limit checking of the absolute value, see Figure 7.1b. Trend checking is, for example, applied for oil pressures and vibrations of oil bearings of turbines or for wear measures of machines.

Limit checking of absolute values and trends can also be combined. This requires, however, a mutual coordination of the thresholds. One possibility is to make the threshold of the absolute values dependent on the trend, e.g. $Y_{max} = f(\dot{Y} > 0)$ and $Y_{min} = f(\dot{Y} < 0)$ in order to detect fast developing deviations early and to avoid false alarms for small trends, if the value is far away from the threshold possibility to be exceeded, compare Figure 7.2, [7.17]. In some applications it is advantageous to make the thresholds adaptive, i.e. a function of other variables. For example, the thresholds for residuals are set in dependence of the input excitation of the process, see Section 7.5.

A further improvement of limit checking can be realized by applying *signal prediction*. By using polynomial regression models

$$Y(k) = a_0 + a_1 k + a_2 k^2 + \dots \qquad (7.3)$$

$k = t/T_0$ discrete time, and recursive least squares parameter estimation, a prediction of the signals can be made for N samples ahead. This is then possibly a better prediction than the trend only. The predicted signal then shows relatively early the danger of exceeding a threshold or avoids a false alarm if the signal returns to the normal zone without action. The prediction can also be made by using stochastic difference equations for the randomly changing monitored signal, [7.17], [7.19].

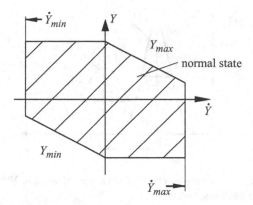

Fig. 7.2. Combination of limit checking for absolute values and trends. The thresholds Y_{max} and Y_{min} are a function of the trend $\dot{Y}$

7.3 Change detection with binary thresholds

7.3.1 Estimation of mean and variance

The monitored variables are usually stochastic variables $Y_i(t)$ with a certain probability density function $p(Y_i)$, mean value and variance

$$\mu_i = E\{Y_i(t)\}; \quad \bar{\sigma}_i^2 = E\left\{[Y_i(t) - \mu_i]^2\right\} \tag{7.4}$$

as nominal values for the non-faulty process. Changes are then expressed by

$$\Delta Y_i = E\{Y_i(t) - \mu_i\} \text{ and } \Delta\sigma^2 = E\left\{[\sigma_i(t) - \bar{\sigma}_i]^2\right\} \tag{7.5}$$

for $t > t_F$, where t_F is the time of fault occurrence, which is unknown.

If the mean and standard deviations before the change caused by a fault are described by μ_0 and σ_0 and after the change has appeared by μ_1 and σ_1, the change detection problem is depicted by Figure 7.3, assuming a normal probability distribution of the variable $Y(t)$. Then the following cases of changes can be distinguished:

1) the mean changes $\mu_1 = \mu_0 + \Delta\mu$; standard deviation $\sigma_1 = \sigma_0$ remains constant;
2) the mean does not change $\mu_1 = \mu_0$; standard deviation changes $\sigma_1 = \sigma_0 + \Delta\sigma$;
3) both, mean and standard deviation change.

As an example now case (1) is considered. If the probability densities do not significantly overlap, one can use a fixed threshold.

$$\Delta Y_{tol} = \kappa \, \sigma_0 \tag{7.6}$$

with, e.g. $\kappa \geq 2$, to detect the change just by observing the average $\mu(Y, t)$. In selecting the threshold, a comparison has to be made between the detection of relatively small changes and false alarms.

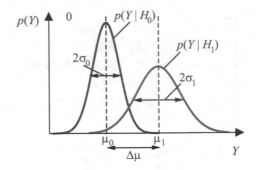

Fig. 7.3. Normal probability density functions of the observed variable Y for the nominal state (index 0) and changed (faulty) state (index 1)

However, the detection problem becomes more involved if the change of the mean

$$\Delta\mu = \mu_1 - \mu_0 \qquad (7.7)$$

is small compared to the standard deviation, say $\kappa \le 1$. Then *statistical tests* have to be applied.

The detection of changes of the random variable $Y(k)$ can be performed off-line or on-line in real-time. For *off-line change detection* within a sample length N it has to be tested where at some unknown time t_F a change in $Y(k)$ occurred from Y_0 to Y_1. This can only be made after storing all data. For fault detection in *real-time* the *on-line change detection* is of more interest. Here at every time k it has to be decided if a change from Y_0 to Y_1 has happened. This means that especially sequential or recursive tests are of interest for fault detection. The first case is easier to decide, because more measurements are available.

a) Statistics of the observation

The decision on a change of the observed variable can now be brought into a statistical context. This is of interest if small changes have to be detected in a noisy environment. It is assumed that the observed variable $Y(t)$ is a scalar function of time and is sampled with discrete time $t = kT_0$, where T_0 is the sampling time. It is further assumed that $Y(k)$ is a random variable with probability distribution $p(Y)$, for example, a Gaussian or normal distribution, see Figure 7.3,

$$p(Y) = \frac{1}{\sqrt{2\pi}\sigma_y}\, e^{-\frac{(Y-\mu_Y)^2}{2\sigma_Y^2}} \qquad (7.8)$$

with the mean (first moment)

$$\mu_Y = E\{Y(k)\} \qquad (7.9)$$

and variance (second or central moment)

$$\sigma_Y^2 = E\left\{(Y(k) - \mu_Y)^2\right\} \tag{7.10}$$

Then the probability that $Y(k)$ lies within certain areas is

$$\left. \begin{array}{l} P(\mu_Y - \sigma \leq Y \leq \mu_Y + \sigma) = 68.3\% \\ P(\mu_Y - 2\sigma \leq Y \leq \mu_Y + 2\sigma) = 95.4\% \end{array} \right\} \tag{7.11}$$

The considered variable can also be described by

$$Y(k) = \mu_Y + n(k) \tag{7.12}$$

where $n(k)$ is a zero mean random variable (noise) with $\sigma_n = \sigma_Y$.

b) Estimation of mean and variance

For the estimation of the mean and variance it can usually be assumed that the signal is ergodic. Therefore the estimates can be formulated for one sample function $Y(k)$ dependent on the time

$$\hat{\mu}_Y(N) = \frac{1}{N} \sum_{k=1}^{N} Y(k) \tag{7.13}$$

$$\hat{\sigma}_Y^2(N) = \frac{1}{N-1} \sum_{k=1}^{N} (Y(k) - \mu_Y)^2 \tag{7.14}$$

For on-line application *recursive forms* are of interest which are obtained by subtracting, e.g. $\hat{\mu}_Y(N)$ from $\hat{\mu}_Y(N + 1)$, resulting in

$$\hat{\mu}_Y(k) = \hat{\mu}_Y(k - 1) + \frac{1}{k} \left[Y(k) - \hat{\mu}_Y(k - 1) \right] \qquad (1 \leq k \leq N) \tag{7.15}$$

$$\begin{aligned} \hat{\sigma}_y^2(k) &= \hat{\sigma}_Y^2(k - 1) + \frac{1}{k} \left[[(Y(k) - \hat{\mu}_Y(k - 1)]^2 - \frac{k}{k-1} \hat{\sigma}_Y^2(k - 1) \right] \\ &= \frac{k-2}{k-1} \hat{\sigma}_Y^2(k - 1) + \frac{1}{k} [Y(k) - \hat{\mu}_Y(k - 1)]^2 \qquad (2 \leq k \leq N) \end{aligned} \tag{7.16}$$

A further possibility is to limit the averaging over a *time window* of length w. The mean then becomes

$$\hat{\mu}_Y(N) = \frac{1}{w} \sum_{k=N-w+1}^{N} Y(k) \tag{7.17}$$

and in recursive form

$$\hat{\mu}_Y(N) = \hat{\mu}_Y(k - 1) + \frac{1}{w} [Y(k) - Y(k - w)] \tag{7.18}$$

Correspondingly the window estimate of the variance yields

$$\hat{\sigma}_Y^2(N) = \frac{1}{w-1} \sum_{k=N-w+1}^{N} (Y(k) - \mu_Y)^2 \tag{7.19}$$

and recursively

$$\hat{\sigma}_Y^2(k) = \hat{\sigma}_Y^2(k-1) + \frac{1}{w-1} \left[\hat{\sigma}_Y^2(k) - \hat{\sigma}_Y^2(k-w) \right.$$
$$\left. -2\, \gamma(k) \tfrac{w-1}{w}\, \hat{\mu}_Y(k-1) - \tfrac{w-1}{w^2}\, \gamma^2(k) \right] \qquad (7.20)$$
$$\gamma(k) = \hat{\mu}(k) - \hat{\mu}_Y(k-w)$$

Still another way is averaging with *exponential forgetting* and forgetting factor $\lambda <$ 1, (e.g. $\lambda = 0.95$)

$$\hat{\mu}_Y(k) = \lambda\, \hat{\mu}_Y(k-1) + (1-\lambda)Y(k) \qquad (7.21)$$

which corresponds to a frozen, i.e. constant $k = k_1 = 1(1-\lambda)$ in (7.15), see [7.18]. This leads then to

$$\hat{\sigma}_Y^2(k) = \frac{2\lambda-1}{\lambda}\, \hat{\sigma}_Y^2(k-1) + (1-\lambda) \left[Y(k) - \hat{\mu}_Y(k-1) \right]^2 \qquad (7.22)$$

(7.21) corresponds to a discrete-time low pass filter of first order for determining $E\{Y(k)\}$

$$\hat{\mu}_Y(k) = \mu_{Yf}(k) = -a_1\, \mu_{Yf}(k-1) + b_0 Y(k)$$
$$\text{with } a_1 = -\lambda = e^{-T_0/T}\; ;\; b_0 = (1+a_1) \qquad (7.23)$$

This leads to a signal flow as in Figure 7.4. Then the variance can also be determined via

$$\hat{\sigma}_Y^2(k) = E\left\{ [Y(k) - \mu_Y(k)]^2 \right\}$$
$$= E\left\{ Y^2(k) \right\} - E\left\{ \mu_Y^2(k) \right\} \qquad (7.24)$$

where $E\left\{ Y^2(k) \right\}$ is also determined by a first order low pass filter

$$Y_f^2(k) = -a_1\, Y_f^2(k-1) + b_0\, Y^2(k) \qquad (7.25)$$

see [7.15], [7.8].

Also higher order low pass filters may be used, however, on cost of detection time.

If *different variables* $Y_i(k)$ have to be compared the variation coefficient can be applied, [7.6]

$$v_i(N) = \frac{\sigma_i(N)}{\mu_i(N)} \qquad (7.26)$$

7.3.2 Statistical tests for change detection

a) Hypothesis testing

For change detection *hypothesis tests* can be applied known from the theory of statistics, [7.23], [7.24], [7.4], [7.26], [7.12]. In hypothesis testing one tests a hypothesis H_0 against one or more alternate hypothesis $H_1, H_2, \ldots$ that are specified. For fault detection mainly of interest is the

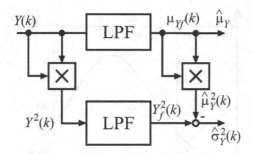

Fig. 7.4. Determination of mean and variance by low pass filters (LPF)

- null hypothesis (no fault)

$$H_0 : Y \in Y_0;$$ (7.27)

- change hypothesis (fault)

$$H_1 : Y \in Y_1$$ (7.28)

where Y_0 is the nominal value of the considered variable, which is assumed to be random. This means a decision has to be made as follows. Based on the assumption that the null hypothesis is true if no fault occurs, the null hypothesis is rejected and the alternate hypothesis H_1 is accepted, if the sample of the random variable Y falls outside the region of acceptance. Otherwise, H_0 is accepted and H_1 is rejected.

If the probability density function $p(Y)$ is known, the tasks of hypothesis testing is illustrated in Figure 7.5. Assuming that the true hypothesis H_0 is $Y = \mu_0$, the question is how much must the estimate $\hat{\mu}_0$ differ from μ_0 for rejecting the hypothesis. If the $\hat{\mu}_0$ can be any value for acceptance, the probability is

$$P(\hat{\mu}_0) = \int_{-\infty}^{\infty} p(\hat{Y}) \, dY = 1$$ (7.29)

If the region of acceptance is limited to $\mu > \mu_{\frac{\alpha}{2}}$ the probability of rejection is

$$P(\hat{\mu}_0 \leq \mu_{\frac{\alpha}{2}}) = \int_{-\infty}^{\mu_{\frac{\alpha}{2}}} p(\hat{Y}) \, dY = \frac{\alpha}{2}$$ (7.30)

and if acceptance is limited to $\mu < \mu_{1-\frac{\alpha}{2}}$ the probability of rejection is

$$P(\hat{\mu}_0 > \mu_{1-\frac{\alpha}{2}}) = \int_{\mu_{1-\frac{\alpha}{2}}}^{\infty} p(\hat{Y}) \, dY = \frac{\alpha}{2}$$ (7.31)

because of symmetry. Herewith α is called the level of significance, which is usually a small number $\alpha = 0.05$ or 0.01. Thus, the range of acceptance of the hypothesis is

- null hypothesis $H_0 : \mu_{\frac{\alpha}{2}} \leq \hat{\mu}_0 \leq \mu_{1-\frac{\alpha}{2}}$

and the range of rejection

- change hypothesis $H_1 : \hat{\mu}_0 \leq \mu_{\frac{\alpha}{2}}$ and $\hat{\mu}_0 \geq \mu_{1-\frac{\alpha}{2}}$

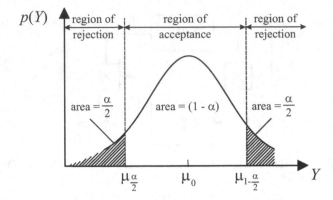

Fig. 7.5. Regions of rejection and acceptance for a symmetric hypothesis test of a random variable Y

The test described above is a two-sided test.

For hypothesis testing many different test methods were developed. Only some of them will be considered here. A first class assumes that the probability distribution of the random variable is normal and is called parametric tests. A second class does not make assumptions on the probability distribution and is called distribution-free or nonparametric tests, [7.12]. In the following mainly parametric tests will be considered, like t-test, likelihood and ratio tests. As a nonparametric test, the F-test is described.

b) Test quantities

The theory of statistical hypothesis testing distinguishes special test quantities, which themselves possess certain probability distributions. These are mainly only known for normal distribution $N(\mu, \sigma)$ of the investigated variables. Usually the mean μ and the standard deviation σ have to be tested. For testing of the mean the test quantity

$$t = \frac{\hat{\mu}(N) - \mu}{\hat{\sigma}/\sqrt{N}} = \frac{\frac{1}{N}\sum_{k=1}^{N} Y(k) - \mu_Y}{\hat{\sigma}_Y(N)} \tag{7.32}$$

is used, [7.14], [7.26], which is t-distributed with degree of freedom $f = N - 1$ (N is the number of variables).

Testing of the variance, the test quantity

$$\chi^2 = \frac{(N-1)\hat{\sigma}^2(N)}{\sigma^2} = \frac{1}{\sigma_Y^2}\sum_{k=1}^{N} (Y(k) - \mu_Y)^2 \tag{7.33}$$

is applied, which is χ^2 distributed, because $\sum \Delta Y^2(k)$ shows a χ^2 distribution with N degrees of freedom. If $\chi^2 > \chi^2_{N-1, 1-\alpha}$ the null hypothesis is rejected, i.e. the variance has increased with probability $(1 - \alpha)$.

Example 7.1: Testing of the mean

If a change of $\Delta\mu = \hat{\mu}(N) - \mu_0$ has to be detected for a variable $Y(k)$ with standard deviation $\hat{\sigma} = 1$ then $t \geq 1.96$ for large N and the significance level $\alpha = 0.05$ (probability 95%) has to be reached, which follows from a t-distribution table. The number of required samples N then yields from (7.32)

$$N \geq \left(\frac{\hat{\sigma}\, t_{\infty,\,0.05}}{\Delta\mu}\right)^2 = \left(\frac{\hat{\sigma}}{\Delta\mu}\right)^2 3.84$$

One obtains then for $\Delta\mu = 0.1$ $N \geq 384$, for $\Delta\mu = 0.2$ $N \geq 96$, for $\Delta\mu = 0.3$ $N \geq 43$ samples to detect the change with 95% probability. □

Example 7.2: Testing of the variance

The task consists of testing if the variance $\hat{\sigma}_1^2(N)$ compared to the previous value $\sigma_0^2 = 1$ has increased to 1.2. The χ^2 distribution table shows for the significance level $\alpha = 0.05$ (95 % probability) and $N = 5, 10, 30$

$$\chi^2(5) = 11.1; \quad \chi^2(10) = 18.3; \quad \chi^2(30) = 43.8$$

(7.33) yields

$$\chi^2(5) = \frac{4 \cdot 1.21^2}{1} = 5.76; \quad \chi^2(10) = 12.96; \quad \chi^2(30) = 41.8$$

For $N = 30$ is holds $\chi^2 > \chi^2_{30,\,0.05}$. This means about 30 samples are required to detect if the standard deviation σ_1 has increased by about 10 %. □

c) Detection of changes in the mean: t-test

The variable $Y(k)$ is assumed to have the mean Y_0 before the change and is observed with N_0 samples. After the change the mean is Y_1 and N_1 samples are available.

A classical test for the comparison of two mean values is the t-test. Here it is assumed that the probability distribution is normal and that the variances before and after the change are not known but equal. The test quantity is then, [7.4],

$$t = \frac{\hat{\mu}_{Y0} - \hat{\mu}_{Y1}}{\sqrt{(N_0 - 1)\hat{\sigma}_{Y0}^2 + (N_1 - 1)\hat{\sigma}_{Y1}^2}} \sqrt{\frac{N_0 N_1 (N_0 + N_1 - 2)}{N_0 + N_1}} \quad (7.34)$$

If $|t| > t_{\alpha,N_0+N_1-2}$ then a change has occurred, where t_{α,N_0+N_1-2} is taken from a t-distribution table for the significance level α.

If the number of new measurements N_1 is assumed to be small compared to previous measurements N_0, i.e. $N_1 \ll N_0$ then

$$t = \frac{\hat{\mu}_{Y0} - \hat{\mu}_{Y1}}{\hat{\sigma}_0^2 / \sqrt{N_1}} \tag{7.35}$$

which is identical with the test quantity (7.32).

d) Run-Sum tests for detection of changes in the mean

Especially in the area of quality control, relatively simple run tests were developed also known as control charts, see, e.g. [7.22], [7.13]. They test the change of the mean, provided the mean μ_0 and standard deviation σ_0 before the change are known and the standard deviation does not change, $\sigma_1 = \sigma_0$. A simple one-sided *run test* is, for example, if one value is larger than 3 times the standard deviation or if $n = 7$ values successively are above the mean, [7.9]. The *run sum tests* make use of the *cumulative sum* (CUSUM) of a random variable. Examples are:

$$RS_1(N) = \sum_{k=1}^{N} \left(\bar{Y}(k) - \mu_0 \right) \quad \text{deviation from a reference} \tag{7.36}$$

$$RS_2(N) = \sum_{k=1}^{N} \left(Y(k) - Y(k-1) \right) \quad \text{successive differences} \tag{7.37}$$

One of the run sum tests classifies the measured variable or estimate of the mean into deviation-bands as multiples v of the standard deviation, Table 7.1. If the observed variable, e.g. enters the band $\mu_0 + 2\sigma < \bar{Y} < \mu_0 + 2\sigma$ the score $sc_j = 2$ is assigned. The run sum is then the sum of the scores

$$RS_3(N) = \sum_{k=1}^{N} sc_j(k) \tag{7.38}$$

If a threshold RS_{th} is exceeded, a change is detected. A run is terminated when a value of Y falls in the opposite side of the mean value, i.e. the sign of the deviation from the mean changes.

e) Detection of changes in the variance: F-test

A further classical test for the comparison of two variances is the F-test. It is assumed that both samples are normally distributed. The mean values must not be known, they can even be different. The test quantity is

$$F(N_1, N_2) = \frac{\hat{\sigma}_{Y1}^2(N_1)}{\hat{\sigma}_{Y2}^2(N_2)} \tag{7.39}$$

Table 7.1. Assigned scores for a two-sided run-sum test

Deviation bands Y	Score sc_j
$> \mu_0 + 3\sigma$	$+3$
$> \mu_0 + 2\sigma$	$+2$
$> \mu_0 + 1\sigma$	$+1$
$> \mu_0 + 0\sigma$	$+0$
$< \mu_0 - 0\sigma$	-0
$< \mu_0 - 1\sigma$	-1
$< \mu_0 - 2\sigma$	-2
$< \mu_0 - 3\sigma$	-3

As F follows a F-distribution with (m_1, m_2) degrees of freedom ($m_1 = N_1 - 1$; $m_2 = N_2 - 1$), $F(m_1, m_2)$ is taken from a F-distribution table for a significance level α. If

$$F(N_1, N_2) > F(m_1, m_2, \alpha) \tag{7.40}$$

then $\hat{\sigma}_{Y1}^2 > \hat{\sigma}_{Y2}^2$ with probability $(1 - \alpha)$.

The statistical tests discussed so far are applicable if as well the measurements N_0 before the change and N_1 after the change are relatively large. However, large N_1 contradicts the early detection of changes. Therefore the classical statistical tests are in general not directly recommended for fast change detection in real-time.

f) Likelihood ratio test for jump detection

It is assumed that $Y(k)$ is statistically independent and that the type of probability distribution density $p_Y(Y)$ is known, e.g. normal distribution and that the parameters of this distribution density, e.g. μ_Y has to be estimated. $p_Y(\mu, \sigma)$ is now called a *likelihood function*, because it is a probability function in which the observations are regarded as fixed numbers and the parameters as the variables.

To test if the observed variable $Y(k)$ is more likely to be Y_0 or Y_1 one can assume two probability densities $p_Y(Y|H_0)$ and $p_Y(Y|H_1)$, compare Figure 7.3. Then one may determine the *likelihood-ratio*

$$\Lambda(Y) = \frac{p_Y(Y|H_1)}{p_Y(Y|H_0)} \tag{7.41}$$

in order to compare both probability densities.

If $\Lambda(Y) > \Lambda_{th}$ H_1 is decided, if $\Lambda(Y) < \Lambda_{th}$ H_0 is true. As the logarithm is a monotonic function also the log-likelihood ratio can be applied

$$ln\ \Lambda(Y) = ln\ p_{Y1}(Y) - ln\ p_{Y0}(Y) \tag{7.42}$$

Then computations become simpler and the decision for H_1 is just

$$ln\ \Lambda(Y) > ln\ \Lambda_{th} \quad \text{or} \quad ln\ \Lambda(Y) - ln\Lambda_{th} > 0 \tag{7.43}$$

i.e. a sign change determines of H_1 or H_0 is true. This is the *sequential probability ratio test* (SPRT), according to [7.25], [7.10]. This test can now be further evaluated for different assumptions about the probability density function, [7.11].

If the probability density is normal, the means μ_0 and μ_1 are known and it is assumed that the variance does not change, i.e. $\sigma_1^2 = \sigma_0^2 = \sigma^2$, then (7.42) leads to

$$
ln\,\Lambda(Y) = \frac{-(Y(k)-\mu_1)^2}{2\sigma^2} + \frac{(Y(k)-\mu_0)^2}{2\sigma^2} = \frac{\mu_1-\mu_0}{\sigma^2}\left[Y(k)-\frac{\mu_0+\mu_1}{2}\right]
$$
(7.44)

[7.1]. If the observations $Y(k)$ are statistically independent, the likelihood-ratio becomes after jump at time k'

$$
\Lambda(Y)|_{k'}^{N} = \prod_{k=k'}^{N} \frac{p_Y(Y(k)|H_1)}{p_Y(Y(k)|H_0)}
$$
(7.45)

Assumption of normal distribution leads with the known jump size $\Delta\mu = \mu_1 - \mu_0$ to

$$
\begin{aligned}
ln\,\Lambda(Y)|_{k'}^{N} &= \frac{\Delta\mu}{\sigma^2}\sum_{k=k'}^{N}\left[Y(k)-\mu_0-\frac{\Delta\mu}{2}\right]\\
&= \frac{1}{\sigma^2}S_{k'}^{N}(\mu,\Delta\mu)
\end{aligned}
$$
(7.46)

Therefore the test quantity

$$
S_i^N(\mu,\Delta\mu) = \Delta\mu\sum_{k=i}^{N}\left[Y(k)-\mu_0-\frac{\Delta\mu}{2}\right]
$$
(7.47)

can be computed for a time window of length $(N-i)$. With $s = S/\Delta\mu$ a recursive form becomes

$$
\begin{aligned}
s_i(k) &= s_i(k-1) + Y(k) - \mu_0 - \frac{\Delta\mu}{2}\\
s_i(k) &= s_i(k-1) + Y(k) - \frac{\mu_0-\mu_1}{2}
\end{aligned}
$$
(7.48)

This change detection algorithm becomes

$$
\begin{aligned}
s_i(k) &= s_i(k-1) = -\frac{\Delta\mu}{2}\ \text{for}\ Y(k) = \mu_0\\
s_i(k) &= s_i(k-1) = \frac{\Delta\mu}{2}\ \ \text{for}\ Y(k) = \mu_1
\end{aligned}
$$

The incremental quantity

$$
\Delta s_i(k) = s_i(k) - s_i(k-1)
$$
(7.49)

changes its sign after a jump has arisen in $Y(k)$. The described detection method is also known as Page-Hinkley stopping rule, [7.21] or cumulative sum algorithm, see also [7.3]. Note, that different cumulative sum algorithms exist, see Section d).

The assumption of a known jump size and direction is rather unrealistic. For unknown jump magnitude one can run two tests in parallel, assuming that μ_0 is known: a minimal jump magnitude can be assumed with two signs, i.e. $+\Delta\mu$ and $-\Delta\mu$, [7.2].

The statistical tests described in this section have shown that generally relative limiting assumptions had to be made, which are mostly not satisfied by experimentally generated residuals. In many cases these residuals are not statistical independent, not normal distributed, non-stationary and with unknown changes of the means and variance. Therefore one should first try the relatively simple tests from Section 7.3.1 and adaptive thresholds, (see Section 7.5) and if possible, to obtain strong deflections of the features by proper design of the fault detection method.

7.4 Change detection with fuzzy thresholds

In many cases the binary decision between "normal" and "faulty" is somewhat artificial, because there is seldom a sharp boarder between both states, Figure 7.6a. Therefore fuzzy thresholds are a more realistic alternative for change detection. The fluctuating variable $Y_i(t)$ can also be described by a fuzzy set $\mu_n(Y)$ for the normal state, see Figure 7.6b. (Here the usual variable μ is used for a membership function which should not be confounded with the mean value μ of the previous sections). If the fuzzy set is selected as a triangle, the center describes the mean and the lower width is $\Delta = \kappa\sigma_i$, $\kappa = 2, 3, \dots$ A membership function μ_{Y+} for "increased" is selected, to obtain a gradual measure for exceeding a threshold, [7.20]. Matching the changed fuzzy set $\mu'(Y)$ with the fuzzy threshold μ_{Y+} leads to the exceeding degree

$$\mu_Y = max_Y \left[\min\left(\mu'(Y), \mu_{Y+}(Y)\right)\right] \qquad (7.50)$$

As result then, e.g. $\mu_Y = 0.6$ is obtained which means that the threshold is reached to 60 %.

Depending on the selection of the thresholds membership functions, this gradual information gives more information on the severity of an unnormal state than binary thresholds.

7.5 Adaptive thresholds

Process model-based fault-detection methods described in the next chapters use process models which do not fully agree with real processes due to model uncertainties. Then, the generated residuals deviate from zero even without faults. These deviations depend then frequently on the amplitude and frequencies of the input excitation. Therefore the residuals may contain a static part which is proportional to the input $U(t)$ and a dynamic part dependent, e.g. on $\dot{U}(t)$. To cope with this problem, [7.16] has introduced an adaptive threshold which uses a first order high pass filter (HPF) for enlarging the threshold, Figure 7.7. A proportional enlargement may be added by a constant c_2, [7.8]. A low pass filter (LPF) is used to smooth the thresholds. The time constants T_1 and T_3 are selected according to the dominating time constant of the process. T_2/T_1 depends on the model uncertainty of the dynamics.

Figure 7.8 shows an example for the time behavior of an adaptive threshold. Adaptive thresholds were also proposed by [7.5], [7.7].

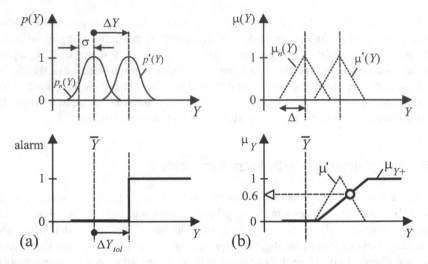

Fig. 7.6. Change detection for stochastic variables: (a) stochastic variables $Y(t)$ with probability density function $p(Y)$. $p_n(Y)$ is the normal state. $p'(Y)$ is the changed state. ΔY_{tol} marks the (binary) threshold; (b) stochastic variable $Y(t)$ as fuzzy set $\mu(Y)$. $\mu_n(Y)$ is the normal state. $\mu'(Y)$ is the changed state. μ_{Y+} is the fuzzy threshold for "increased" and $\mu_Y = 0.6$ is degree of exceeding the threshold "increased"

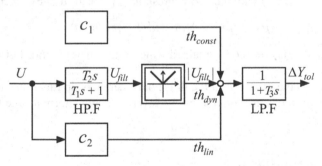

Fig. 7.7. Generation of an adaptive threshold dependent on process input excitation. The constant threshold is $th_{const} = c_1$

7.6 Plausibility checks

A rough supervision of measured variables is sometimes performed by checking the plausibility of its indicated values. This means that the measurements are evaluated with regard to credible, convincing values and their compatibility among each other. Therefore, a *single measurement* is examined whether the sign is correct and the value is within certain limits. This is also a limit checking, however, with usually wide tolerances. If *several measurements* are available for the same process then the measurements can be related to each other with regard to their normal ranges by using logic rules, like

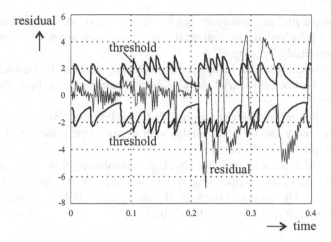

Fig. 7.8. Example for a residual r with adaptive threshold

$$\text{IF } [Y_{1min} < Y_1(t) < Y_{1max}] \text{ THEN } [Y_{2min} < Y_2(t) < Y_{2max}] \qquad (7.51)$$

For example, one expects for a circulation pump with rotating speed n and pressure p

$$\text{IF } [1000 \text{ rpm} < n < 3000 \text{ rpm}] \text{ THEN } [3 \text{ bar} < p < 8 \text{ bar}]$$

The plausibility check can also be made dependent on the operating condition, like

$$\text{IF [Operating condition 1] THEN } [Y_{3min} < Y_3(t) < Y_{3max}] \qquad (7.52)$$

One example is the oil pressure p_{oil} of a combustion engine with speed n and cooling water temperature ϑ_{H20}

$$\text{IF } [n < 1500 \text{ rpm}] \text{ AND } [\vartheta_{H20} < 50° \text{C}] \text{ THEN } [3 \text{ bar} < p_{oil} < 5 \text{ bar}] \quad (7.53)$$

Hence, plausibility checks may be formulated by using rules with binary logic connections like AND, OR. These rules and ranges of the measurements allow a rough description of the expected behavior of the process under normal conditions. If these rules are not satisfied either the process or the measurements are faulty. Then, one needs further testing to localize the fault and its cause.

These plausibility checks presuppose the ranges of measured process variables under certain operating conditions and represent rough process models. If the ranges of the variables are increasingly made smaller, many rules would be required to describe the process behavior. Then, it is better to use mathematical process models in form of equations to detect abnormalities. Therefore, plausibility tests can be seen as a first step towards model-based fault-detection methods.

7.7 Problems

1) What are the advantages in combining limit checking for absolute values and for trends?

2) The pressure $p(t)$ in a water network has to be supervised at the end of a branch. Write the equations for estimating the mean and variance with a time window and exponential forgetting.

3) What statistical tests can be applied for detecting changes in the mean and the variance of a stochastic variable $Y(k)$? Which parameters of the signals have to be known?

4) Which of the test methods are applicable for on-line real-time fault detection?

5) Compare binary and fuzzy thresholds with regard to the selection of thresholds and interpretation of the result.

6) State the rules for a plausibility check of a pneumatic flow valve if the measured variables are: manipulating air pressure p_{air}, valve position U_{valve} and flow $\dot{m}_{flow}$. The ranges of the variables are: $p_{air} = 0.2 \dots 1.0$ bar; $U_{valve} = 1\% \dots 100\%$; $\dot{m}_{flow} = 0.1 \dots 10 m^3/h$.

8

Fault detection with signal models

Many measured signals of processes show oscillations that are either of harmonic or stochastic nature, or both. If changes of these signals are related to faults in the actuators, the process and sensors, signal model-based fault-detection methods can be applied. Especially for machine vibration, the measurement of position, speed or acceleration allows to detect, for example, imbalance or bearing faults (turbo machines), knocking (gasoline engines) chattering (metal grinding machines). But also signals from many other sensors like electrical current, position, speed, force, flow and pressure, contain frequently oscillations with a variety of higher frequencies than the process dynamics.

The task of fault detection by the analysis of signal models is summarized in Figure 8.1. By assuming special mathematical models for the measured signal, see Chapter 6, suitable features are calculated as, for example, amplitudes, phases, spectrum frequencies and correlation functions for a certain frequency band width $\omega \leq \omega \leq \omega_{max}$ of the signal. A comparison with the observed features for normal behavior provide changes of the features which then are considered as *analytical symptoms*.

The signal models can be divided in nonparametric models, like frequency spectra or correlation functions, or parametric models, like amplitudes for distinct frequencies or ARMA type models. The following sections describe some signal-analysis methods for harmonic oscillations, stochastic signals and instationary signals, compare the survey presented in Figure 8.2. For more detail see the special books on digital signal analysis, like [8.46], [8.12], [8.40], [8.29], [8.23], [8.28], [8.55].

8.1 Analysis of periodic signals

It is now assumed that periodic signals $y(t)$ are superimposed on a steady state signal value Y_{00} such that the measured absolute signal is, see Section 6.1,

$$Y(t) = Y_{00} + y(t) \tag{8.1}$$

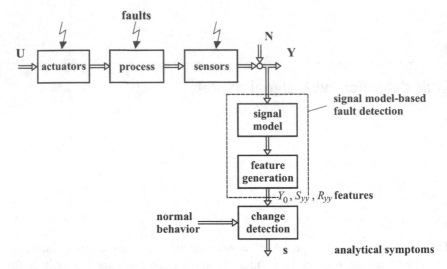

Fig. 8.1. Scheme for the fault detection with signal models

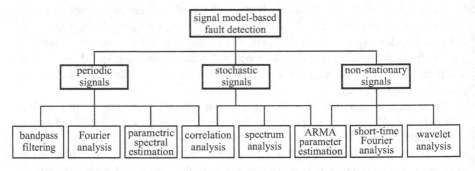

Fig. 8.2. Survey of signal-analysis methods for signal model-based fault detection

If the steady state value is removed, the signal $y(t)$ only has to be analyzed. Usually the periodic signal $y(t)$ is composed of a usable signal part $y_u(t)$ and a noise part $n(t)$

$$y(t) = y_u(t) + n(t) \tag{8.2}$$

The usable signal $y_u(t)$ contains the dynamic signal components which have to be analyzed. The noise $n(t)$ is assumed to have zero mean and is uncorrelated with $y_u(t)$.

According to the theory of Fourier series, each periodic signal can be described by the superposition of harmonic components

$$y_u(t) = \sum_{v=1}^{N} y_{0v}\, e^{-d_v t} \sin\left(\omega_v t + \varphi_v\right) \tag{8.3}$$

Each component is determined by the amplitude y_{0v}, the frequency ω_v, the phase angle φ_v and the damping factor d_v. These parameters have now be determined by a signal analysis method. In many cases it is sufficient for fault detection to determine ω_ν and $y_{0\nu}$.

8.1.1 Bandpass filtering

The classical method of obtaining the amplitudes of the harmonic components in dependence on the frequency, i.e. the *frequency spectrum*, it to pass the signal through a number of analog bandpass filters with different center frequencies, see Figure 8.3, or a filter where center frequency is moved over a frequency range. The bandpass filters have a certain bandwidth like the width of transmission between three dB frequencies and the steepness of the flanks, expressed, e.g. in the frequency band for 60 dB attenuation. The filter may be realized analog or digital. In the last case the signals are converted in an A/D converter and then operated by recursive filter algorithms, allowing real-time filtering. The bandpass filters may have a constant absolute bandwidth or a constant relative (percentage) bandwidth. Constant bandwidth gives a uniform frequency resolution on a linear frequency scale. This is used if the considered frequency range is limited, as for example, for two decades. Constant percentage bandwidth gives uniform solution on a logarithmic scale and is used for wide frequency ranges of three or more decades.

Commercially available signal analyzers allow to send the filtered signal to a detector if one is interested in the power spectrum. The signal is then squared and integrated over a certain time to obtain an average power value. If the square root of this mean square value is taken, a root mean square (RMS) amplitude is obtained.

Figure 8.3 shows a scheme for a discrete-stepped bandpass filter analyzer. Here, the bandpass filters have different (stepped) center frequencies. The detector is connected sequentially to the filter outputs and measure the signal amplitudes or power in each frequency band. The output is then amplified and recorded by a pen recorder or printer.

For narrow-band analysis it is more appropriate to use a single filter with tunable center frequency. The filter can have constant bandwidth or constant percentage bandwidth. The frequency spectrum then results continuously in frequency. For more details see, e.g. [8.46], [8.43], [8.35], [8.49], [8.14].

8.1.2 Fourier analysis

If the sinusoidal phase shifted oscillations (8.3) are developed into a Fourier series, it holds

$$y(t) = \frac{a_0}{2} + \sum_{\nu=1}^{N} a_\nu \cos \nu \, \omega_0 t + \sum_{\nu=1}^{N} b_\nu \sin \nu \, \omega_0 t \qquad (8.4)$$

If the frequencies $\nu \, \omega_O$ are known, the amplitudes can be determined by the *Fourier coefficients*

$$a_\nu = \frac{2}{T_p} \int_0^{T_p} y(t) \cos (\nu \omega_0 t) \, dt$$

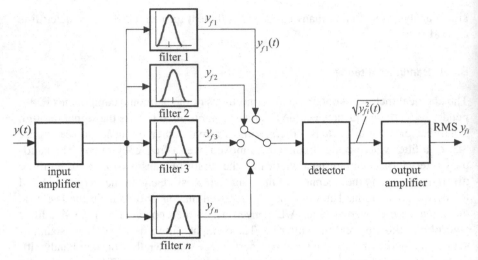

Fig. 8.3. Scheme for bandpass filtering with stepped filters

$$b_v = \frac{2}{T_p} \int_0^{T_p} y(t) \sin(v\omega_0 t) \, dt \tag{8.5}$$

The Fourier series can be put in the complex form

$$y(t) = c_0 + \sum_{v=1}^{N} c_v e^{iv\omega_0 t} + \sum_{v=1}^{N} c_{-v} e^{-iv\omega_0 t}$$

$$= \sum_{v=-\infty}^{\infty} c_v e^{iv\omega_0 t} \tag{8.6}$$

with the complex Fourier coefficients

$$c_v (iv\omega_0) = \frac{1}{T_p} \int_0^{T_p} y(t) e^{-iv\omega_0 t} \, dt \tag{8.7}$$

For $T_p \to \infty$ and thus $\omega_0 \to d\omega$ and $v\omega_0 = \omega$, the periodic function becomes a non-periodic function resulting in the Fourier transform $y(i\omega)$, (8.13).

8.1.3 Correlation Functions

The autocorrelation function (ACF), respectively auto-covariance function, for stationary zero mean disturbing and periodic signals is given by the general form

$$R_{yy}(\tau) = \lim_{T \to \infty} \frac{1}{T} \int_0^T y(t) \, y(t + \tau) \, dt \tag{8.8}$$

For periodic signals, one has to average over integer periods. Thus applies

$$R_{yy}(\tau) = \lim_{n \to \infty} \frac{1}{nT_{pv}} \int_0^{nT_{pv}} y(t)\, y(t + \tau)\, dt \tag{8.9}$$

For a sinusoidal phase-shifted oscillation with $\omega_v = 2\pi/T_{pv}$

$$y_u(t) = y_{0v} \sin(\omega_v t + \varphi_v) + n(t) \tag{8.10}$$

the ACF becomes, [8.17],

$$R_{yy}(\tau) = \frac{y_{0v}^2}{2} \cos\omega_v \tau \tag{8.11}$$

and thus again a periodic function. The result is independent of the phase shift φ_v. The stationary disturbing signal components $n(t)$ and oscillations with $\omega \neq \omega_v$ for $n \to \infty$ have no influence on the ACF. Thus, the ACF is suitable for analyzing periodic signals with stochastic disturbing signal components.

For the cross-correlation function between the input signal $u(t) = u_0 \sin \omega t$ of a linear system and the output signal $y(t)$, it holds

$$R_{yu}(\tau) = \lim_{n \to \infty} \frac{u_0}{nT_{pv}} \int_0^{nT_{pv}} y(t) \sin \omega_v (t + \tau)\, dt \tag{8.12}$$

only oscillation components of $y(t)$ with $\omega = \omega_v$ have an influence. One may consider the similarity to the Fourier coefficient b_v, (8.5).

8.1.4 Fourier transformation

The Fourier transform of a non-periodic signal $y(t)$ is defined as

$$y(i\omega) = \mathcal{F}\{y(t)\} = \int_{-\infty}^{\infty} y(t)e^{-i\omega t}\, dt \tag{8.13}$$

To ensure its convergence, the following condition must be fulfilled

$$\int_{-\infty}^{\infty} |y(t)|\, dt < \infty \tag{8.14}$$

Figure 8.4 shows the amplitude densities $|y(i\omega)|$ for some examples of finite periodic signals, for which the convergence condition is fulfilled, [8.37]. If the Fourier transform is applied to an oscillation of finite duration T, one obtains a peak at $\omega = \omega_v$, Figure 8.4c. The longer the duration T, the higher and more narrow the amplitude density $|y(i\omega)|$ around $\omega = \omega_v$. A steady state oscillation with $T \to \infty$ yields $|y(i\omega)|$. In this case, the convergence condition is no longer fulfilled.

Sampling the continuous-time signal $y(t)$ with the sampling time T_0 results, from (8.12) in the case of $y(t) = 0$ for $t < 0$, approximately in

$$y(i\omega) \approx T_0 \sum_{k=0}^{\infty} y(kT_0)\, e^{-i\omega kT_0} \tag{8.15}$$

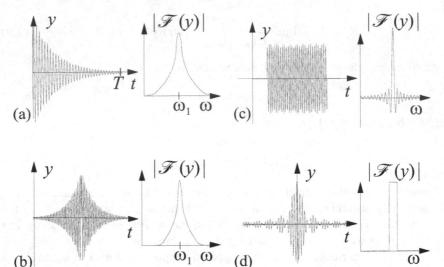

Fig. 8.4. Amplitude densities of some signals of finite duration: (a) decaying oscillation; (b) growing and decaying oscillation; (c) finite periodic signal; (d) time course of a rectangular Fourier transform

Omitting the constant T_0 yields the discrete Fourier transform (DFT)

$$y_D(i\omega) = \sum_{k=0}^{\infty} y(kT_0)\, e^{-i\omega kT_0} \qquad (8.16)$$

Restricting its application only to a finite measuring interval $0 \le k \le N - 1$, then applies

$$\hat{y}_D(i\omega) = \sum_{k=0}^{N-1} y(kT_0)\, e^{-i\omega kT_0}$$

$$= \sum_{k=0}^{N-1} y(kT_0)\, \cos \omega kT_0 - i \sum_{k=0}^{N-1} y(kT_0)\, \sin \omega kT_0$$

$$= Re\,(y_D(i\omega)) + i\, Im\,(y_D(i\omega)) \qquad (8.17)$$

whereas the discrete amplitude spectrum can be calculated from

$$|y_D(i\omega)| = \left[Re^2(y_D(i\omega)) + Im^2(y_D(i\omega)) \right]^{\frac{1}{2}} \qquad (8.18)$$

and the discrete phase spectrum from

$$\alpha_D(i\omega) = arc\, tg\, [Im(y_D(i\omega))/Re(y_D(i\omega))] \qquad (8.19)$$

Introducing the abbreviation

$$z = e^{i\omega T_0} \tag{8.20}$$

one obtains the z-transform

$$\hat{y}_D(z) = \sum_{k=0}^{N-1} y(kT_0) \, z^{-k} \tag{8.21}$$

For each angular frequency ω, $2N$ multiplications and $2(N-1)$ additions are necessary. Therefore, the computing effort is relatively high.

8.1.5 Fast Fourier transformation (FFT)

A reduction of the long computing time of the DFT can be achieved by utilizing the cyclic properties of the oscillation pointers in the complex plane, [8.7], [8.5]. The analysis is then restricted to re-sorting data and multiplications with precalculated sine and cosine values. The angular frequencies are as well discretized with a spacing of $\Delta\omega$. With a finite measuring duration $T = NT_0$, the lowest describable angular frequency is $\omega_{min} = 2\pi/T = 2\pi/NT_0$ and the highest angular frequency is $\omega_{max} = \pi/T_0$ according to the Shannon sampling theorem. The frequency resolution becomes

$$\omega = n\Delta\omega \text{ with } \Delta\omega = \frac{2\pi}{NT_0} \tag{8.22}$$

Using (8.16) gives the series

$$\hat{y}_D(in\Delta\omega) = \sum_{k=0}^{N-1} y(kT_0) \, e^{-ikn\Delta\omega T_0} = \sum_{k=0}^{N-1} y(kT_0) W_N^{kn} \tag{8.23}$$

with constant data set-independent complex factors W_N

$$W_N = e^{-i\Delta\omega T_0} = e^{i2\pi/N} = \text{const.} \tag{8.24}$$

The analysis of several data of equal length permits the factors W_N^{kn} to be precalculated and stored as sine/cosine tables for different lengths of data sets N in order to save computing time, e.g. for real-time processing. In addition, the following symmetry characteristic is valid

$$W_N = e^{-i2\pi/N} = \left\{ e^{-i2\pi/(N/2)} \right\}^{1/2} = W_{N/2}^{1/2} \tag{8.25}$$

A decomposition of the sample set $y(k)$ into two parts, one containing only even-numbered samples and the second containing odd-numbered samples

$$y_{even} = y(2k) \, ; \, y_{odd} = y(2k+1), \, k = 0\ldots(\frac{N}{2}-1) \tag{8.26}$$

yields two sub-sequences from (8.23) and (8.24)

$$\hat{y}_{even}(n) = \sum_{k=0}^{N/2-1} y(2k)\, W_{N/2}^{2kn} = \sum_{k=0}^{N/2-1} y(2k)\, W_N^{2kn}$$

$$\hat{y}_{odd}(n) = \sum_{k=0}^{N/2-1} y(2k+1)\, W_{N/2}^{2kn} = \sum_{k=0}^{N/2-1} y(2k+1)\, W_N^{2kn} \qquad (8.27)$$

Thus, the entire series of the signal $y(k)$ can be denoted

$$\hat{y}_D(n) = \sum_{k=0}^{N/2-1} \left\{ y(2k)\, W_N^{2kn} + y(2k+1)\, W_N^{(2k+1)\,n} \right\}$$

$$= \hat{y}_{even}(n) + W_N^n\, \hat{y}_{odd}(n) \qquad (8.28)$$

in such a way as to permit a calculation by formation of two sub-sequences, each with the half data set length. According to (8.28), the decomposition can be proceeded as long as the length of the sub-sequences is even. A perfect utilization of the calculation symmetry is possible if the length of the data set N represents a power of two, $N = 2^\nu$. In this case, the evaluation of (8.22) degenerates into a rearrangement of $y(k)$ as well as into multiplications of the precalculable complex values W_N. This corresponds to a computing effort of each $4N\ lgN/lg2$ real multiplications and additions for this FFT.

According to the order of the operations, one distinguishes between Cooley-Tukey algorithms (re-sorting with subsequent multiplication) and Sande-Tukey algorithms (multiplication of the data with sine/cosine values and subsequent re-sorting, [8.41]).

A comparison of the computing effort between the DFT and FFT with fixed data set length N shows the strong increase of the DFT versus the FFT with increasing length N, Table 8.1. Due to the large savings in computing time, the disadvantage of a specific data set length (power of two) required for the FFT is usually accepted.

Table 8.1. Comparison of required calculations for DFT and FFT

Data set length	Required calculations for		
N	DFT	FFT	
128	33282	3584	multiplications
	256	3584	additions
1024	2101250	40960	multiplications
	2048	40960	additions
4096	33570818	196608	multiplications
	8192	196608	additions

Example 8.1:

The signal processor DSP32C (AT&T) needs a computing time of 80 ns with a clock frequency of $f_c = 50$ MHz ($T_c = 20$ ns) for one calculation step (floating-point addition and/or multiplication). Due to its specific architecture (multiplier-adder cascade), both an addition and a multiplication operation can be executed together within one calculation step. Regarding the calculation of the FFT, with the same number of both operations half of the computing time can be saved. However, with the DFT only minor effects are noticeable, see Table 8.1.

Table 8.2. Comparison of the required computing time of a signal processor for DFT and FFT

N	DFT	FFT	Factor DFT/FFT
128	26.8 ms	2.85 ms	94
1024	1.68 s	32.75 ms	513
4096	26.86 s	0.1575 s	1705

Table 8.2 shows that the factor of time-saving becomes larger with increasing data set length.

□

Regarding the successive utilization of symmetries with the FFT, the data set length N is limited to a power of two numbers ($N = 2^v$). Often, $N = 1024 = 2^{10}$ is used. Whenever a data set does not meet these requirements, it must be either truncated or padded with zeros ("zero padding") up to the next power of two numbers, which means a corruption of the raw data.

Another disadvantage of the DFT and FFT of a discrete spectrum arises from the selected representation of a discrete spectrum. Due to its finite curvature behavior, sharp spectral lines ("Peaks"), which may arise from discrete sinusoidal oscillation components, cannot be modelled accurately. A finite data set length in the time domain ($k = 0 \dots N - 1$) can be generated from an infinitely long data set length by a rectangular windowing function f_{rec}, with

$$f_{rec}(kT_0) = \begin{cases} 1 \text{ for } 0 \leq k \leq N \\ 0 \text{ for else} \end{cases} \qquad (8.29)$$

see Figure 8.5. Instead of a sequence $y(k)$ of the real measured signal, a sequence of finite duration $y_{tr}(k)$ is transformed into the frequency domain, which results from the multiplication of the signal and the windowing function

$$y_{tr}(k) = f_{rec}(k)y(k) \qquad (8.30)$$

In the frequency domain, this operation leads to a convolution of the spectrum of the measured signal and the Fourier transform of the rectangular window, which forms the sinc-function dependent on the data set length N. The amplitudes of all discrete frequency components smear over by such a convolution performed over the entire frequency range *(leakage effect)*, see, e.g. [8.23].

A reduction of the leakage effect can often be achieved by introducing specific windowing functions $f_{win}(k)$, which are multiplied with the measured signal prior to the transformation, as shown in Figure 8.6

$$y_{tr}(k) = f_{win}(k)y(k) \tag{8.31}$$

These windowing functions $f_{win}(k)$ continuously change their values from 1 in the middle of the measuring interval to 0 on the edges. Besides the reduction of the leakage effect, however, the disadvantage of a diminished resolution of the resulting spectrum due to the corruption of the signal values on the edges occurs.

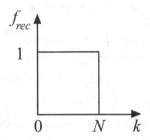

Fig. 8.5. Rectangular windowing function $f_{rec}(k)$

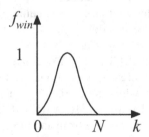

Fig. 8.6. Generalized windowing function $f_{win}(k)$

Example 8.2:

The sinusoidal oscillation

$$y(t) = 1V \sin (2\pi \cdot 1Hz \cdot t)$$

is sampled ($T_0 = 200ms$) within the measuring interval $k = 0...N - 1$. The amplitude spectrum is estimated by means of the FFT from the appropriate sampled sequence $y(k)$:

a) Without additional windowing function f_{win}

$$y(n\Delta\omega) = FFT\{f_{rec}(k)y(k)\}$$

From Figure 8.7, it becomes evident that the limitation of the measuring interval to N samples, particularly for small values of $N(N = 128)$, leads to a significant estimation error of the signal amplitude and signal frequency.

b) With additional windowing function (Hanning-window)

$$y(n \, \Delta\omega) = FFT \{f_{win}(k) f_{rec}(k) \, y(k)\}$$
$$f_{win}(k) = 0.5 \left\{ 1 - \cos\left(\frac{2\pi k}{n}\right) \right\}$$

The additional Hanning-window causes an improvement of the frequency and amplitude estimation for small values $N(N = 128)$, as illustrated in Figure 8.8. The influence of the leakage effect on the estimation results decreases with increasing data set length N.

For complete details on using the FFT, refer, for instance, to [8.1], [8.5], [8.34], [8.47], [8.45], [8.23], [8.28], [8.55].

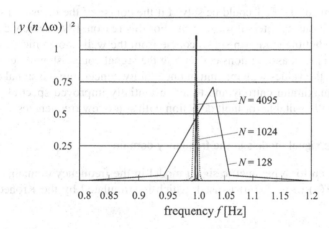

Fig. 8.7. FFT of a sinusoidal oscillation with 1 Hz, $y(k)$ for different data set lengths N

□

If one is interested in the *power spectrum* of the signal $y(k \, T_0)$ based on the discrete Fourier transform, one can first determine the discrete amplitude spectrum $|y_D(i \, \omega)|$, (8.16), by using the FFT, and the using the relation for the periodogram

$$P_y(n \, \Delta\omega) = \frac{1}{N} |\dot{y}_D(n \, \Delta\omega)|^2 \qquad (8.32)$$

[8.24], [8.27], [8.15].

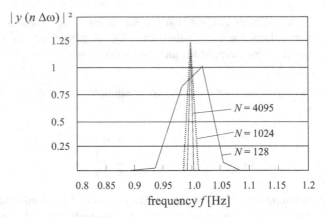

Fig. 8.8. FFT of a sinusoidal oscillation with 1 Hz or different data set length N and additional Hanning-window

8.1.6 Maximum entropy spectral estimation

Most problems of the FFT could be solved if the course of the measured signal outside of the measuring interval was known. For this reason, [8.6] searched for an approach to predict the unknown signal course from the well-known measured values, whereby no a priori assumptions concerning the signal course should be made. This estimation of the values with maximum uncertainty concerning the signal course led to the term maximum entropy and to a substantially improved spectral estimation and is especially suitable for fault detection with some few frequencies.

a) Parametric signal models in the frequency domain

As an approach for a parametric signal model in the frequency domain, a fictitious form filter $F(z)$ resp. $F(i\omega)$ is used, which is stimulated by the Kronecker-Delta pulse

$$\delta(k) = \begin{cases} 1 \text{ for } k = 0 \\ 0 \text{ for else} \end{cases} \quad \delta(z) = 1 \qquad (8.33)$$

to generate steady state oscillations y(k), Figure 8.9 The dynamic system behavior of the form filter is to be determined in such a way that it yields

$$y(z) = F(z)\delta(z) = F(z) \text{ with } \delta(z) = 1 \qquad (8.34)$$

to give an identical frequency response of the form filter $F(z)$ and amplitude spectrum of the measured signal $y(z)$. This applies similarly to the power spectral density

$$S_{yy}(\omega) = |F(i\omega)|^2 S_{\delta\delta}(\omega) = |F(i\omega)|^2 \text{ with } S_{\delta\delta}(\omega) = 1 \; \forall \; \omega \qquad (8.35)$$

Three possible parametric model approaches of such filters can be distinguished, [8.4]. The MA (moving average) model is

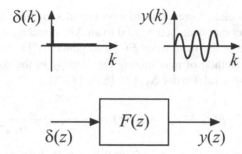

Fig. 8.9. Generation of steady state oscillations $y(k)$ by means of a fictitious form filter $F(z)$ and stimulation with a δ-impulse

$$F_{MA}(z) = \beta_0 + \beta_1\, z^{-1} + \ldots + \beta_n\, z^{-n} \qquad (8.36)$$

Here, the signal spectrum is approximated by a polynomial of limited order n. Thus, the spectrum can only represent limited variations of the amplitude and is unsuitable for the modelling of periodic signals, whose amplitude spectrum only consists of discrete peaks. In the time domain, an MA approach corresponds to the filter-difference equation

$$y(k) = \beta\, \delta\,(k) + \beta_1\, \delta\,(k-1) + \ldots + \beta_n\, \delta\,(k-n) \qquad (8.37)$$

A pure autoregressive (AR) model

$$F_{AR}(z) = \frac{\beta_0}{1 + \alpha_1\, z^{-1} + \ldots + \alpha_n\, z^{-n}} \qquad (8.38)$$

is able to approximate sharp spectral lines of periodic signals according to the poles of the denominator polynomial. Thus, it is particularly suitable for estimating the spectra of harmonic oscillations, [8.36]. The corresponding filter-difference equation is

$$\beta_0\, \delta\,(k) = y(k) + \alpha_1\, y(k-1) + \ldots + \alpha_n\, y(k-n) \qquad (8.39)$$

After a single stimulation by a δ-pulse, the following course of $y(k)$ is only dependent on its passed values $y(k-i)$.

The mixed (ARMA)-filter model approach

$$F_{ARMA}\,(z) = \frac{\beta_0 + \beta_1\, z^{-1} + \ldots + \beta_p\, z^{-p}}{1 + \alpha_1\, z^{-1} + \ldots + \alpha_n\, z^{-n}} \qquad (8.40)$$

is given by the ARMA-difference equation

$$y(k) + \alpha_1\, y(k-1) + \ldots + \alpha_n\, y(k-n) = \beta_0\delta(k) + \ldots + \beta_p\delta(k-p) \qquad (8.41)$$

and is composed of both the MA and AR model approach.

However, strong convergence problems arise with regard to the estimation as the consequence of doubling the number of parameters (β_j, α_i). Regarding this type of model, more complex and more special estimation procedures are described in the literature, see [8.26], which, however, yield worse results for periodic signals (mainly autoregressive signal components) compared to an AR model approach.

A model structure of the form filter $F(z)$ can generally be derived also for periodic signals by the method of maximizing the entropy in the form of a pure AR model for the power spectral density $S_{yy}(z)$, [8.8], [8.48]

$$S_{yy}(z) = F(z) \; F(z^{-1}) \cdot S_{\delta\delta}(z) = \frac{\beta_0^2}{|1 + \sum_{i=1}^n \alpha_i \; z^{-i}|^2} \tag{8.42}$$

By estimating the coefficients α_i and β_0 from the measured signal $y(k)$, one obtains a parametric, autoregressive model in the frequency domain for the power spectral density $S_{yy}(\omega)$, which is characterized by $(n + 1)$ parameters (typically: $n = 4$ ... 30).

b) Determination of the coefficients

A suppression of stochastic signal components with respect to the estimation can be achieved, if instead of the signal $y(t)$

$$y(t) = \sum_{v=1}^m y_{0v} \; e^{-d_v t} \sin(\omega_v \; t + \varphi_v) \tag{8.43}$$

its autocorrelation function $R_{yy}(\tau)$ is used

$$R_{yy}(\tau) = E \; \{y(t) \; y(t + \tau)\} = \sum_{v=1}^m \frac{y_{0v}^2}{2} \; e^{-d_v \tau} \cos(\omega_v \tau) \tag{8.44}$$

Since the autocorrelation function $R_{yy}(\tau)$ of a periodic signal $y(t)$ yields again a periodic function in τ of the type (8.43), for this function a form filter model given by (8.34) can be assumed as well.

$$R_{yy}(z) = F(z) \; \delta(z) \tag{8.45}$$

The eigen-behavior of $y(t)$, represented by its m characteristic frequencies ω_v and damping coefficients d_v, is also contained in $R_{yy}(\tau)$. However, the phase information gets lost and the amplitudes of the ACF become

$$R_{0v} = 0.5 \; y_{0v}^2 \tag{8.46}$$

The approach of a general form filter model of the ARMA-type (8.41) yields a filter-difference equation for the ACF of the measured signal with m eigenfrequencies and thus of the order $n = 2m$

$$\begin{aligned} R_{yy}(\tau) = & -\alpha_1 \; R_{yy}(\tau - 1) - \alpha_2 \; R_{yy}(\tau - 2) - \ldots - \alpha_n \; R_{yy}(\tau - 2m) \\ & + \beta_0 \; R_{\delta\delta}(\tau) + \beta_1 \; R_{\delta\delta}(\tau - 1) + \ldots + \beta_{2m-1} \; R_{\delta\delta}(\tau - 2m + 1) \\ & + R_{nn}(\tau) \end{aligned} \tag{8.47}$$

if an additive-affecting, uncorrelated disturbing signal $n(t)$ of zero-mean value is taken into account. Its autocorrelation function $R_{nn}(\tau)$

$$R_{nn}(\tau) = \begin{cases} n_0 \text{ for } \tau = 0 \\ 0 \text{ else} \end{cases} \tag{8.48}$$

gives only a constant contribution for $\tau = 0$.

The ARMA-signal model (8.47) reaches the steady state after $\tau = 2m$ steps. In the model (8.47), all β-parameters are then omitted. From this time, the ACF $R_{yy}(\tau)$ proceeds in the form of a stationary steady state oscillation and can be exclusively described by the AR-part of (8.47).

In the case of the AR model, the eigen-behavior of the autocorrelation function can be expressed by

$$R_{nn}(\tau) = R_{yy}(\tau) + \alpha_1 \ R_{yy}(\tau-1) + \alpha_2 \ R_{yy}(\tau-2) + \ldots + \alpha_n \ R_{yy}(\tau-2m) \tag{8.49}$$

This relationship yields a system of equations for different shifts τ for the determination of the coefficients α_j and n_0.

$$\begin{bmatrix} R_{yy}(0) & R_{yy}(1) & \ldots & R_{yy}(2m) \\ R_{yy}(1) & R_{yy}(0) & \ldots & R_{yy}(2m-1) \\ \ldots & \ldots & \ldots \ldots & \\ R_{yy}(2m) & R_{yy}(2m-1) & \ldots & R_{yy}(0) \end{bmatrix} \begin{bmatrix} 1 \\ \alpha_1 \\ \ldots \\ \alpha_{2m} \end{bmatrix} = \begin{bmatrix} n_0 \\ 0 \\ \ldots \\ 0 \end{bmatrix} \tag{8.50}$$

The coefficient n_0 is a measure for the mean-square model error

$$R_{nn}(0) = n_0 = E\{n^2(k)\} = E\left\{[y(k) - \hat{y}(k)]^2\right\} \tag{8.51}$$

with $\hat{y}(k)$ as the model prediction for $y(k)$, (8.38). To resolve the system of equation (8.50), estimates of the ACF $R_{yy}(\tau)$ for $\tau = 0\ldots2m$ have to be determined from the measured signal sequence $y(k), k = 0 \ldots N - 1$.

$$R_{yy}(\tau) = \frac{1}{N - |\tau| + 1} \sum_{k=0}^{} y(k) \ y(k + \tau) \tag{8.52}$$

For an efficient solution of the equation system (8.50), the Burg algorithm, [8.41], is recommended. Here, the signal model (8.49) is interpreted as a predictor filter for the unknown autocorrelation values $R_{yy}(\tau)$. Starting from the order of $m = 1$, the coefficients and values of the autocorrelation function are estimated alternately up to the final order of $2m$, [8.6].

The eigenfrequencies of significant discrete oscillation components in $y(t)$ are calculated by pole decomposition of the denominator polynomial in the AR-signal model (8.38)

$$A^*(z) = z^{2m} \ A(z^{-1}) = z^{2m} \left\{1 + \alpha_1 \ z^{-1} + \alpha_2 \ z^{-2} + \ldots + \alpha_{2m} \ z^{-2m}\right\}$$

$$= \prod_{\nu=1}^{m} (1 + \alpha_{1\nu} \ z + \alpha_{2\nu} z^2) \tag{8.53}$$

or by searching the maximum in $A^*(z)$. The resulting conjugate-complex poles z_ν permit a factorization into square parts, (8.53)

$$\alpha_{2\nu} z^2 + \alpha_{1\nu}\, z + 1 = \alpha_{2\nu}\, (z - z_{\nu 1})\, (z - z_{\nu 2}) = 0 \qquad (8.54)$$

From a corresponding table of the z-transform, e.g. [8.16], one obtains for each conjugate-complex pair of poles ($z_{\nu 1}$, $z_{\nu 2}$ from (8.53)) the angular frequency ω_ν of the appropriate sinusoidal partial oscillation $y_\nu(t)$ in $y(t)$

$$\omega_\nu = \frac{1}{T_0}\, \arccos\left[\frac{-\alpha_{1\nu}}{2\sqrt{\alpha_{2\nu}}}\right] \qquad (8.55)$$

In this way all significant partial oscillation frequencies ω_ν of the measured signal $y(t)$ are calculable.

c) Estimation of the amplitudes

A determination of the amplitudes from the AR-signal model (8.49), (8.50) is only inaccurately possible. The amplitudes $y_{0\nu}$ of each partial oscillation of a significant eigenfrequency z_ν result with (8.42) from

$$S_{yy}(z_\nu) = \frac{\beta_0^2}{|A^*(z_\nu)|^2} \qquad (8.56)$$

with $A(z_\nu) = A^*(z_\nu)$ given by (8.53). They are dependent on the denominator coefficient α_i and the constant numerator coefficient β_0. Here, the slightest estimation errors of the coefficients result in large variations of the amplitudes. For this reason, a second estimation level is particularly performed to determine the amplitudes, [8.33], [8.31].

In the autocorrelation function of periodic oscillations

$$R_{yy}(\tau) = E\ \{y(t)\, y(t+\tau)\} = \sum_{\nu=1}^{m} \frac{y_{0\nu}^2}{2}\, e^{-d_\nu \tau}\, \cos\,(\omega_\nu \tau) \qquad (8.57)$$

the damping term can be neglected for small damping values

$$R_{yy}(\tau) = E\ \{y(t)\, y(t+\tau)\} = \sum_{\nu=1}^{m} \frac{y_{0\nu}^2}{2}\, \cos(\omega_\nu \tau) \qquad (8.58)$$

without obtaining a noticeable influence on the accuracy of the estimation result. With known eigenfrequencies ω_ν provided by the first estimation level, one obtains a system of equations for determination of the appropriately demanded amplitudes of the autocorrelation $R_{0\nu}$

$$
\begin{bmatrix} R_{yy}(1) \\ R_{yy}(2) \\ \dots \\ R_{yy}(m) \end{bmatrix}
\begin{bmatrix} \cos(\omega_1 T_0) & \cos(\omega_2 T_0) & \dots & \cos(\omega_m T_0) \\ \cos(\omega_1 2T_0) & \cos(\omega_2 2T_0) & \dots & \cos(\omega_m 2T_0) \\ \dots & \dots & \dots & \dots\ \dots \\ \cos(\omega_1 m T_0) & \cos(\omega_2 m T_0) & \dots & \cos(\omega_m m T_0) \end{bmatrix}
\begin{bmatrix} R_{01} \\ R_{02} \\ \dots \\ R_{0m} \end{bmatrix}
\qquad (8.59)
$$

In consideration of (8.46), the signal amplitudes can be evaluated by means of the amplitudes of autocorrelation with

$$y_{0\nu} = \sqrt{2\ R_{0\nu}} \qquad (8.60)$$

by inverting the (non-singular) matrix (8.59).

Thus, a parametric model representation for the demanded power spectral density $S_{yy}(\omega)$ of the measured signal $y(kT_0)$ was found, which represents the spectrum by a parametric AR model respectively by frequencies of significant sinusoidal oscillation components with appropriate amplitudes. For applications only the sampling time T_0, the data set length N and the order m of the expected significant partial oscillations must be given.

Figure 8.10 shows the amplitude spectrum of the current of an asynchronous motor of a hacksawing machine for nine frequencies. A worn-out saw blade leads to higher frequencies of the motor current. Further examples are given in [8.32].

The advantage of this modified maximum entropy spectral estimation is that the amplitudes $y_{0\nu}$ of the oscillations and the frequencies are directly estimated. This means that instead of a distribution of many amplitudes over the frequency range, like obtained with a FFT, only some (e.g. 5) distinct frequencies are determined precisely and their amplitudes are given. Therefore, this method is advantageous for fault detection based on periodic signals. Applications for grinding machines are given in [8.22] and [8.21], see also [8.18].

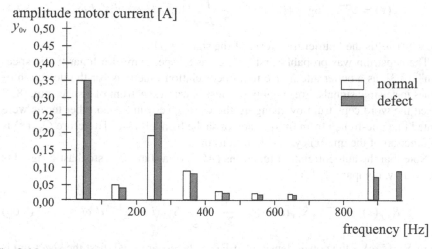

Fig. 8.10. Estimated amplitudes $y_{0\nu}$ of the current of an asynchronous motor of a hacksawing machine for an intact and worn-out saw blade, $m = 0, N = 100$

8.1.7 Cepstrum analysis

For the detection of harmonic signals with small amplitudes among harmonics with larger amplitudes the cepstrum may be used. There exist different definitions for the cepstrum, [8.43]. In the case of periodic signals $y(t)$ the *power cepstrum* is defined as the inverse Fourier transform of the logarithm of the power spectrum of the signal $y(t)$.

$$C_{yy}(\tau) = \mathcal{F}^{-1} \{\log P(\omega)\} = \frac{1}{2\pi} \int_{-\infty}^{\infty} \log P(\omega) e^{i\omega\tau} d\omega \qquad (8.61)$$

The power spectrum results from the representation of periodic functions as Fourier series with the Fourier coefficient at discrete frequencies $\nu \omega_0$, (8.7)

$$c_\nu(i \nu\omega_0) = \frac{1}{T_p} \int_0^{T_p} y(t) e^{-i\nu\omega_0 t} dt \qquad \nu = 0, 1, \dots, N \qquad (8.62)$$

with the period T_p (e.g. $T_p = 1/\omega_0$). Therefore periodic signals result in a discrete amplitude spectrum with dimension [amplitude] and phase spectrum. To obtain the power content of each harmonic, one has to take the square of the amplitude of the Fourier coefficient

$$P(\omega)_{\nu=\nu\,\omega_0} = |c_\nu(i \nu \omega_0)|^2 \qquad (8.63)$$

now with dimension [(amplitude)^2]. The evaluation of (8.61) is then performed by sampling $P(\omega)$.

The *complex cepstrum* is defined as

$$C_y(\tau) = \mathcal{F}^{-1} \{\log y(i \omega)\} = \frac{1}{2\pi} \int_{-\infty}^{\infty} \log y(i \omega) e^{i\omega\tau} d\omega \qquad (8.64)$$

where $y(i \omega)$ is the Fourier transform of the signal $y(t)$.

The cepstrum was probably first defined as a "spectrum of a logarithmic spectrum", [8.3], as a better alternative to autocorrelation functions for the detection of echoes in seismic signals. Presumably because is was a spectrum of a spectrum, [8.3] coined the word ceps-trum by changing the word spec-trum. Also other terms were created like que-frency from fre-que-ncy or saphe from ph-as-e. Therefore $C_{yy}(\tau)$ is is a function of the quefrency τ with dimension [s].

Note that the autocorrelation functions (ACF) of a stationary stochastic signal is obtained by, compare (6.21),

$$R_{yy}(\tau) = \mathcal{F}^{-1} \{S_{yy}(i \omega)\} = \frac{1}{2\pi} \int_{-\infty}^{\infty} S_{yy}(i \omega) e^{i\omega\tau} d\omega \qquad (8.65)$$

where $S_{yy}(i \omega)$ is the power density. If FFT-analyzers are used, first the spectrum is determined and then by the inverse Fourier transform the ACF instead of operating in the time domain, [8.43], [8.35].

The advantages of a cepstrum with respect to an autocorrelation function are to be seen if a comparison is made between the *power spectrum* with dimension [(amplitude)]^2 of a signal in linear scale and logarithmic scale. The power spectrum in the logarithmic scale shows much more peaks compared to the linear scale, where only the main harmonics dominate. This is because the logarithm attenuates small values in comparison to large values. Therefore the cepstrum shows the harmonics with their periods in [s] for small amplitudes better, as shown in an example for a ball bearing fault in [8.43]. Hence, the cepstrum is suitable for detecting periodic effects in the logarithmic spectrum, like families of harmonics, sidebands or echoes. It is used, e.g. in speech analysis, seismic analysis and machine fault detection, see also [8.25].

The *complex cepstrum* is mentioned to be more powerful than the power spectrum, but is more difficult to deal with, [8.43]. It contains also a phase information and it is possible to return to the original time domain. Applications are, for example, echo removal and speech synthesis.

8.2 Analysis of non-stationary periodic signals

Many signals have no constant frequency spectrum but change their frequency contents over time. These non-stationary signals should not be analyzed with the conventional Fourier transform, because then only averaged results are produced which are not associated to particular time instants. Two approaches for the analysis of these non-stationary periodic signals are considered in this section the *short-time Fourier transform* and the *wavelet transform*.

8.2.1 Short-time Fourier transform

A straightforward way to obtain the frequency spectrum of a time-varying periodic signal $y(t)$ is to apply the short-time Fourier transform (STFT)

$$y_{STFT}(i\omega, \pi) = \int_{-\infty}^{\infty} y(t)\, f(t - \tau)\, e^{-i\omega t}\, dt \tag{8.66}$$

where $f(t - \tau)$ is a window function around the time τ of interest. The STFT calculates the similarity between the signal $y(t)$ and the function $f(t - \tau)\exp(i\omega t)$. The function $f(t - \tau)$ has usually a short-time duration. By changing τ the STFT describes how the frequency content evolves over time. In selecting the length of the window function $f(t - \tau)$, always a compromise between the resolution Δt of the signal and the resolution $\Delta\omega$ of the spectrum has to be made, because of the uncertainty condition, [8.42]. If, e.g. a small resolution of the spectrum is desired the window length must be long.

8.2.2 Wavelet transform

The STFT determines the similarity between the investigated signal and a windowed harmonic signal. In order to obtain a better approximation of short-time signal changes with sharp transients, also the similarity with a short-time prototype function of finite duration can be calculated. Such prototype or basis functions which show some damped oscillating behavior are wavelets which origin from a mother-wavelet

$\Psi(t)$, [8.42], [8.2], [8.52]. Figure 8.11 shows some typical mother-wavelets. These mother-wavelets can now be time-scaled (dilatation) by the factor a and time-shifted by τ (translation) and leads to

$$\Psi^*(t, a, \tau) = \frac{1}{\sqrt{a}} \Psi\left(\frac{t-\tau}{a}\right) \tag{8.67}$$

(The factor $1/\sqrt{a}$ is introduced in order to reach a correct scaling of the power-density spectrum, [8.2]). If the mean frequency of the wavelet is ω_0, the scaling the wavelet by t/a results in the scaled mean frequency ω_0/a.

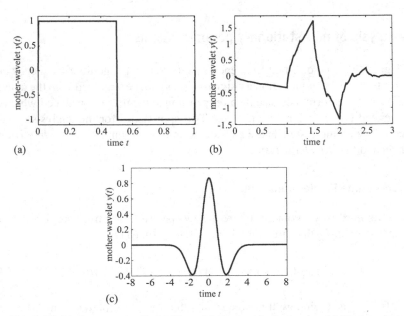

(a)

(b)

(c)

Fig. 8.11. Mother-wavelet Ψ: (a) Haar; (b) Daubechie 2nd order; (c) Mexican hat

The continuous-time wavelet transformation (CWT) then becomes

$$\text{CWT}(a, \tau) = \frac{1}{\sqrt{a}} \int_{-\infty}^{\infty} y(t)\Psi\left(\frac{t-\tau}{a}\right) d\tau \tag{8.68}$$

which is a real function for real $y(t)$ and $\Psi(t)$. Note, that the STFT is usually a complex function. Examples for wavelet functions are

1) *Morlet-Wavelet*

$$\Psi(t) = e^{-t^2/T^2} e^{2\pi f_0 i t}$$

2) *Mexican-hat wavelet*

$$\Psi(t) = (1 - t^2) e^{-t^2/2}$$

3) *One-cycle-sine wavelet*

$$\Psi(t) = \begin{cases} \sin(t) & |t| < \tau \\ 0 & \text{else} \end{cases}$$

The advantages of the wavelet transform stem from the signal adapted basis function and the better resolution in time and frequency. The wavelet functions correspond to certain band pass filters, where, for example, by a reduction of the mean frequency through the scale factor also a reduction of the bandwidth is achieved, compared to STFT where the bandwidth stays constant, [8.50].

Example 8.3:

A periodic signal with changing frequency is considered and some overlap between

$$y(t) = \begin{cases} \cos(2\pi f_1 t) & t/T_0 \le 400 \\ \cos(2\pi f_1 t) + \cos(2\pi f_2 t) & 400 < t/T_0 \le 724 \\ \cos(2\pi f_2 t) & t/T_0 > 724 \end{cases}$$

$$f_1 = 80\,\text{Hz}; \quad f_2 = 240\,\text{Hz}; \quad T_0 = 125\mu s$$

For the wavelet transform the wavelet Mexican hat was used, [8.50].
Figure 8.12b shows the time-shifted wavelets for $a = 9$ and $a = 25$. Figure 8.12c presents the results of the CWT, indicating the amplitude for different a and τ. Figure 8.12d and e show the calculated wavelet coefficients $\Psi(a, \tau)$ in dependence on the time for $a = 9$ and 25. The maximum values of $\Psi(a, \tau)$ are marked with a square and circle and determine the best agreement between the analyzed signal $y(t)$ and the wavelets. Hence, this example shows that the changed frequencies can immediately be detected by taking the maximal values of the CWT.

□

Sudden changes in the amplitude or frequency of periodic signals can therefore be detected after the CWT exceeds certain thresholds. An application of the wavelet transform for the misfire detection of a 6 cyl. gasoline engine with measured exhaust gas pressure and wavelet analysis was demonstrated by [8.50].

8.3 Analysis of stochastic signals

The analysis of stochastic signals applies the signal model equations described in Section 6.2.

8.3.1 Correlation analysis

For single signals the autocorrelation function (ACF) is used, mostly for sampled signals in form of

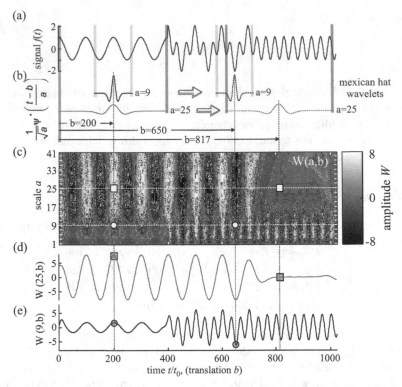

Fig. 8.12. Application of wavelet transform to periodic signals: (a) Signal to be analyzed; (b) Scaled and time-shifted Mexican hat wavelets; (c) Scalogram $\Psi(a,\tau)$ for $a = 1...41$ and $\tau = 0...1000$ (bright means large amplitude) (d) Time behavior of wavelet coefficient $\Psi(a,\tau)$ for $a = 25$; (e) Time behavior of wavelet coefficient $\Psi(a,\tau)$ for $a = 9$

$$\hat{\phi}_{yy}(\tau) = \frac{1}{N} \sum_{k=0}^{N-1} y(k) \, y(k+\tau) \tag{8.69}$$

or if the mean $\bar{x}$ is known in form of the auto-covariance function

$$\hat{R}_{yy}(\tau) = \frac{1}{N} \sum_{k=0}^{N-1} y(k) \, y(k+\tau) - \bar{y}^2 \tag{8.70}$$

For the following discussion it is assumed that $\bar{y} = 0$. As for finite N only $N - |\tau|$ products exist (8.70) leads to a bias b

$$E\left\{\hat{R}_{yy}(\tau)\right\} = \left[1 - \frac{|\tau|}{N}\right] R_{yy}(\tau) = R_{yy}(\tau) + b \tag{8.71}$$

which vanishes for $N \to \infty$, such that the estimate is consistent. An alternative would be to use $N - |\tau|$ instead of N in the nominator. Then, however, the variance of the estimates increase, see [8.17]. Therefore, (8.70) is preferred and N should be sufficiently large compared to $|\tau|$.

For on-line application in real time, the ACF can also be turned into a recursive correlation estimator

$$\hat{R}_{yy}(\tau, k) = R_{yy}(\tau, k-1) + \frac{1}{k+1} \, [y(k-\tau)y(k) - \hat{R}_{yy}(\tau, k-1)]$$

$$\begin{array}{cccccc} \text{new} & \text{old} & \text{correction} & \text{new} & - & \text{old} \qquad (8.72) \\ \text{estimate} & \text{estimate} & \text{factor} & \text{product} & & \text{product} \end{array}$$

see [8.17].

8.3.2 Spectrum analysis

As the power spectral density is defined as Fourier transform of the auto-covariance function, the result of Section 8.1.4 can be applied. Instead of signal $y(kT_0)$ the auto-covariance function $R_{yy}(\tau)$ has to be used. Hence the power spectral density becomes

$$S_{yy}(\omega) = \sum_{\tau=-\infty}^{\infty} R_{yy}(\tau)e^{-i\omega\tau T_0} \qquad (8.73)$$

which is the two-sided discrete Fourier transform, compare (6.21). As $R_{yy}(\tau)$ is symmetric for large measurement from N also

$$S_{yy}(\omega) = 2 \sum_{\tau=0}^{\infty} R_{yy}(\tau) \, e^{-i\omega\tau T_0} \qquad (8.74)$$

can be used. Then, the fast Fourier transform (FFT), Section 8.1.4 can be applied directly.

8.3.3 Signal parameter estimation with ARMA-models

If the autoregressive-moving average (ARMA) model (6.30)

$$\begin{aligned} y(k) + c_1 y(k-1) + \ldots + c_n y(k-p) \\ = d_0 v(k) + d_1 v(k-1) + \ldots + d_m v(k-p) \end{aligned} \qquad (8.75)$$

is applied, the parameters c_i and d_j have to be estimated. Assuming the order p as known, the ARMA model can be written as

$$y(k) = \boldsymbol{\psi}^T(k)\hat{\boldsymbol{\theta}}(k-1) + v(k) \qquad (8.76)$$

where

$$\boldsymbol{\psi}^T(k) = [-y(k-1)\ldots - y(k-p)v(k-1)\ldots v(k-p)] \qquad (8.77)$$

$$\hat{\boldsymbol{\theta}}^T = \left[\hat{c}_1 \ldots \hat{c}_p \ \hat{d}_1 \ldots \hat{d}_p\right] \qquad (8.78)$$

If the white noise values $v(k-1)\ldots v(k-p)$ were known, the recursive least squares method (RLS) could be used, see Section 8.1. $v(k)$ can be interpreted as equation error, which is statistically independent by definition.

For the time of the measured $y(k)$ the values $y(k-1),\ldots,y(k-p)$ are known. Assuming that the estimates $\hat{v}(k-1),\ldots,\hat{v}(k-p)$ and $\hat{\theta}(k-1)$ are known, the most recent signal $\hat{v}(k)$ can be estimated using (8.76)

$$\hat{v}(k) = y(k) - \boldsymbol{\psi}^T(k)\hat{\boldsymbol{\theta}}(k-1) \tag{8.79}$$

where

$$\hat{\boldsymbol{\psi}}^T(k) = [-y(k-1)\ldots - y(k-p)\hat{v}(k-1)\ldots\hat{v}(k-p)] \tag{8.80}$$

Then

$$\hat{\boldsymbol{\psi}}^T(k+1) = [-y(k)\ldots - y(k-p+1)\hat{v}(k)\ldots\hat{v}(k-p+1)] \tag{8.81}$$

is also determined, such that the recursive LS parameter estimation algorithms, Section 8.1, can be used. For more details see [8.19]. This method can be applied for stationary signals and also for special types of non-stationary signals.

This chapter on fault detection with signal models has shown how certain features of measured signals can be generated. According to Figure 8.1 the observed features are compared with the normal features. If the differences exceed a threshold, the exceeding values represent symptoms. The exceeding of these thresholds can be detected by the methods of limit checking and change detection, described in Chapter 7.

8.4 Vibration analysis of machines

8.4.1 Vibrations of rotating machines

Many machines contain drive systems with motors, clutches, gears, shafts, belts or chains, and different ball/rolling or oil bearings. Vibrations are usually generated by

- inherent machine oscillations (e.g. piston-crankshaft, toothed machine tool cutting, axial piston pumps, induction motors);
- shaft oscillations with radial or axial displacement;
- irregular speed of the shaft (e.g. Kardan joint or excentric gears);
- torsional shaft oscillation;
- impulsewise excitation (e.g. through backlash, cracks, pittings, broken gear teeth).

Some of the vibrations indicate a normal state of the machines. However, changes of these vibrations and additional appearing ones may be caused by faults. Therefore vibration analysis is a well established field in machine or monitoring supervision, [8.43], [8.54], [8.53], [8.25], [8.13].

Machine vibrations are usually measured as acceleration $a(t)$ with lateral accelerometers in one, two or three orthogonal directions (horizontal, vertical and axial or rotational accelerometers) at the casing of the machines. Therefore, there is a machine transfer behavior between the source of the vibrations and the measurement location. This transfer behavior, expressed by a frequency response $G_m(i\omega)$, may contain one or several resonance frequencies $\omega_{res,i}$ of the machine structure, resulting from different mass-spring-damper systems.

The measurement principle of the accelerometer is based, e.g. on the measurement of forces, as for piezoeletric force sensors, or the measurement of the displacement of a seismic mass, as with inductive sensors. Usually a highpass filter follows the accelerometer in order to damp low-frequency disturbancies with a cut off frequency of, e.g. 100-200 Hz.

Instead of the acceleration $a(t)$ also the vibration speed $v(t)$ or the vibration displacement $d(t)$ can be measured. If, for example, a sinusoidal oscillation of the displacement

$$d(t) = d_0 \sin \omega t$$

is considered, the other signals are related by

$$v(t) = \dot{d}(t) = d_0 \omega \cos \omega t \tag{8.82}$$
$$a(t) = \ddot{d}(t) = -d_0 \omega^2 \sin \omega t \tag{8.83}$$

Therefore, higher frequent components are better represented by the measurement of the acceleration. Acceleration is also more easily measured as speed or displacement, which both need a non-oscillating reference point. However, also high frequent noise is amplified which may reduce the signal-to-noise ratio. This requires appropriate low pass filtering for the very high frequencies, see Figure 8.13.

In the following, first the *modelling of vibration* signals is considered, especially the resulting frequencies caused by bearing and gear defects. Then some applicable *vibration analysis methods* are considered, based on the earlier sections of this chapter. In general, the analyzed vibration signal is denoted by $y(t)$, which stands for $a(t)$, $v(t)$ or $d(t)$.

8.4.2 Vibration signal models

Faults in rotating machines may generate additional stationary harmonic signals or pulsewise signals. *Harmonic signals* arise because of *linearly superimposed effects* like unbalance, inexact alignment or centering or deformed shafts, tooth faults, ball bearing faults, electrical flux differences in electrical motors, or by changes of the machine's periodic operation. The resulting harmonic signals may then appear as additive vibrations, like

$$y(t) = y_1(t) + y_2(t) + \cdots + y_n(t) = \sum_{i=1}^{n} y_i(t) \tag{8.84}$$

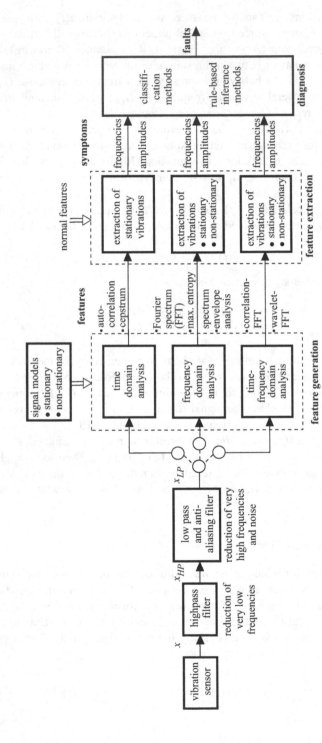

Fig. 8.13. Vibration analysis methods for detection and diagnosis of rotating machinery faults (bearings, gears)

Typical frequencies in machines with gears and ball or roller bearings are:

ω_s basic rotational shaft angular frequency
ω_t tooth contact frequency $\omega_t = z\,\omega_s$ (z: teeth number)
ω_{or} outer race frequency (frequency through rolling over an uneven point of the outer ring)
ω_{ir} inner race frequency
ω_c cage frequency
ω roller or ball spin frequency

The ball or roller frequencies can be calculated from geometrical data, see, e.g. [8.53], [8.13], [8.9], under the assumption that no slip occurs. However, with significant thrust loads and internal preloads these frequencies change because of other contact angles and slip, [8.54].

As shown is Section 6.1, additional frequencies appear also through *nonlinear effects* like nonlinear characteristics, backlash, mechanical looseness of the machines itself or its parts because of foundation cracks or broken mounting or by hysteresis through dry friction and slip-stick effects.

Another reason for additional frequencies are *amplitude modulations* of a basic waveform. A typical example is a pair of gears where one gear is not centered precisely. If the non-centered gear rotates with angular frequency ω_1 and has 22 teeth, the teeth contact frequency is 22 impacts per revolution, resulting in angular frequency $\omega_2 = 22\,\omega_1$ and amplitude y_{02}. Because of the non-centered gear, the amplitude of the tooth impacts oscillate with the speed ω_1 as the gear moves closer and farther away from the second gear resulting in oscillating contact forces. Hence, the frequency with which the level of the impacts are modulated (changed) is the rotation frequency ω_1 of the non-centered gear with amplitude x_{01}, see Figure 8.14. This yields as resulting oscillations

$$
\begin{aligned}
y(t) &= y_{02} \sin \omega_2\, t \,[1 + y_{01} \sin (\omega_1 t + \varphi_1)] \\
&= y_{02} \sin \omega_2\, t + y_{01}\, y_{02}\, \tfrac{1}{2}\, [\sin\,((\omega_2 - \omega_1)\,t - \varphi_1) \\
&\quad + \sin\,((\omega_2 + \omega_1) + \varphi_1)]
\end{aligned}
\tag{8.85}
$$

Then the tooth contact frequency ω_2 can be observed and the two sideband frequencies $\omega_2 - \omega_1$ and $\omega_2 + \omega_1$, see Figure 8.14d. The rotation frequency ω_1 does not appear.

In addition the non-centered gear may also cause a *frequency modulation*, because its effective radius changes as it moves closer and farther from the other gear. Therefore the frequency ω_2 of the tooth contact oscillates, resulting in

$$
y(t) = y_{02} \sin\,[\omega_2 + y_{01} \sin \omega_1\, t]\, t
\tag{8.86}
$$

Therefore also here sidebands with same frequencies as for amplitude modulation arise, [8.10]. Similar effects with sideband effects are observed for rolling element bearings s_1 and electrical motor-bar defects.

Another form of vibrations appear as *periodic impulses*, for example, in ball or rolling bearings. The impact pulse is generated every time a ball or roller hits a defect

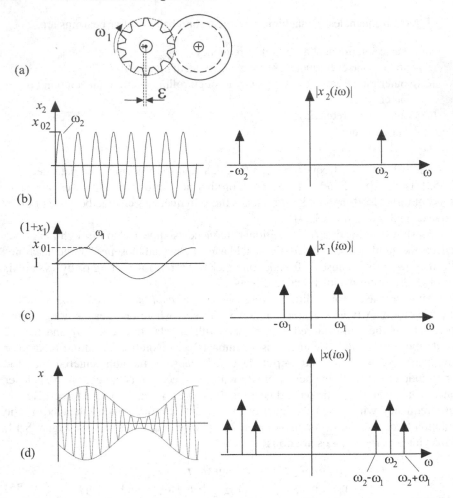

Fig. 8.14. Vibration signals and Fourier spectrum for a non-centered gear by excentricity ϵ: (a) scheme of a one-stage gear with excentricity; (b) gear with teeth contact frequency ω_2; (c) non-centered gear with rotation frequency ω_1; (d) amplitude-modulated teeth contact frequency

in the raceway or every time a defect in a ball hits the raceway. Each such impulse excites a short transient vibration in the bearings and the mechanical structures at its natural frequencies, for example, rigid body frequencies, see Figure 8.15a. Considering one dominating mass-spring-damper system with eigenfrequency ω_{0e}, the momentum of each impact generates a damped impulse response of the position $d(t)$ at the measurement location

$$d_0(t) = a_0 \, e^{-\delta t} \sin \omega_{0e} t \qquad (8.87)$$

This damped vibration with decaying constant δ repeats with period T_{imp} or angular impact frequency $\omega_{imp} = 2\pi / T_{imp}$. The resulting signal is then described by

$$d(t) = \sum_{\nu} d_0 \left(t - \nu \, T_{imp}\right) \qquad (8.88)$$

and its Fourier transform becomes periodic

$$\mathcal{F}(i\omega) = \mathcal{F}\left(i \left(\omega \pm \nu \, \omega_{imp}\right)\right), \; \nu = 0, 1, 2, \dots$$

as known from sampled-data systems. Hence, a frequency spectrum shows peaks at frequencies $\nu \, \omega_{imp}$.

The described signal models for rotating machinery are valid for single ball/rolling bearings and single gear stages. If several different bearings and gears act together the vibration effects are added and create several distinct frequencies by linear superposition and nonlinear effects, also resulting in several sideband frequencies. Hence, the frequency spectra become increasingly complex and it may not be straightforward to isolate the effects and to diagnose the faults.

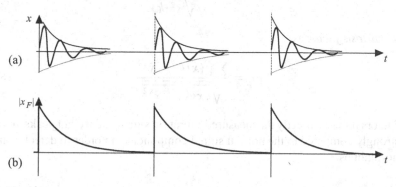

Fig. 8.15. (a) Vibration $x(t)$ caused by periodic impulses; (b) Envelopes $|x_F(t)|$ after low pass filtering of $x(t)$ and rectification

8.4.3 Vibration analysis methods

The goal of vibration analysis is to extract features from the measurements $x(t)$ or $x(kT_0)$, T_0 sampling time, to be used for fault detection and diagnosis. The sampling frequency $\omega_0 = 2\pi/T_0$ has according to Shannon's sampling theorem to be selected with $\omega_0 > 2\,\omega_{max}$ where ω_{max} is the highest frequency of interest. In order to avoid anti-aliasing effects anti-aliasing low pass filters have to be applied with $\omega_{cut} = \omega_0/2 > \omega_{max}$. The corresponding analysis methods operate in the time domain, the frequency domain or in both the time and frequency domain, see, e.g. [8.54], [8.53], [8.25], [8.13], [8.9] and Figure 8.13.

a) Time domain methods

The application of *autocorrelation functions* (ACF) according to (8.9) or *autocovariance functions* allows to separate the vibration signal from noise. The vibration signals are represented as periodic functions $R_{yy}(\tau + v\ T_{pi})$, $v = 0, 1, 2, \ldots$, for the harmonics i. However, interpretation by inspection is only possible for very few harmonics. Otherwise the ACF can be used as data preprocessing for frequency analysis methods.

The *power cepstrum* according to (8.61) is defined as a Fourier-backtransformation of the logarithmic power spectrum with angular frequency $\omega[1/s]$ into the time domain. It is used if harmonics with small amplitudes among dominating harmonics have to be detected, like for ball bearing faults, [8.43], [8.25]. Impulses from defects show as distinct peaks in the distance of one rotation time.

For the detection of peaks and outliers the measured signal $x(k)$ can be used for the calculation of a crest and kurtosis factor.

A *crest factor* is defined as

$$cr = \max \frac{|x(k)|}{\sqrt{x^2(k)}} \tag{8.89}$$

and the *kurtosis factor* as

$$kur = \frac{\sum_k (x(k) - \bar{x})^4}{\sqrt{(x(k) - \bar{x})^2}} \tag{8.90}$$

Both factors process directly the measured vibration signal and weight peaks and outliers strongly compared to the normal signal components, in order to detect changes or abnormalities.

b) Frequency domain methods

The classical way to analyze stationary harmonic vibration signals is the *Fourier analysis* described in Section 8.1.2, especially in the algorithmic efficient form of *Fast Fourier Transform* (FFT). The resulting peaks allow an intuitive way to extract changes in the frequency peaks caused by faults. Comparison with expected frequencies for, e.g. ball bearings and gears discussed earlier in this section then may allow to isolate respective faults. However, the observed frequencies are not always uniquely related to distinct faults, [8.53].

The *maximum entropy spectral estimation*, Section 8.1.6, is recommended for automatically finding some (2 to 5) distinct frequencies on cost of computations. Good results have been obtained for a grinding machine, [8.20] and hacksawing machine, [8.31].

For the extract of damped impulse responses as result of impact impulses caused by ball/rolling bearings or toothed gears, the *envelope analysis method* is suitable. Here, the eigenfrequencies $\omega_{0e,v}$ of the machine modes are suppressed by a low pass or band pass filter which include the eigenfrequencies. After determining the magnitude $|x_F(t)|$ of the low pass filtered signal, only the positive part of the envelopes remain, see Figure 8.15b. This signal is then analyzed by a Fast Fourier Transform

(FFT). By this way the impact frequency ω_{imp} and its higher harmonics are much better represented as in the FFT of the original signal $x(t)$, where it may not be recognized because of the higher contributions of the machine modes.

c) Time-frequency domain analysis

Time-frequency analysis methods first perform some data preprocessing in the time-domain, to improve the signal-to-noise ratio or to improve the signal contents with regard to certain frequency ranges.

A first way is to apply *autocorrelation* to remove noise effects or *cross-correlation* if the relation to another frequency is of interest and to analyze the resulting time-dependent periodic signals with FFT.

For non-stationary periodic signals the short-time Fourier transform or the Wave-lettransform can be applied, described in Section 8.2.

Vibration analysis can be used to detect faults in rotor systems. Through the identification of fault models with equivalent forces and suitable locations for position and acceleration measurements. it was shown by using a laboratory testing that imbalances and larger axle rents can be detected, [8.39], [8.38].

This summary of some basic methods for vibration analysis of machines, as depicted in Figure 8.13, shows that the design of the various filters, the selection of sampling time and length of measurements, the applied analysis methods and their appropriate combinations together with the selection of the measurement equipment has to be performed with good knowledge of the individual machine properties. Examples of successful applications are given in [8.53], [8.25]. A comparison of different methods in [8.9] shows relatively similar results and gives recommendation for practical cases.

8.4.4 Speed signal analysis of combustion engines[1]

Increasing demands on economy, reliability and particularly the reduction of exhaust gas emissions is forcing vehicle manufactures to develop suitable detection and diagnosis functions for combustion engines. Legal regulations, such as the On Board Diagnosis II (OBD II) introduced by the California Air Resources Board (CARB) in 1996 (California's OBDII Requirements) or the European On Board Diagnosis (EOBD) introduced by the European Union in 1998, have promoted the development of supervision methods of all components in a passenger car that cause an increase of exhaust gas emissions in the case of faults. Statistics, [8.30], have shown that an increase in exhaust gas emissions and decrease in engine performance is, in most cases, caused by faults in the injection or mixture preparation. Regarding spark-ignition engines, misfire detection is a very demanding task. When a cylinder misfires, e.g. due to faults in the mixture preparation or ignition system with the effect that no combustion or incomplete combustion occurs, unburned fuel enters the exhaust system, which then burns in the hot catalytic converter. The released heat

---

[1] compiled by Frank Kimmich

may damage or destroy the catalytic converter by thermal overloading. If a given misfire ratio is exceeded, the fuel supply for the misfiring cylinders can be cut off in order to protect the catalytic converter from damage and to avoid exceeding the emission standard. One way to detect misfiring cylinders is to evaluate the engine speed signal at the engine flywheel.

The signal characteristics of a combustion engine are determined by the batch behavior of the combustion, which depends on the crankshaft angle CA. Each cylinder of a four-stroke engine fires every 720°CA. This corresponds to one working cycle and specifies the engine base period. All relevant signal components are multiples of this base frequency. During a working cycle, each cylinder fires one time so that a combustion every 180°CA results for a four-cylinder engine. If the engine angular speed, measured at the flywheel, is denoted by ω_E, the frequency of this oscillation corresponds to the ignition frequency f_I by

$$f_1 = \frac{\omega_E}{4\pi} i_c, \quad i_c : \text{number of cylinders} \tag{8.91}$$

In Figure 8.16, a typical engine speed signal of a spark-ignition (SI) engine measured at idle speed without misfiring is depicted showing speed oscillations with the ignition frequency around the engine speed mean value (approx. 800 rpm).

If misfires or faults in the injection mass occur, the engine speed decreases significantly. Figure 8.17 shows the measured engine speed of a four-cylinder engine in the case of continuous misfiring of one cylinder. Then, additional low-frequency oscillations arise, as can be clearly seen from the low-pass filtered engine speed signal. The appearing frequency components are harmonics of the engine base frequency. Depending on the misfiring cylinders, different frequency patterns result.

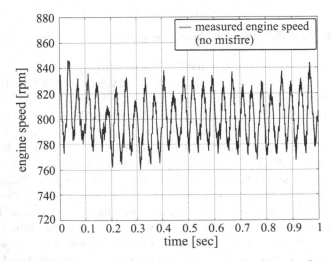

Fig. 8.16. Measured engine speed signal at idle speed of an SI engine (no misfires)

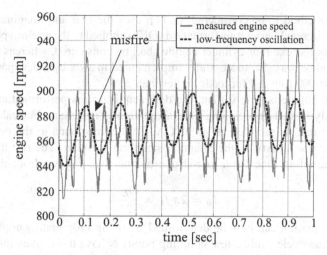

Fig. 8.17. Measured engine speed signal and low-pass filtered signal at idle speed with misfires in cylinder 1

In the past few years, methods have been investigated using the Fourier and the fast Fourier analysis to evaluate these frequency components, see [8.44]. Figure 8.18 shows the Fourier transforms of both engine speed signals (no misfire and misfire in cylinder 1). Without misfire, the ignition frequency means that only the fourth engine harmonic appears in the spectrum. In the case of misfires, additional frequency components arise. Evaluating these frequency components means that not only misfires, but also the misfiring cylinder, can be detected and located.

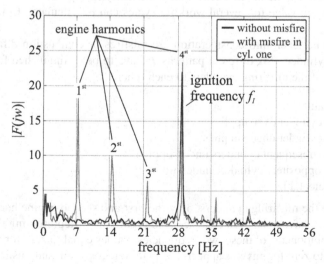

Fig. 8.18. Fourier transform of the engine speed signal without and with misfires in cylinder 1

Another method to be considered, [8.11], uses the real and imaginary components of the discrete Fourier transformation (DFT) applied to the engine speed signal. Thereby, a four-stroke four-cylinder engine shall be considered, whereas the principle of the method was also successfully implemented in a six-cylinder spark-ignition engine up to 6000 rpm and for loads higher than 20%.

Since the engine is time-variant, the data acquisition is performed crank angle synchronously so that no sampling time adaptation is necessary. For calculation of the DFT, the data is sampled all at 90°CA. This corresponds to the double ignition frequency, so satisfying the Shannon sampling theorem. The resulting speed-dependent sampling time for a four-stroke engine then follows from the ignition frequency:

$$T_o = 2\pi f_1 = \frac{\omega_E}{2} \tag{8.92}$$

The DFT evaluation can now be determined by using only eight sampling points per combustion cycle. Only a few sampling points N have to be taken into account, which is an easy real-time application. To compute the DFT, the amplitudes and the phase angle can be calculated as follows:

$$A_m = \sqrt{\left(\sum_{i=1}^{N-1} \omega_i \cos\left(\frac{2\pi mi}{N}\right)\right)^2 + \left(\sum_{i=1}^{N-1} \omega_i \sin\left(\frac{2\pi mi}{N}\right)\right)^2} \tag{8.93}$$

$$\varphi_m = \arctan \frac{\sum_{i=1}^{N-1} \omega_i \sin\left(\frac{2\pi mi}{N}\right)}{\sum_{i=1}^{N-1} \omega_i \cos\left(\frac{2\pi mi}{N}\right)} \tag{8.94}$$

whereas m denotes the order. Because of the usually non-cyclic combustion variations, an average value for several working cycles can be calculated from the measured data.

Faults to be taken into consideration are misfires or combustion differences in one or two cylinders. Six different patterns P have to be distinguished for the relative location of the misfiring cylinders to each other:

P0: no fault
P1: one cylinder oversupplies
P2: one cylinder undersupplies
P3: two subsequent cylinders undersupply
P4: two oppositely cylinders undersupply
PX: undetectable.

To locate the misfiring cylinders, only the first and second engine harmonics ($m = 1$ and $m = 2$) have to be evaluated, see also Figure 8.19. Representing the real and imaginary components of these two frequencies, values equal to zero for no misfires and unequal to zero for misfires appear. For pattern recognition and misfire detection respectively, comparisons of the amplitude values and the real and imaginary components have to be performed. Also, two thresholds T1, T2, which are dependent on engine speed and load, have to be determined. The flowchart in Figure 8.19 shows the signal flow of monitoring and diagnosis of the possible fault patterns. Depending on the fault case, different vector patterns arise, with which the defective cylinders can be detected. Thus, the fault diagnosis is executed by a pattern recognition method of the amplitudes and phases of the DFT.

The performance of the proposed method is, on the one hand, limited by faults in the data acquisition (for example, error in measurements) and, on the other hand, by overlaid disturbances on the measured signal. It can be used for misfire detection as well as for monitoring smooth engine operation for the whole operation area, except for too-low loads and high speeds. A similar approach was developed by [8.51] for measuring the exhaust gas pressure.

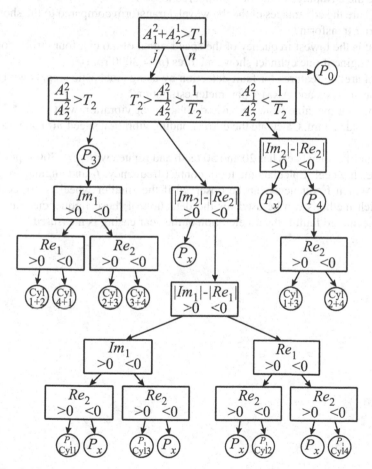

Fig. 8.19. Scheme for detection of misfires and diagnosis of the faulty cylinders

8.5 Problems

1) State three typical tasks for fault detection with signal models in the areas of machine tools, turbo engines and car suspensions. Which measurements should be made? Which signal-analysis methods can be applied?

2) A passenger car shows strong vibrations in the steering wheel around 80 km/h? What can be the fault? Which variables should be measured to diagnose the fault?

3) The seat of a truck drive is provided with an accelerometer. Which signal-analysis methods can be used to determine the sources of the observed fluctuations?

4) State the advantage of the FFT compared to DFT.

5) What are the advantages of the the wavelet transform compared to the short time Fourier transform?

6) What is the lowest frequency of the speed signal of a 6 cyl. four-stroke combustion engine if one cylinder shows misfires ($n = 3000$ rpm)?

7) What are the features for fault detection by using stochastic signals with correlation analysis and ARMA-parameter estimation?

8) A rotating machine shows a basic displacement vibration with $f = 5$ Hz and amplitude 2 mm. Calculate the corresponding vibration speed $v(t)$ and acceleration $a(t)$.

9) A pair of gear-wheels has 20 and 50 teeth and rotates with $n = 3000$ rpm for the 20 teeth wheel. Calculate the tooth contact frequency f_t and angular frequency ω_t. Which frequencies can be observed if the smaller wheel is not centered? Which methods can be used for the detection of these frequencies due to the non-centered fault if the acceleration at the gear casing is measured?

9

Fault detection with process-identification methods

As described in Chapter 5 mathematical process models describe the relationship between input signals $u(k)$ and output signals $y(k)$ and are fundamental for model-based fault detection. In many cases the process models are not known at all or some parameters are unknown. Furtheron, the models have to be rather precise in order to express deviations as result of process faults. Therefore, process-identification methods have to be applied frequently before applying any model-based fault-detection method. But also the identification method itself may be a source to gain information on, e.g. process parameters which change under the influence of faults. First publications known on fault detection with identification methods are [9.24], [9.4], [9.26], [9.27], [9.28], [9.14] and [9.13].

The chapter gives a brief introduction into the most important identification methods for linear and nonlinear processes with single-input single-output which are relevant for fault detection. Table 9.1 shows a survey of the most important identification methods.

In Figure 9.1 those methods are extracted which can be applied for a wide range of processes and excitation signals at the input. Especially for dynamic processes the input signals have to change periodically, stochastically or by special test signals. These signals may be the normal operating signals (like in servo systems, actuators or driving vehicles) or may be artificially introduced for testing (after fault detection with other methods or for quality control during manufacturing or maintenance). A considerable advantage of identification methods is that with only one input and one output signal several parameters (up to about six) can be estimated, which give a detailed picture on internal process quantities. The generated features for fault detection are then impulse response values in the case of correlation methods or parameter estimates. Other identification methods like step response model matching or frequency response measurement are only in special cases suitable for automatic fault detection because of the special excitation and evaluation procedure.

The identification methods are first described for discrete-time and continuous-time linear processes, both in open loop and in closed loop and then for nonlinear processes.

Table 9.1. Survey of important identification methods TVS: time-variant systems; MIMO: multi-input multi-output systems, NLS: non-linear systems

input signal	model	output signal	identification method	used device	allowable disturbances	digital computer coupling		data processing		reachable accuracy	extendibility			application example
						off-line	on-line	one shot	real-time		TVS	MIMO	NLS	
	$\frac{K}{(1+Ts)^n}$ param.		determination of characteristic valves	recorder	very small	-	-	-	-	small	-	-	-	rough model, controller tuning
	$G(i\,\omega_v)$ nonparam.		Fourier analysis	recorder	small	X	-	X	-	medium	-	X	-	verification of theoretical models
	$G(i\,\omega_v)$ nonparam.		frequency response measurement	recorder, F. resp. meas. device	medium	X	X	-	-	very large	-	X	-	verification of theoretical models
	$g(t)$ nonparam.		correlation	correlator / process computer	large	- / X	- / X	- / X	X / X	large	X	X	X	detection of signal relations, time delay identification
			model adjustment	analog computer	small	-	-	-	X	medium	X	-	-	analog-adaptive control
	$\frac{b_0+b_1 s+\ldots}{1+a_1 s+\ldots}$ param.		parameter estimation	process computer	large	X	X	X	X	large	X	X	X	design of advanced controllers, adaptive control, fault detection
	param.		neural net	process computer	small/ medium	X	-	X	-	very large	-	X	X	design of non-linear controllers, learning controllers, fault detection

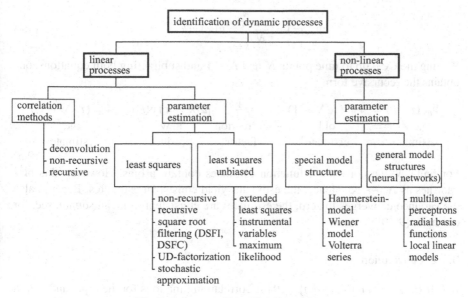

Fig. 9.1. Survey of identification methods

9.1 Identification with correlation functions

If stationary stochastic signals act on a linear process in open loop, then the impulse response $g(v)$ can be determined if the autocorrelation function of the input signal and the cross-correlation function of the input and output signal are known.

9.1.1 Estimation of correlation functions

The autocorrelation function (ACF) is defined by

$$\Phi_{uu}(\tau) = E\{u(k)u(k+\tau)\} = \lim_{N\to\infty} \frac{1}{N} \sum_{k=0}^{N-1} u(k)u(k+\tau) \qquad (9.1)$$

For finite samples of measured signals an estimate is given by

$$\hat{\Phi}_{uu}(\tau) = \frac{1}{N} \sum_{k=0}^{N-1} u(k)u(k+\tau) \qquad (9.2)$$

For the cross-correlation function (CCF)

$$\Phi_{uy}(\tau) = E\{u(k)y(k+\tau)\} = \lim_{N\to\infty} \frac{1}{N} \sum_{k=0}^{N-1} u(k)y(k+\tau) \qquad (9.3)$$

the estimate is

$$\hat{\Phi}_{uy}(\tau) = \frac{1}{N} \sum_{k=0}^{N-1} u(k)y(k+\tau) \qquad (9.4)$$

Writing the CCF up to time points N and $N-1$ and subtracting both equations, one obtains the recursive form

$$\hat{\Phi}_{uy}(\tau, N) = \hat{\Phi}_{uy}(\tau, N-1) + \tfrac{1}{N+1} \, [u(k-\tau)y(k) - \hat{\Phi}_{uy}(\tau, N-1)]$$

$$\begin{array}{ccccccc} \text{new} & = & \text{old} & + \text{correction} & \text{new} & - & \text{old} \\ \text{estimate} & & \text{estimate} & \text{factor} & \text{product} & & \text{estimate} \end{array}$$

$$(9.5)$$

For finite N the correlation function estimates contain a bias. However, this bias vanishes as $N \to \infty$. Hence, the estimates yield consistent estimates. Because also the variance converges to zero, the estimates are consistent in mean square, see, for example, [9.30].

9.1.2 Convolution

If $E\{u(k)\} = 0$ and $E\{y(k)\} = 0$ the correlation functions for the input and output signals of a linear process are related by the convolution sum

$$\Phi_{uy}(\tau) = \sum_{v=0}^{\infty} g(v)\Phi_{uu}(\tau - v) \qquad (9.6)$$

If $l+1$ values of $g(v)$ have to be determined and the convolution sum is truncated for $v > l$, then

$$\Phi_{uy}(\tau) \approx \Phi_{uu}^{T}\mathbf{g} \qquad (9.7)$$

where

$$\Phi_{uu}^{T} = [\Phi_{uu}(\tau)\Phi_{uu}(\tau-1)\ldots\Phi_{uu}(\tau-l)]$$
$$\mathbf{g}^{T} = [g(0)\,g(1)\ldots g(l)]$$

Now $l+1$ equations are required to get a unique solution. Therefore τ is varied within

$$-P < \tau < -P + 2l$$

and the equation system becomes

$$\Phi_{uy} \approx \hat{\Phi}_{uu} \cdot \hat{\mathbf{g}} \qquad (9.8)$$

As $\hat{\Phi}_{uu}$ is a $(l+1) \times (l+1)$ square matrix the impulse response estimates result from the deconvolution equation

$$\hat{\mathbf{g}} \approx \hat{\Phi}_{uu}^{-1} \, \Phi_{uy} \qquad (9.9)$$

where, of course, $\hat{\Phi}_{uu}$ can be inverted only if

$$\det\hat{\Phi}_{uu} \neq 0 \qquad (9.10)$$

which is an identifiability condition. In other words, the process must be persistently excited (of order $l + 1$).

If the input signal $u(k)$ is white noise with ACF

$$\Phi_{uu}(\tau) = \sigma_u^2 \, \delta \, (\tau) = \Phi_{uu}(0)\delta(\tau)$$

if follows from (9.6) that

$$\hat{g}(\tau) = \frac{1}{\hat{\Phi}_{uu}(0)} \hat{\Phi}_{uu}(\tau)$$

The impulse response is then proportional to the CCF.

If stochastic, stationary noise $n(k)$ acts on the process output, the necessary conditions for the consistent estimation of $\hat{g}(\tau)$ in mean square are the following:

- $n(k)$ and $y(k)$ are stationary;
- $E\{u(k)\} = 0$;
- $u(k)$ is persistently exciting;
- $n(k)$ is not correlated with $u(k)$.

For more details see [9.12], [9.30].

9.2 Parameter estimation for linear processes

It is assumed that the process can be described by the linear difference equation

$$\begin{aligned} y_u(k) &+ a_1 y_u(k-1) + \ldots + a_m \, y_u(k-m) \\ &= b_1 \, u(k-d-1) + \ldots + b_m u \, (k-d-m) \end{aligned} \tag{9.11}$$

Here,

$$\begin{aligned} u(k) &= U(k) - U_{00} \\ y_u(k) &= Y_u(k) - Y_{00} \end{aligned} \tag{9.12}$$

are the deviations of the absolute signals $U(k)$ and $Y_u(k)$ from the operating point described by U_{00} and Y_{00}, k is the discrete time $k = t/T_0 = 0, 1, 2, \ldots, T_0$ is the sampling time and $d = T_t/T_0 = 0, 1, 2, \ldots$ is the discrete dead-time of the process. The corresponding transfer function in the z-domain is

$$\begin{aligned} G_P(z) &= \frac{y_u(z)}{u(z)} = \frac{B(z^{-1})}{A(z^{-1})} z^{-d} \\ &= \frac{b_1 \, z^{-1} + \ldots + b_m z^{-m}}{1 + a_1 \, z^{-1} + \ldots + a_m z^{-m}} z^{-d} \end{aligned} \tag{9.13}$$

The measured signal contains a stationary, stochastic disturbance

$$y(k) = y_u \, (k) + n \, (k) \quad \text{with} \ E\{n(k)\} = 0 \tag{9.14}$$

The task is to determine the unknown parameters a_i and b_i from N measured input and output signal data points, see Figure 9.2.

9.2.1 Method of least squares (LS)

a) Equation error methods

Let the model parameters obtained from the data up to the sample $(k-1)$ be denoted by $\hat{a}_i$ and $\hat{b}_i$. Then, (9.11) becomes in the presence of a disturbed output signal

$$
\begin{aligned}
& y(k) + \hat{a}_1\, y(k-1) \ldots + \hat{a}_m\, y(k-m) \\
& -\hat{b}_1\, u(k-d-1) - \ldots - \hat{b}_m u(k-d-m) = e(k)
\end{aligned}
\tag{9.15}
$$

where the *equation error* (residual) $e(t)$ is introduced instead of "0". This error corresponds to a generalized error, see Figure 9.2. This can be seen by rewriting (9.15), compare Figure 9.2a.

$$
\hat{A}\,(z^{-1})y(z) - \hat{B}(z^{-1})\, z^{-d}\, u(z) = e\,(z)
\tag{9.16}
$$

e is linearly dependent on the parameters sought for (linear in the parameters).

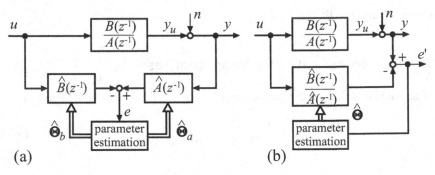

Fig. 9.2. Model structures for parameter estimation: (a) equation error; (b) output error

From (9.15), $\hat{y}(k|k-1)$ can be interpreted as the one-step-ahead prediction, based on the measurements up to sample $(k-1)$

$$
\hat{y}(k|k-1) = \boldsymbol{\psi}^T(k)\hat{\boldsymbol{\theta}}
\tag{9.17}
$$

with the data vector

$$
\boldsymbol{\psi}^T(k) = [-y\,(k-1)\ldots - y\,(k-m)\,|\, u(k-d-1)\ldots \\
u(k-d-m)]
\tag{9.18}
$$

and the parameter vector

$$
\hat{\boldsymbol{\theta}} = [\hat{a}_1 \ldots \hat{a}_m \,|\, \hat{b}_1 \ldots \hat{b}_m]^T
\tag{9.19}
$$

Consequently, (9.15) can be written as

$$
y(k) = \boldsymbol{\psi}^T(k)\hat{\boldsymbol{\theta}} + e(k)
\tag{9.20}
$$

The measured signals for $k = m + d, \ldots, m + d + N$ are written in vectors, e.g.

$$\mathbf{y}^T (m + d + N) = [y(m + d) \ldots y(m + d + N)] \tag{9.21}$$

Then,

$$y(m + d + n) = \mathbf{\Psi}(m + d + N)\hat{\boldsymbol{\theta}} + e(m + d + N) \tag{9.22}$$

where $\mathbf{\Psi}$ is a $((N + 1) \times 2m)$-data matrix. Minimizing the sum of errors squared

$$V = \sum_{k=m+d}^{m+d+N} e^2 (k) = \mathbf{e}^T (m + d + N)\, \mathbf{e}\, (m + d + N) \tag{9.23}$$

yields

$$\left. \frac{dV}{d\boldsymbol{\theta}} \right|_{\boldsymbol{\theta}=\hat{\boldsymbol{\theta}}} = -2\mathbf{\Psi}^T [\mathbf{y} - \mathbf{\Psi}\hat{\boldsymbol{\theta}}] = \mathbf{0} \tag{9.24}$$

for the unknown parameters. From this, the (nonrecursive) estimation equation of the least squares (LS) method can be obtained

$$\hat{\boldsymbol{\theta}} = [\mathbf{\Psi}^T \mathbf{\Psi}]^{-1} \mathbf{\Psi}^T \mathbf{y} \tag{9.25}$$

The matrix

$$\mathbf{P} = [\mathbf{\Psi}^T \mathbf{\Psi}]^{-1} \tag{9.26}$$

has the dimension $(2m, 2m)$. The inverse exists if and only if

$$\det [\mathbf{\Psi}^T \mathbf{\Psi}] = \det \mathbf{P}^{-1} \neq 0 \tag{9.27}$$

Also,

$$\frac{\partial^2 V}{\partial \boldsymbol{\theta} \, \partial \boldsymbol{\theta}^T} = \mathbf{\Psi}^T \mathbf{\Psi} \tag{9.28}$$

has to be positive-definite such that the loss function V has a minimum. Both requirements are satisfied if and only if

$$\det [\mathbf{\Psi}^T \mathbf{\Psi}] = \det \mathbf{P}^{-1} > 0 \tag{9.29}$$

This condition also includes that the input signal is persistently exciting the process and that the process is *stable*.

From parameter estimation methods, it is usually required that the estimate is not biased for a finite number of data samples N

$$E \{\boldsymbol{\theta} \, (N)\} = \boldsymbol{\theta}_0 \tag{9.30}$$

($\boldsymbol{\theta}_0$ denotes the true parameters) and is consistent in the quadratic mean

$$\lim_{N \to \infty} E \, \hat{\boldsymbol{\theta}} \, (N) = \boldsymbol{\theta}_0 \tag{9.31}$$

$$\lim_{N \to \infty} E \, [\hat{\boldsymbol{\theta}} \, (N) - \boldsymbol{\theta}_0] \, [\boldsymbol{\theta} \, (N) - \boldsymbol{\theta}_0]^T = \mathbf{0} \tag{9.32}$$

For the least squares method, (9.31) becomes, by substituting (9.22) into (9.25)

$$E\left\{\hat{\theta}\left(N\right)\right\} = \theta_0 + E\left\{[\Psi^T\ \Psi]^{-1}\Psi^T\mathbf{e}\right\}$$
$$= \theta_0 + \mathbf{b} \tag{9.33}$$

In order to have a vanishing bias (systematic estimation error) $\mathbf{b}$, Ψ^T and $\mathbf{e}$ must be uncorrelated. Consequently, $e(k)$ must not be correlated and $E\{e(k)\} = 0$. The estimation is unbiased if the disturbance signal $n(k)$ is generated by the disturbance filter

$$G_v(z) = \frac{n(z)}{v(z)} = \frac{1}{A(z^{-1})} \tag{9.34}$$

where $v(k)$ is discrete white noise, see Figure 9.3. Since this filter does not exist in practice, the least squares estimation, in general, yields biased estimates. These systematic estimation errors are the larger the greater the variance σ_n^2 of the disturbance signal is compared to the output signal σ_{yu}^2.

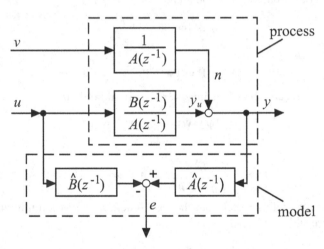

Fig. 9.3. Model configuration for the least squares method with equation error (generalized error)

For the covariance matrix, the following is true if $\hat{\theta} = \theta_0$ (which means $e = 0$)

$$\mathrm{cov}\left[\Delta\ \theta\right] = E\left\{\left[\hat{\theta} - \theta_0\right]\left[\hat{\theta} - \theta_0\right]^T\right\}\ \sigma_e^2\ E\ \{\mathbf{P}\}$$
$$= \sigma_e^2\ E\left\{\left[\frac{1}{N+1}\Psi^T\ \Psi\right]^{-1}\right\}\frac{1}{N+1}$$
$$= \sigma_e^2\left\{\Phi^{-1}(N+1)\right\}\frac{1}{N+1} \tag{9.35}$$

σ_e^2 is the variance of $e(k)$. Ψ is a matrix whose elements are correlation functions. For $N \to \infty$, (9.32) is satisfied. $E\{\mathbf{P}\}$ is proportional to the covariance matrix of the parameter estimation errors.

Because of the biased estimates for the least squares algorithm, this method can only be used for processes with no or only small disturbance signals. A big advantage of the least squares algorithm, however, is that the parameter vector $\hat{\theta}$ can be determined in one batch calculation and no iterative methods are necessary. This is possible since the employed error measure is linear in the parameters.

b) Output error methods

Instead of the equation error the output error

$$e'(k) = y(k) - y_M\left(\hat{\theta}, k\right) \tag{9.36}$$

can be used, where

$$y_M\left(\hat{\theta}, z\right) = \frac{\hat{B}(z^{-1})}{\hat{A}(z^{-1})} u(z) \tag{9.37}$$

is the model equation output, see Figure 9.2b. But then no direct calculation of the parameter estimates $\hat{\theta}$ is possible because $e'(k)$ is nonlinear in the parameters. Therefore the loss function (9.23) is minimized by a numerical optimization method, e.g. downhill-simplex. The computational effort is then larger, and on-line real-time application, in general, not possible. However, relative precise parameter estimates may be obtained, [9.10], [9.50].

c) Recursive least squares (RLS) methods

Writing the nonrecursive estimation equations for $\hat{\theta}(k+1)$ and $\hat{\theta}(k)$ and subtracting one from the other, results in the recursive parameter estimation algorithm

$$
\underset{\substack{\text{new} \\ \text{estimate}}}{\hat{\theta}(k+1)} = \underset{\substack{\text{old} \\ \text{estimate}}}{\hat{\theta}(k)} + \underset{\substack{\text{correction} \\ \text{factor}}}{\gamma(k)} \; [\underset{\substack{\text{new} \\ \text{measurement}}}{y(k+1)} - \underset{\substack{\text{one - step - ahead} \\ \text{prediction of the new} \\ \text{measurement}}}{\psi^T(k+1)\hat{\theta}(k)}] \tag{9.38}
$$

The correcting vector is given by

$$\gamma(k) = \mathbf{P}(k+1)\,\psi(k+1)$$

$$= \frac{1}{\psi^T(k+1)\,\mathbf{P}(k)\,\psi(k+1)+1}\,\mathbf{P}(k)\,\psi(k+1) \tag{9.39}$$

and

$$\mathbf{P}(k+1) = [\mathbf{I} - \gamma(k)\,\psi^T(k+1)]\,\mathbf{P}(k) \tag{9.40}$$

To start the recursive algorithm one sets

$$\hat{\theta}(0) = 0$$
$$P(0) = \alpha I \qquad (9.41)$$

with α large ($\alpha = 100, \ldots, 1000$). The expectation of the matrix P is proportional to the covariance matrix of the parameter estimates

$$E\{P(k+1)\} = \frac{1}{\sigma_e^2} \operatorname{cov}[\Delta\theta(k)] \qquad (9.42)$$

with

$$\sigma_e^2 = E\{e^T e\} \qquad (9.43)$$

and the parameter error

$$\Delta\theta(k) = \hat{\theta}(k) - \theta_0 \qquad (9.44)$$

Hence, the recursive algorithm contains the variances of the parameter estimates (diagonal elements of covariance matrix). (9.38) can also be written as

$$\hat{\theta}(k+1) = \hat{\theta}(k) + \gamma(k)\, e(k+1) \qquad (9.45)$$

To improve the numerical properties of the basic RLS algorithms, modified versions are recommended, see Section 9.2.3.

DC value estimation

As for process parameter estimation the variations of $u(k)$ and $y(k)$ of the measured signals $U(k)$ and $Y(k)$ have to be used. The DC (direct current or steady-state) values U_{00} and Y_{00} either have also to be estimated or have to be removed. The following methods are available.

Differencing
The easiest way to to obtain the variations without knowing the DC values is just to take the differences

$$U(k) - U(k-1) = u(k) - u(k-1) = \Delta u(k)$$
$$Y(k) - Y(k-1) = y(k) - y(k-1) = \Delta y(k) \qquad (9.46)$$

Instead of $u(z)$ and $y(z)$, the signals $\Delta u(z) = u(z)(1 - z^{-1})$ and $\Delta y(z) = y(z)(1 - z^{-1})$ are then used for the parameter estimation. As this special high-pass filtering is applied to both the process input and output, the process parameters can be estimated in the same way as in the case of measuring $u(k)$ and $y(k)$. In the parameter estimation algorithms $u(k)$ and $y(k)$ have to be replaced by $\Delta u(k)$ and $\Delta y(k)$. However, if the DC values are required explicitly, other methods habe to be used.

Averaging
The DC values can be estimated simply by averaging:

$$\hat{Y}_{00} = \frac{1}{M} \sum_{k=1}^{M} Y(k) \tag{9.47}$$

The recursive version of this is

$$\hat{Y}_{00} = \hat{Y}_{00}(k-1) + \frac{1}{k} \left[Y(k) - \hat{Y}_{00}(k-1) \right] \tag{9.48}$$

For slowly time-varying DC values, recursive averaging with exponential forgetting leads to

$$\hat{Y}_{00} = \hat{Y}_{00}(k-1) + (1-\lambda) Y(k) \tag{9.49}$$

with $\lambda < 1$. A similar argument applies for U_{00}. The variations $u(k)$ and $y(k)$ can be determined by (9.12).

Implicit estimation of a constant
The estimation of the DC values U_{00} and Y_{00} can also be included into the parameter estimation. Substituting (9.12) into (9.11) results in

$$\begin{aligned} Y(k) &= -a_1 Y(k-1) - \ldots - a_m Y(k-m) + b_1 U(k-d-1) \\ &\quad + \ldots + b_m U(k-d-m) + C \end{aligned} \tag{9.50}$$

where

$$C = (1 + a_1 + \ldots + a_m) Y_{00} - (b_1 + \ldots + b_m) U_{00} \tag{9.51}$$

Extending the parameter vector $\hat{\theta}$ by including the element C and the data vector $\psi^T(k)$ by adding the number 1, the measured $Y(k)$ and $U(k)$ can be directly used for the estimation and C also can be estimated. Then, for one given DC value the other can be calculated, using (9.51). For closed-loop identification it is convenient to use

$$Y_{00} = W(k) \tag{9.52}$$

Explicit estimation of a constant
The parameters $\hat{a}_i$ and $\hat{b}_i$ for the dynamic behavior and the DC constant C can also be estimated separately. First the dynamic parameters are estimated using the differencing method above. Then with

$$\begin{aligned} L(k) &= Y(k) + \hat{a}_1 Y(k-1) + \ldots + \hat{a}_m Y(k-m) \\ &\quad - \hat{b}_1 U(k-d-1) - \ldots - \hat{b}_m U(k-d-m) \end{aligned} \tag{9.53}$$

the equation becomes

$$e(k) = L(k) - C \tag{9.54}$$

and after applying the LS method

$$C(m + d + N) = \frac{1}{N + 1} \sum_{k=m+d}^{m+d+N} L(k) \tag{9.55}$$

For large N one obtains

$$\hat{C} \approx \left[1 + \sum_{i=1}^{m} \hat{a}_i\right] \hat{Y}_{00} - \left[\sum_{i=1}^{m} \hat{b}_i\right] \hat{U}_{00} \tag{9.56}$$

If the $\hat{Y}_{00}$ is of interest and U_{00} is known, it can be calculated from (9.56) using the estimate $\hat{C}$.

In this case $\hat{\theta}$ and $\hat{C}$ are only coupled in one direction, as $\hat{\theta}$ does not depend on $\hat{C}$. A disadvantage can be the worse noise-to-signal ratio through differencing.

The final selection of the DC method depends on the particular application.

9.2.2 Extended least squares (ELS) method

If instead of the LS method

$$A(z^{-1})y(z) - B(z^{-1}) z^{-d} u(z) = \varepsilon(z) \tag{9.57}$$

with an correlated error signal $\varepsilon(z)$ the ARMAX model

$$A(z^{-1}) y(z) - B(z^{-1}) z^{-d} u(z) = D(z^(- 1)) e(z) \tag{9.58}$$

with a correlated signal $\varepsilon(z) = D(z^{-1}) e(z)$ is used, the recursive methods for dynamical processes and for stochastic signals can be combined to form an extended least squares method, [9.61], [9.47]. Based on

$$y(k) = \psi^T(k) \, \hat{\theta}(k - 1) + e(k) \tag{9.59}$$

the following extended vectors are introduced:

$$\psi^T(k) = [-y(k - 1) \ldots - y(k - m) \, u(k - d - 1) \ldots$$
$$u(k - d - m) \, \hat{v}(k - 1) \ldots \hat{v}(k - p)] \tag{9.60}$$

$$\hat{\theta}^T = [\hat{a}_1 \ldots \hat{a}_m \, \hat{b}_1 \ldots \hat{b}_m \, \hat{d}_1 \ldots \hat{d}_m] \tag{9.61}$$

The recursive version is especially suited to parameter estimation. The parameters are then obtained by

$$\hat{\theta}(k + 1) = \hat{\theta}(k) + \gamma(k) [y(k + 1) - \hat{\psi}^T(k + 1)\hat{\theta}(k)] \tag{9.62}$$

and equations corresponding to (9.39) – (9.40). The signal values $\hat{v}(k) = e(k)$ in $\hat{\psi}^T(k + 1)$ are calculated recursively. Therefore the roots of $D(z) = 0$ must lie within the unit circle of the z-plane. The parameter estimation is unbiased and consistent in mean square if the convergence conditions of the LS method are transferred to the model equation (9.59). That means that (9.58) has to be valid. In addition,

$$H(z) = 1/\left[D(z)\right] - 1/2 \tag{9.63}$$

must be positive real. Besides this ELS method several methods exist, for example, those of instrumental variables or maximum likelihood, see, e.g. [9.12], [9.37], [9.62] and [9.30].

9.2.3 Modifications of basic recursive estimators

To improve some of the properties of basic parameter estimation methods, the corresponding algorithms can be modified. These modifications serve to improve the numerical properties in digital computers, to give access to intermediate results and to diminish the influence of starting values. The numerical properties are important if the word length is relatively small, as with 8-bit or 16-bit microcomputers, or if the changes of input signals becomes small, as in adaptive control or fault detection. In both cases ill-conditioned equation systems result.

The numerical conditions can now be improved by not calculating $\mathbf{P}$ as intermediate value, in which the squares of signals appear, but square roots of $\mathbf{P}$. This leads to *square-root filtering methods* or *factorization methods*, see, for example, [9.6]. By this means, forms can be distinguished which start from the covariance matrix $\mathbf{P}$ or the information matrix $\mathbf{P}^{-1}$, [9.33], [9.6], [9.34]. The following treatment leans on [9.31].

a) Factorization methods for P

For *discrete square-root filtering in covariance form* (DSFC) the symmetric matrix $\mathbf{P}$ is decomposed in two triangular matrices

$$\mathbf{P} = \mathbf{S}\mathbf{S}^T \tag{9.64}$$

where $\mathbf{S}$ is called the *square root* or the *Cholesky factor* of $\mathbf{P}$. For the RLS method the resulting algorithm then becomes:

$$
\begin{aligned}
\hat{\boldsymbol{\theta}}(k+1) &= \hat{\boldsymbol{\theta}}(k) + \boldsymbol{\gamma}(k)e(k+1) \\
\boldsymbol{\gamma}(k) &= a(k)\mathbf{S}(k)\mathbf{f}(k) \\
\mathbf{f}(k) &= \mathbf{S}^T(k)\boldsymbol{\psi}(k+1) \\
\mathbf{S}(k+1) &= [\mathbf{S}(k) - g(k)\boldsymbol{\gamma}(k)\mathbf{f}^T(k)]/\sqrt{\lambda(k)} \\
1/[a(k)] &= \mathbf{f}^T(k)\mathbf{f}(k) + \lambda(k) \\
g(k) &= 1/[1 + \sqrt{\lambda(k)a(k)}]
\end{aligned}
\tag{9.65}
$$

The starting values are $\mathbf{S}(0) = \sqrt{\alpha}\,\mathbf{I}$ and $\hat{\boldsymbol{\theta}}(0) = \mathbf{0}$. A disadvantage is the calculation of the square roots for each recursion.

Another method has been proposed by [9.6], the so-called *U-D factorization* (DUDC). Here, the covariance matrix is factorized by

$$\mathbf{P} = \mathbf{U}\mathbf{D}\mathbf{U}^T \tag{9.66}$$

where $\mathbf{D}$ is diagonal and $\mathbf{U}$ is an upper triangular matrix with ones in the diagonal. Then the recursions for the covariance matrix are

$$\mathbf{U}(k+1)\,\mathbf{D}(k+1)\,\mathbf{U}^T(k+1)$$
$$= \frac{1}{\lambda}[\mathbf{U}(k)\,\mathbf{D}(k)\,\mathbf{U}^T(k) - \gamma(k)\,\boldsymbol{\psi}^T(k+1)\,\mathbf{U}(k)\,\mathbf{D}(k)\,\mathbf{U}^T(k)] \quad (9.67)$$

After substitution of (9.39) and (9.99) the right-hand sides becomes

$$\mathbf{UDU}^T = \frac{1}{\lambda}\mathbf{U}(k)\left[\mathbf{D}(k) - \frac{1}{\alpha(k)}\mathbf{v}(k)\,\mathbf{f}^T(k)\,\mathbf{D}(k)\right]\mathbf{U}^T(k)$$
$$= \frac{1}{\lambda}\mathbf{U}(k)\left[\mathbf{D}(k) - \frac{1}{\alpha(k)}\mathbf{v}(k)\,\mathbf{v}^T(k)\right]\mathbf{U}^T(k) \quad (9.68)$$

where

$$\mathbf{f}(k) = \mathbf{U}^T(k)\,\boldsymbol{\psi}(k+1)$$
$$\mathbf{v}(k) = \mathbf{D}(k)\,\mathbf{f}(k)$$
$$\alpha(k) = \lambda + \mathbf{f}^T(k)\,\mathbf{v}(k) \quad (9.69)$$

The correcting vector then yields

$$\gamma(k) = \frac{1}{\alpha(k)}\,\mathbf{U}(k)\mathbf{v}(k) \quad (9.70)$$

If the term $(\mathbf{D} - \alpha^{-1}\mathbf{vv}^T)$ in (9.68) is again factorized the recursion for the elements $\mathbf{U}$, $\mathbf{P}$ and λ become

$$\left.\begin{array}{rcl}\alpha_j &=& \alpha_{j-1} + v_f\,f_j \\ d_j(k+1) &=& d_j(k)\alpha(j-1)/(\alpha_j - \lambda) \\ b_j &=& v_j \\ v_j &=& f_j/\alpha_{j-1}\end{array}\right\} \quad j = 2,\dots,2m \quad (9.71)$$

[9.6] with the initial values

$$\begin{array}{rcl}\alpha_1 &=& \lambda + v_1\,f_1; \qquad d_1(k+1) = \dfrac{d_1(k)}{\alpha_1\lambda} \\ b_1 &=& v_1\end{array} \quad (9.72)$$

For each j the following expressions hold for the elements of $\mathbf{U}$

$$\left.\begin{array}{rcl}u_{ij}(k+1) &=& u_{ij}(k) + r_j\,b_i \\ b_i &=& b_i + u_{ij}\,v_j\end{array}\right\} \quad i = 1,\dots,j \quad (9.73)$$

$$\gamma(k) = \frac{1}{\alpha_{2m}}\,\mathbf{b} \quad (9.74)$$

The parameters are finally obtained from (9.38)

$$\begin{aligned}
\hat{\boldsymbol{\theta}}(k+1) &= \hat{\boldsymbol{\theta}}(k) + \boldsymbol{\gamma}(k)\, e(k+1) \\
e(k+1) &= \mathbf{y}(k+1) - \boldsymbol{\psi}^T(k+1)\, \hat{\boldsymbol{\theta}}(k)
\end{aligned} \tag{9.75}$$

(9.74), (9.71) and (9.73) are calculated instead of (9.39) and (9.40). As compared to DSFC, here no routines are required for square-root calculations. The computational expense is comparable to that of RLS. The numerical properties are similar to those of DSFC, only the matrix elements of $\mathbf{U}$ and $\mathbf{D}$ may become larger than those of $\mathbf{S}$.

To reduce the calculations after each sampling, invariance properties of the matrices, [9.36], may be used to generate fast algorithms. A saving of calculation time only results for order $m > 5$, but at the cost of greater storage requirements and higher sensitivity for starting values.

b) Factorization methods for $\mathbf{P}^{-1}$[1]

Discrete square-root filtering in information form (DSFI) results from the nonrecursive LS method of the form

$$\mathbf{P}^{-1}(k+1)\, \hat{\boldsymbol{\theta}}(k+1) = \boldsymbol{\Psi}^T(k+1)\, \mathbf{y}(k+1) = \mathbf{f}(k+1) \tag{9.76}$$

with

$$\begin{aligned}
\mathbf{P}^{-1}(k+1) &= \lambda\, \mathbf{P}^{-1}(k) + \boldsymbol{\psi}(k+1)\boldsymbol{\psi}^T(k+1) \\
\mathbf{f}(k+1) &= \lambda\, \mathbf{f}(k) + \boldsymbol{\psi}(k+1)y(k+1)
\end{aligned} \tag{9.77}$$

The information matrix $\mathbf{P}^{-1}$ is now split into upper triangular matrices $\mathbf{R}$:

$$\mathbf{P}^{-1} = \mathbf{R}^T\mathbf{R} \tag{9.78}$$

Note that $\mathbf{R} = \mathbf{S}^{-1}$, cf. (9.64). Then $\hat{\boldsymbol{\theta}}(k+1)$ is calculated from (9.76) by back-substitution from

$$\mathbf{R}(k+1)\, \hat{\boldsymbol{\theta}}(k+1) = \mathbf{b}(k+1) \tag{9.79}$$

This equation follows from (9.76) introducing an orthonormal transformation matrix $\mathbf{Q}$ (with $\mathbf{Q}^T\mathbf{Q} = \mathbf{I}$) such that

$$\boldsymbol{\Psi}^T\mathbf{Q}^T\mathbf{Q}\boldsymbol{\Psi}\hat{\boldsymbol{\theta}} = \boldsymbol{\Psi}^T\mathbf{Q}^T\mathbf{Q}\mathbf{y} \tag{9.80}$$

Here

$$\mathbf{Q}\boldsymbol{\Psi} = \begin{bmatrix} \mathbf{R} \\ \mathbf{0} \end{bmatrix} \tag{9.81}$$

possesses an upper triangular from, and the equation

$$\mathbf{Q}\mathbf{y} = \begin{bmatrix} \mathbf{b} \\ \mathbf{w} \end{bmatrix} \tag{9.82}$$

holds. With equation (9.80) it follows that

---

[1] compiled by Michael Vogt

$$\mathbf{Q}(k+1)\mathbf{\Psi}(k+1)\hat{\boldsymbol{\theta}}(k+1) = \mathbf{Q}(k+1)\mathbf{y}(k+1) \qquad (9.83)$$

Actually, DSFI uses a different idea to minimize the sum of errors squared

$$V = \sum e^2(k) = \|\mathbf{e}\|_2^2 = \|\mathbf{\Psi}\hat{\boldsymbol{\theta}} - \mathbf{y}\|_2^2 \qquad (9.84)$$

Whereas the LS method solves the *normal equations* $\nabla V = \mathbf{0}$, here the *QR factorization* $\mathbf{Q}\mathbf{\Psi} = \binom{\mathbf{R}}{\mathbf{0}}$ is used to simplify (9.84). This relies on the fact that the multiplication with an orthonormal matrix $\mathbf{Q}$ does not change the norm of a vector:

$$V = \|\mathbf{\Psi}\hat{\boldsymbol{\theta}} - \mathbf{y}\|_2^2 = \|\mathbf{Q}\mathbf{\Psi}\hat{\boldsymbol{\theta}} - \mathbf{Q}\mathbf{y}\|_2^2 = \|\begin{pmatrix}\mathbf{R}\\\mathbf{0}\end{pmatrix}\hat{\boldsymbol{\theta}} - \begin{pmatrix}\mathbf{b}\\\mathbf{w}\end{pmatrix}\|_2^2$$

$$= \|\begin{pmatrix}\mathbf{R}\hat{\boldsymbol{\theta}} - \mathbf{b}\\\mathbf{0} - \mathbf{w}\end{pmatrix}\|_2^2 = \|\mathbf{R}\hat{\boldsymbol{\theta}} - \mathbf{b}\|_2^2 + \|\mathbf{w}\|_2^2 = \min_{\hat{\boldsymbol{\theta}}} \qquad (9.85)$$

As already stated in (9.79), the parameters $\hat{\boldsymbol{\theta}}$ are determined by solving the system $\mathbf{R}\hat{\boldsymbol{\theta}} - \mathbf{b} = \mathbf{0}$, whereas $\|\mathbf{w}\|_2^2$ is the remaining residual, i.e. the sum of errors squared for the optimal parameters $\hat{\boldsymbol{\theta}}$. The advantage of this orthogonalization method can be seen from the error sensitivity of the system that determines the parameters, see [9.15]. If the normal equations (9.76) are directly solved by the LS method, the parameter error is bounded by

$$\frac{\|\Delta\hat{\boldsymbol{\theta}}\|}{\|\hat{\boldsymbol{\theta}}\|} \leq \text{cond}(\mathbf{P}^{-1}) \frac{\|\Delta\mathbf{y}\|}{\|\mathbf{y}\|} = \text{cond}^2(\mathbf{\Psi}) \frac{\|\Delta\mathbf{y}\|}{\|\mathbf{y}\|}, \qquad (9.86)$$

where $\text{cond}(\cdot)$ is the *condition number* measuring the error sensitivity of the solution with respect to the error $\Delta\mathbf{y}$ in the process output signal. A similar bound can also be found for the input signal. However, if the orthogonalization approach is used, the upper bound for the parameter errors is given by

$$\frac{\|\Delta\hat{\boldsymbol{\theta}}\|}{\|\hat{\boldsymbol{\theta}}\|} \leq \text{cond}(\mathbf{R}) \frac{\|\Delta\mathbf{b}\|}{\|\mathbf{b}\|} = \text{cond}(\mathbf{\Psi}) \frac{\|\Delta\mathbf{y}\|}{\|\mathbf{y}\|}, \qquad (9.87)$$

i.e. the system (9.79) is much less sensitive to measurement errors then the normal equations (9.76) themselves.

The main effort of the method described above is the computation of $\mathbf{R}$ and $\mathbf{b}$. This is usually done by applying *Householder transformations* to the matrix $(\mathbf{\Psi}, \mathbf{y})$, see [9.15], so that $\mathbf{Q}$ does not need to be computed. However, DSFI computes $\mathbf{R}$ and $\mathbf{b}$ *recursively*. Assuming that in each step one row is appended to $(\mathbf{\Psi}, \mathbf{y})$, (9.83) is now transferred to a recursive form, [9.33],

$$\begin{bmatrix}\mathbf{R}(k+1)\\\mathbf{0}^T\end{bmatrix} = \mathbf{Q}(k+1)\begin{bmatrix}\sqrt{\lambda}\,\mathbf{R}(k)\\\boldsymbol{\psi}^T(k+1)\end{bmatrix} \qquad (9.88)$$

$$\begin{bmatrix}\mathbf{b}(k+1)\\w(k+1)\end{bmatrix} = \mathbf{Q}(k+1)\begin{bmatrix}\sqrt{\lambda}\,\mathbf{b}(k)\\y(k+1)\end{bmatrix} \qquad (9.89)$$

Then $\mathbf{R}(k + 1)$ and $\mathbf{b}(k + 1)$ are used to calculate $\hat{\boldsymbol{\theta}}(k + 1)$ with (9.79), whereas $w(k + 1)$ is the current residual. This form is partially nonrecursive and partially recursive and has the advantage that no starting values have to be assumed and that $\mathbf{R}(0) = \mathbf{0}$. The method is especially suitable if the parameters are not required for each sampling time. Then only $\mathbf{R}$ and $\mathbf{b}$ have to be calculated recursively. This is done by applying *Givens rotations* to the right hand sides of (9.88) and (9.89). The Givens rotation

$$\mathbf{G} = \begin{bmatrix} \gamma & \sigma \\ -\sigma & \gamma \end{bmatrix} \tag{9.90}$$

is applied to a $2 \times \mu$ matrix $\mathbf{M}$ in order to eliminate the element m'_{21} in the transformed matrix $\mathbf{M}' = \mathbf{GM}$, i.e. to introduce a zero in the matrix

$$\begin{bmatrix} \gamma & \sigma \\ -\sigma & \gamma \end{bmatrix} \begin{bmatrix} m_{11} & m_{12} & \cdots \\ m_{21} & m_{22} & \cdots \end{bmatrix} = \begin{bmatrix} m'_{11} & m'_{12} & \cdots \\ 0 & m'_{22} & \cdots \end{bmatrix} \tag{9.91}$$

The two conditions

$$\det(\mathbf{G}) = \gamma^2 + \sigma^2 = 1 \qquad \text{(normalization)} \tag{9.92}$$
$$m'_{21} = -\sigma m_{11} + \gamma m_{21} = 0 \qquad \text{(elimination of } m'_{21}) \tag{9.93}$$

yield the rotation parameters

$$\gamma = \frac{m_{11}}{\sqrt{m_{11}^2 + m_{21}^2}} \qquad \text{and} \qquad \sigma = \frac{m_{21}}{\sqrt{m_{11}^2 + m_{21}^2}}$$

This transformation is now sequentially applied to $\boldsymbol{\psi}^T(k + 1)$ and the rows of $\sqrt{\lambda}\mathbf{R}$ in (9.88), where $\mathbf{G}$ is now interpreted as a $(n + 1) \times (n + 1)$ matrix

$$\begin{bmatrix} * & * & * \\ 0 & * & * \\ 0 & 0 & * \\ * & * & * \end{bmatrix} \xrightarrow{\mathbf{G}_1} \begin{bmatrix} \bullet & \bullet & \bullet \\ 0 & * & * \\ 0 & 0 & * \\ 0 & * & * \end{bmatrix} \xrightarrow{\mathbf{G}_2} \begin{bmatrix} \bullet & \bullet & \bullet \\ 0 & \bullet & \bullet \\ 0 & 0 & * \\ 0 & 0 & * \end{bmatrix} \xrightarrow{\mathbf{G}_3} \begin{bmatrix} \bullet & \bullet & \bullet \\ 0 & \bullet & \bullet \\ 0 & 0 & \bullet \\ 0 & 0 & 0 \end{bmatrix}$$

The product of the Givens matrices is the transformation matrix $\mathbf{Q}(k + 1)$:

$$\begin{bmatrix} \mathbf{R}(k + 1) \\ \mathbf{0}^T \end{bmatrix} = \underbrace{\mathbf{G}_n(k + 1) \dots \mathbf{G}_1(k + 1)}_{\mathbf{Q}(k + 1)} \begin{bmatrix} \sqrt{\lambda}\,\mathbf{R}(k) \\ \boldsymbol{\psi}^T(k + 1) \end{bmatrix} \tag{9.94}$$

that produces $\mathbf{R}(k + 1)$. The same method is used to compute $\mathbf{b}(k + 1)$. A complete DSFI update step can be described as follows:

Compute for $i = 1, \ldots, n$:

$$r_{ii}(k+1) = \sqrt{\lambda r_{ii}^2(k) + (\psi_i^{(i)}(k+1))^2}$$
$$\gamma = r_{ii}(k)/r_{ii}(k+1)$$
$$\sigma = \psi_i^{(i)}(k+1)/r_{ii}(k+1)$$
$$r_{ij}(k+1) = \sqrt{\lambda}\gamma r_{ij}(k) + \sigma\psi_j^{(i)}(k+1) \left.\begin{array}{r}\\ \\\end{array}\right\}$$
$$\psi_j^{(i+1)}(k+1) = -\sqrt{\lambda}\sigma r_{ij}(k) + \gamma\psi_j^{(i)}(k+1) \quad j = i+1, \ldots, n$$
$$b_i(k+1) = \sqrt{\lambda}\gamma b_i(k) + \sigma y^{(i)}(k+1)$$
$$y^{(i+1)}(k+1) = -\sqrt{\lambda}\sigma b_i(k) + \gamma y^{(i)}(k+1)$$

$$(9.95)$$

No essential differences in the numerical properties can be observed for DSFC and DSFI. Therefore, also DSFI requires the computation of n square-roots in each step. There are also factorizations for $\mathbf{P}^{-1}$ that do not require square-roots, just like the U-D factorization for $\mathbf{P}$. These techniques replace the Givens rotations by *fast Givens rotations*, see [9.15], or employ recursive forms of the *Gram Schmidt orthogonalization*. These fast orthogonalization methods show the same error sensitivity, but their matrix elements may become larger than those of DSFI.

Further discussion of square-root filtering may be found in [9.49], [9.16] and [9.55].

9.2.4 Parameter estimation of time-varying processes

For many processes the parameters of the models are not constant. They change because of internal or external influence with time. Frequently the case arises that the dynamic behavior is linearized around the operating point for small signal changes. After changes of the operating point the real nonlinear behavior becomes effective. If the nonlinear behavior is not very strong and the operating point changes slowly, useful results can be obtained with linear difference equations and time-varying parameters.

The following treatment is for RLS.

a) Exponential weighting with constant forgetting

Up to now it has been assumed that the process parameters to be estimated are constant and therefore the measured signals $u(k)$ and $y(k)$ and the equation error $e(k)$ are weighted equally over the measuring time $k = 0, \ldots, N$. If the recursive estimation algorithms are to be able to follow slowly time-varying process parameters, more recent measurements must be weighted more strongly than old measurements. Therefore the estimation algorithms should have a *fading memory*. This can be incorporated into the least-squares method by time-depending weighting of the squared errors, the method of weighted least squares, see, e.g. [9.29]:

$$V(k) = \sum_{i=m+d}^{m+d+N} w(i)\, e^2(i) \tag{9.96}$$

By choice of

$$w(k) = \lambda^{(m+d+N)-k} = \lambda^{N'-k}, \quad 0 < \lambda < 1 \tag{9.97}$$

the errors $e(k)$ are weighted as shown in Table 9.2 for $N' = 50$. The weighting then increases exponentially to 1 for N'.

Table 9.2. Weighting factors due to (9.97) for $N' = 50$

k	1	10	20	30	40	47	48	49	50
$\lambda = 0.99$	0.61	0.67	0.73	0.82	0.90	0.97	0.98	0.99	1
$\lambda = 0.95$	0.08	0.13	0.21	0.35	0.60	0.85	0.90	0.95	1

This leads to an exponential forgetting memory. The recursive estimation algorithms are then

$$\hat{\boldsymbol\theta}(k+1) = \hat{\boldsymbol\theta}(k) + \boldsymbol\gamma(k)\left[y(k+1) - \boldsymbol\psi^T(k+1)\hat{\boldsymbol\theta}(k)\right] \tag{9.98}$$

$$\boldsymbol\gamma(k) = \frac{1}{\boldsymbol\psi^T(k+1)\,\mathbf{P}(k)\,\boldsymbol\psi(k+1) + \lambda}\mathbf{P}(k)\,\boldsymbol\psi(k+1) \tag{9.99}$$

$$\mathbf{P}(k+1) = \left[\mathbf{I} - \boldsymbol\gamma(k)\,\boldsymbol\psi^T(k+1)\right]\mathbf{P}(k)\frac{1}{\lambda} \tag{9.100}$$

The influence of the forgetting factor λ can be recognized directly from the inverse of the covariance matrix

$$\mathbf{P}^{-1}(k+1) = \lambda\,\mathbf{P}^{-1}(k) + \boldsymbol\psi(k+1)\,\boldsymbol\psi^T(k+1) \tag{9.101}$$

$\mathbf{P}^{-1}$ is proportional to the information matrix $\mathbf{J}$ given by

$$\mathbf{J} = \frac{1}{\sigma_e^2}\,E\left\{\boldsymbol\Psi^T\boldsymbol\Psi\right\} = \frac{1}{\sigma_e^2}\,E\left\{\mathbf{P}^{-1}\right\} \tag{9.102}$$

see, [9.12], [9.29].

By taking $\lambda < 1$, the information on the last step is diminished or the covariances are increased. This means worse estimates are obtained, such that the new measurements get more weight.

For $\lambda = 1$ we have

$$\lim_{k\to\infty} E\{\mathbf{P}(k)\} = \mathbf{0} \tag{9.103}$$

$$\lim_{k\to\infty} E\{\boldsymbol\gamma(k)\} = \lim_{k\to\infty} E\{\mathbf{P}(k+1)\,\boldsymbol\psi(k+1)\} = \mathbf{0} \tag{9.104}$$

For large k the measurements then have practically no influence on $\hat{\boldsymbol\theta}(k+1)$. Then the elements $\mathbf{P}^{-1}(k+1)$ tend to infinity, (9.101).

If, however, $\lambda < 1$ then, from (9.101)

$$\mathbf{P}^{-1}(k) = \lambda^k \, \mathbf{P}^{-1}(0) + \sum_{i=0}^{k} \lambda^{k-i} \, \boldsymbol{\psi}(i) \, \boldsymbol{\psi}^T(i) \tag{9.105}$$

For large values α of the starting matrix $\mathbf{P}(0) = \alpha \mathbf{I}$ the first term vanishes. As for $\lambda < 1$

$$\lim_{k \to \infty} \sum_{i=1}^{k} \lambda^{k-i} = \lim_{k \to \infty} \sum_{i=0}^{k-1} \lambda^i < \infty \tag{9.106}$$

(convergent series with positive elements) $\mathbf{P}^{-1}(k)$ converges to fixed values

$$\lim_{k \to \infty} E\left\{\mathbf{P}^{-1}(k)\right\} = \mathbf{P}^{-1}(\infty) \tag{9.107}$$

and does not approach infinity. Hence,

$$\lim_{k \to \infty} E\{\mathbf{P}(k)\} = \mathbf{P}(\infty) \tag{9.108}$$

and

$$\lim_{k \to \infty} E\{\boldsymbol{\gamma}(k)\} = \boldsymbol{\gamma}(\infty) \tag{9.109}$$

are finite and nonzero. Therefore the new measurements get a constant weight for large k and the estimator remains sensible for parameter changes. Because of the smaller effective averaging time the noise influence increases and so do the variances. Examples are shown in [9.30].

The forgetting factor λ has to be selected as follows:

- λ small, if the speed of parameter changes is large (say $\lambda = 0.90$). Then only small noise is allowed;
- λ large, if the speed of parameter changes is small (say $\lambda = 0.98$). Then the noise can be larger.

As the RML and RELS methods converge more slowly during the starting phase, the convergence can be accelerated by smaller weights at the beginning.

Parameter estimation algorithms with *constant forgetting factor* are suited for processes with small parameter changes and persistent input excitation. Also if the process parameters are constant, good results are obtainable if the noise with regard to the memory length $M = 1/(1 - \lambda)$ is not too large. However, problems may arise if with constant forgetting factor $\lambda < 1$ the input is not sufficiently exciting. Then the values $\mathbf{P}^{-1}(k + 1)$ decrease because $\boldsymbol{\psi}(k + 1) \approx \mathbf{0}$, see (9.101), or the elements of $\mathbf{P}(k + 1)$ increase continuously (covariance matrix blows up). As the correcting vector is

$$\boldsymbol{\gamma}(k) = \mathbf{P}(k + 1) \, \boldsymbol{\psi}(k + 1) \tag{9.110}$$

the estimator becomes more and more sensitive. Then a small disturbance or a numerical error may suffice to generate sudden hight parameter estimate changes. The estimator then becomes unstable. This situation can be observed with adaptive control systems. Therefore, the input excitation has to be monitored or the forgetting factor has to be time-variant.

9.2.5 Parameter estimation for continuous-time signals

Parameter estimation methods for dynamic processes were first developed for process models in discrete time in combination with digital control systems. For some applications, e.g. the validation of theoretical models or for fault diagnosis, however, parameter estimation methods for models with continuous-time signals are needed.

Method of least squares

A stable process with lumped parameters is considered, which can be described by the linear, time-invariant differential equation

$$a_n y_u^{(n)}(t) + a_{n-1} y_u^{(n-1)}(t) + \ldots + a_1 y_u^{(1)}(t) + y_u(t)$$
$$= b_m u^{(m)}(t) + b_{m-1} u^{(m-1)}(t) + \ldots + b_1 u^{(1)}(t) + b_0 u(t) \quad m < n \tag{9.111}$$

It is assumed that the derivatives of the output signal

$$y^{(j)}(t) = d^j y(t)/dt^j, \quad j = 1, 2, \ldots, n \tag{9.112}$$

and of the input signal for $j = 1, 2, \ldots, m$ exist. $u(t)$ and $y(t)$ are the deviations

$$u(t) = U(t) - U_{00}$$
$$y(t) = Y(t) - Y_{00} \tag{9.113}$$

of the absolute signals $U(t)$ and $Y(t)$ from the operating point described by U_{00} and Y_{00}. The transfer function corresponding to (9.111) is

$$G_P(s) = \frac{y_u(s)}{u(s)} = \frac{B(s)}{A(s)}$$
$$= \frac{b_0 + b_1 s + \ldots + b_{m-1} s^{m-1} + b_m s^m}{1 + a_1 s + \ldots + a_{n-1} s^{n-1} + a_n s^n} \tag{9.114}$$

The measurable signal $y(t)$ contains an additional disturbance signal $n(t)$

$$y(t) = y_u(t) + n(t) \tag{9.115}$$

Substituting (9.115) into (9.111) and introducing an equation error $e(t)$ yields

$$y(t) = \boldsymbol{\psi}^T(t) \, \boldsymbol{\theta} + e(t) \tag{9.116}$$

with

$$\boldsymbol{\psi}^T(t) = \left[-y^{(1)}(t) \ldots - y^{(n)}(t) \mid u(t) \ldots u^{(m)}(t) \right] \tag{9.117}$$

$$\boldsymbol{\Theta} = [a_1 \ldots a_n \quad b_0 \ldots b_m]^T \tag{9.118}$$

The input and output signals are measured at discrete time samples $t = k \, T_0$, $k = 0, 1, 2, \ldots, N$ with sampling time T_0 and the derivatives are generated. Based on this, $N + 1$ equations can be written down

$$y(k) = \boldsymbol{\psi}^T(k)\,\hat{\boldsymbol{\theta}} + e(k) \tag{9.119}$$

This system of equations can be written in matrix notation as

$$\mathbf{y} = \boldsymbol{\Psi}\,\hat{\boldsymbol{\theta}} + \mathbf{e} \tag{9.120}$$

Minimizing the loss function

$$V = \mathbf{e}^T(N)\,\mathbf{e}(N) = \sum_{k=0}^{N} e^2(k) \tag{9.121}$$

yields with $dV/d\hat{\boldsymbol{\theta}} = \mathbf{0}$ as previously shown in Section 9.2.1 the vector of parameter estimates for the least squares method

$$\hat{\boldsymbol{\theta}}(N) = \left[\boldsymbol{\Psi}^T\,\boldsymbol{\Psi}\right]^{-1}\,\boldsymbol{\Psi}^T\,\mathbf{y} \tag{9.122}$$

The existence of a unique solution requires that the matrix $\boldsymbol{\Psi}^T\boldsymbol{\Psi}$ is positive-definite. After dividing this matrix by the measurement time, the elements of the resulting matrix are the estimates of the correlation functions $\boldsymbol{\Phi}(\tau)$ of the derivatives of the signals for $\tau = 0$ with no time shift. It can be seen that the form is very similar to the least squares method for models with discrete time signals. Hence, a lot of the derivations can be directly transferred, such as the recursive formulation and the numerically improved versions. However, particular problems arise concerning the convergence and the evaluation of the needed derivatives of the signals.

A *convergence analysis* shows that the estimates for continuous signals are also biased if the error signal $e(k)$ is statistically independent. Hence, the estimates in general are biased for disturbed processes.

If the needed derivatives of the signals are directly measurable (e.g. as for vehicle applications), these values can be written in the data matrix $\boldsymbol{\Psi}$ and the correlation functions in the matrix $[\boldsymbol{\Psi}^T\boldsymbol{\Psi}]/(N+1)$ can be directly calculated. However, if the derivatives are not measurable, the derivatives have to be evaluated from the sampled signals $u(t)$ and $y(t)$. For this, there basically exist the following methods. The *numerical differentiation* in combination with interpolation approaches (splines, Newton's method) is usually not able to suppress noise due to disturbance signals. State variable filters (SVF), see Figure 9.4,

$$F(s) = \frac{y_f(s)}{y(s)} = \frac{1}{f_0 + f_1 s + \ldots + f_{n-1} s^{n-1} + s^n} \tag{9.123}$$

have proven to yield good results. The state variable filter is a low-pass filter that provides the derivatives as well as filters the disturbance signals. With the state variable filter, the input signal $u(t)$ and the output signal $y(t)$ is filtered. The choice of the filter parameters f_i is relatively free. The design of a Butterworth filter is recommended, see [9.48]. A further possibility is the application of finite impulse response filters (FIR), where the derivations of the impulse response of a low-pass filter are convoluted with the signal, [9.46], [9.59], [9.60].

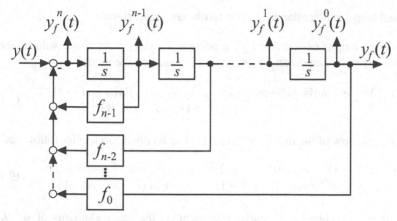

Fig. 9.4. State variable filter

For large signal-to-disturbance ratios, this least squares method has been shown to yield good results. For larger disturbance signals, consistent parameter estimation methods should be employed such as *the instrumental variables method*.

Since these parameter estimation methods are based on discrete time estimation methods, a lot of results and estimation methods for discrete time models can be transferred to models for continuous time.

9.2.6 Parameter estimation in closed loop

For the parameter estimation in closed loop special conditions have to be satisfied. The problem is obvious if correlation analysis is considered. For the convergence of the cross-correlation function between input $u(t)$ and output $y(t)$, the input $u(t)$ must be uncorrelated with the noise $n(k)$, see Figure 9.5. As feedback generates such a correlation, correlation methods cannot be applied directly if no external excitation signals are introduced. In the case of parameter estimation, the situation changes as only the error signal $e(k)$ must be uncorrelated with the elements of the data vector $\psi^T(k)$. A detailed treatment of closed loop identification if given in [9.31] and [9.17]. Now two cases are considered.

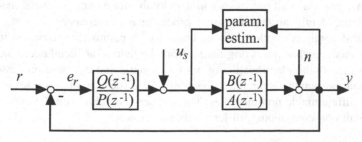

Fig. 9.5. Scheme for process identification in closed loop. u_s: additional perturbation signal

a) Closed loop identification without external excitation signal

If the closed loop is only driven by the noise $n(k)$ and the reference value does not change $r(k) = 0$, then direct LS estimation is based on the model

$$y(k) = \boldsymbol{\psi}^T(k)\boldsymbol{\Theta} = [-y(k-1) \ldots -y(k-m_a)\, u(k-d-1) \qquad (9.124)$$
$$\ldots u(k-d-m_b)]\boldsymbol{\Theta}$$

$\boldsymbol{\psi}^T(k)$ is one row of the matrix $\boldsymbol{\Psi}$ in (9.22). The feedback controller follows as

$$u(k-d-1) = -p_1 u(k-d-2) - \ldots - p_\mu\, u(k-\mu-d-1) \qquad (9.125)$$
$$-q_0 y(k-d-1) - \ldots - q_\nu y(k-\nu-d-1)$$

$u(k-d-1)$ is therefore linearly dependent on the other elements of $\boldsymbol{\psi}^T(k)$ if $\mu \leq m_b - 1$ and $\nu \leq m_a - d - 1$. Only for $\mu \geq m_b$ or $\nu \geq m_a - d$ this linear dependence vanishes. Therefore an identifiability condition is that the order of the controller must be large enough. If, for example, $m_b = 3$ and $m_a = 3$ with $d = 0$, the controller orders must be $\mu \geq 3$ or $\nu \geq 3$, and with $d = 1$ is holds $\mu \geq 3$ or $\nu \geq 2$. If this identifiability condition is satisfied, one-step-ahead prediction error parameter estimation method as LS or ELS can be applied directly on the measured signals $u(k)$ and $y(k)$.

b) Closed loop identification with external perturbation

If the closed loop is perturbed by a sufficiently exciting external input signal $u_s(k)$ or $r(k)$, Figure 9.5, the process input $u(k-d-1)$, (9.125), is no longer linearly dependent on the elements of the data vector $\boldsymbol{\psi}^T(k)$ and the process model can be directly identified by measuring $u(k)$ and $y(k)$ as in open loop. However, the orders of the process model must be known a priori.

9.3 Identification of nonlinear processes

Many processes show a nonlinear static and dynamic behavior, especially if wide areas of operations are considered. Examples are vehicles, aircraft, combustion engines and thermal plants and all processes with Coulomb friction and magnetic hysteresis. Therefore, the identification of nonlinear processes is of increasing interest.

Classical nonlinear models in combination with parameter estimation methods as well as model architectures originating from the field of artificial neural networks are well suited to the identification of nonlinear static and dynamic processes. The next sections present model architectures for the identification of processes with continuously differentiable nonlinearities. The last section deals with the experimental modelling of non-continuously differentiable processes.

9.3.1 Parameter estimation for nonlinear static processes

A process is considered for which the output signal depends nonlinearly from the input signal according to a polynomial

$$Y_u(k) = K_0 + U(k)K_1 + U^2(k)K_2 + \ldots + U^q(k)K_q \tag{9.126}$$

compare Figure 9.6. The output may be contaminated by disturbance n(k)

$$Y(k) = Y_u(k) + n(k) \tag{9.127}$$

Defining the vectors

$$\left.\begin{array}{l} \mathbf{U}(k) = \begin{bmatrix} 1 \; U(k) \; U^2(k) \ldots U^q(k) \end{bmatrix} \\ \mathbf{K}^T = \begin{bmatrix} K_0 \; K_1 \; K_2 \ldots K_q \end{bmatrix} \end{array}\right\} \tag{9.128}$$

leads with $k = 0, \ldots, N - 1$ to the regression model

$$\mathbf{Y} = \mathbf{U}\,\mathbf{K} + \mathbf{n} \tag{9.129}$$

Introducing the equation error

$$\mathbf{e} = \mathbf{Y} - \mathbf{U}\,\mathbf{K} \tag{9.130}$$

where $\mathbf{U}$ and $\mathbf{Y}$ contain the measurements, shows that the equation error is linear in the parameters and the loss function

$$V = \mathbf{e}^T\,\mathbf{e} \tag{9.131}$$

Minimization with

$$\frac{dV}{d\mathbf{K}} = \mathbf{0} \tag{9.132}$$

yields the least squares estimate

$$\hat{\mathbf{K}} = \begin{bmatrix} \mathbf{U}^T\,\mathbf{U} \end{bmatrix}^{-1} \mathbf{U}^T\,\mathbf{Y} \tag{9.133}$$

Existence of this estimate requires

$$\det = \begin{bmatrix} \mathbf{U}^T\,\mathbf{U} \end{bmatrix} \neq 0 \tag{9.134}$$

which means that the input signal $U(k)$ must change during the measurements. If $E\{n(k)\} = 0$, the parameter estimates are consistent in mean square, [9.30].

For fault detection one directly compares the estimated parameters K_i or the nonlinear characteristic curves, e.g. for pumps.

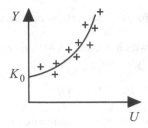

Fig. 9.6. Parameter estimation (regression) for a nonlinear characteristic

9.3.2 Parameter Estimation with Classical Nonlinear Models

Classical methods for the identification of dynamic systems are mostly based on polynomial approximators. One distinguishes between general approaches, e.g. Volterra-series or Kolmogorov-Gabor polynomials, and approaches that involve special structure assumptions such as Hammerstein, Wiener or nonlinear difference equation (NDE) models, [9.12], [9.18], see also [9.30], [9.31].

Static polynomial approximators have the advantage of being linear in the parameters. This advantage can be maintained for certain dynamic polynomial models. This way, computationally expensive iterative optimization methods can be avoided.

In the following, the linear difference equation is written with the shift operator q^{-1}, where $(q^{-1}\,y(k) = y(k-i))$

$$A(q^{-1})\,y(k) = B\,(q^{-1})\,q^{-d};u(k) + D\,(q^{-1})\,v(k) \qquad (9.135)$$

according to (9.58).

The following examples of classical nonlinear dynamic models are based on the representation of the nonlinearity by polynomials.

Generalized Hammerstein model

$$\begin{aligned} A(q^{-1})\,y(k) &= B_1(q^{-1})\,u(k) \\ &+ B_2\,(q^{-1})\,u^2(k) + \ldots + D\,(q^{-1})\,v(k) \end{aligned} \qquad (9.136)$$

Parametric Volterra model

$$\begin{aligned} A(q^{-1})\,y(k) &= B_1\,(q^{-1})\,u(k) \\ &+ \sum_\alpha B_2\,(q^{-1})\,u(k)[q^{-\alpha}\,u(k)] \\ &+ D\,(q^{-1})\,v(k) \end{aligned} \qquad (9.137)$$

Nonlinear Model, [9.35]

$$\begin{aligned} A_1(q^{-1})\,y(k) + \sum_\alpha A_2(q^{-1})\,y(k)\,[q^{-\alpha}\,y(k)] \\ = B\,(q^{-1})\,u(k) + D\,(q^{-1})\,v(k) \end{aligned} \qquad (9.138)$$

In the case of an equation error optimization, these models have the advantage of being linear in the parameters. Therefore, linear parameter estimation methods like LS, RLS and RELS can be applied directly, [9.35], [9.31].

9.3.3 Artificial Neural Networks for Identification

For a general identification approach, methods of interest are those that do not require specific knowledge of the process structure and hence are widely applicable. Artificial neural networks fulfil these requirements. They are composed of mathematically formulated neurons. At first, these neurons were used to describe the behavior of biological neurons, [9.39]. The interconnection of neurons in networks allowed the description of relationships between input and output signals, [9.53], [9.58]. In the sequel, artificial neural networks (ANNs) are considered that map input signals u to output signals y, Figure 9.7. Usually, the adaptable parameters of neural networks are unknown. As a result, they have to be adapted or "trained" or "learned" by processing measured signals u and y, [9.21], [9.20]. This is a typical system identification problem. If inputs and outputs are gathered into groups or clusters, a classification task in connection with, e.g. pattern recognition is given, [9.7]. In the following, the problem of nonlinear system identification is considered (supervised learning). Thereby, the capability of ANNs to approximate nonlinear relationships to any desired degree of accuracy is utilized. Firstly, ANNs for describing static transfer behavior, [9.19], [9.51], will be investigated, which will then be extended to dynamic behavior, [9.2], [9.44], [9.32].

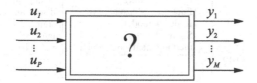

Fig. 9.7. System with P inputs and M outputs, which has to be approximated by an artificial neural network

a) Artificial neural networks for static systems

Neural networks are universal approximators for static nonlinearities and are consequently an alternative to polynomial approaches. Their advantages are the need for only little a priori knowledge about the process structure and the uniform treatment of single-input and multi-input processes. In the following, it is assumed that a nonlinear system with P inputs and M outputs has to be approximated, see Figure 9.7.

Neuron model
Figure 9.8 shows the block diagram of a neuron. In the input operator (synaptic function), a similarity measure between the input vector $\mathbf{u}$ and the (stored) weight vector $\mathbf{w}$ is formed, e.g. by the scalar product

$$x = \mathbf{w}^T \mathbf{u} = \sum_{i=1}^{P} w_i u_i = \left| \mathbf{w}^T \right| |\mathbf{u}| \cos\phi \qquad (9.139)$$

or the Euclidean distance

$$x = ||\mathbf{u} - \mathbf{w}||^2 = \sum_{i=1}^{P} (u_i - w_i)^2 \tag{9.140}$$

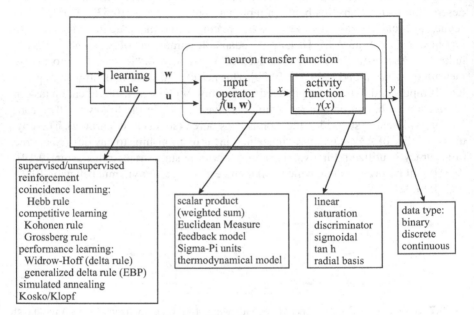

Fig. 9.8. General neuron model

If $\mathbf{w}$ and $\mathbf{u}$ are similar, the resulting scalar quantity x will be large in the first case and small in the second case. The quantity x, also called the activation of the neuron, affects the activation function and consequently the output value

$$y = \gamma(x - c) \tag{9.141}$$

Figure 9.9 shows several examples of those in general nonlinear functions. The threshold c is a constant causing a parallel shift in the x-direction.

Network structure

The single neurons are interconnected to a network structure, Figure 9.10. Hence, one has to distinguish between different layers with parallel arranged neurons: the input layer, the first, second, ... hidden layer and the output layer. Generally, the input layer is used to scale the input signals and is not often counted as a separate layer. Then, the real network structure begins with the first hidden layer. Figure 9.10 shows the most important types of internal links between neurons: feedforward, backward,

(a) hyperbolic tangens (Tangens Hyperbolicus)

$$y = \frac{e^{(x-c)} - e^{-(x-c)}}{e^{(x-c)} + e^{-(x-c)}} = 1 - \frac{2}{1 + e^{2(x-c)}}$$

(b) Sigmoidal function

$$y = \frac{1}{1 + e^{-(x-c)}}$$

(c) limiter

$$y = \begin{cases} 1 & ; \ x - c \geq 1 \\ x - c & ; \ |x - c| < 1 \\ -1 & ; \ x - c \leq -1 \end{cases}$$

(d) neutral zone

$$y = \begin{cases} 0 & ; \ |x - c| \leq 1 \\ x - c - 1 & ; \ x - c > 1 \\ x - c + 1 & ; \ x - c \leq -1 \end{cases}$$

(e) Gauss-functions $\quad y = e^{-(x-c)^2}$

(f) binary function

$$y = \begin{cases} 0 & ; \ x - c < 0 \\ 1 & ; \ x - c > 0 \end{cases}$$

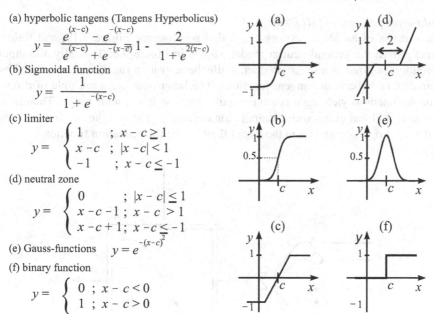

Fig. 9.9. Examples of activation functions

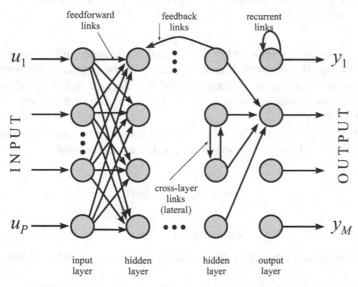

Fig. 9.10. Network structure: layers and links in a neural network

lateral and recurrent. With respect to their range of values, the input signals can be either binary, discrete or continuous. Binary and discrete signals are used especially for classification, while continuous signals are used for identification tasks.

Multi-layer perceptron (MLP) network
The neurons of an MLP network are called perceptrons, Figure 9.11, and follow directly from the general neuron model, shown in Figure 9.8. Typically, the input operator is realized as a scalar product, while the activation functions are realized by sigmoidal or hyperbolic tangent functions. The latter ones are a multiple of differentiable functions yielding a neuron output with $y = 0$ in a wide range. Therefore, they have a global effect with extrapolation capability. The weights w_i are assigned to the input operator and lie in the signal flow before the activation function.

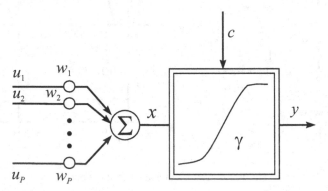

Fig. 9.11. Perceptron neuron with weights w_i, summation of input signals (scalar product) and nonlinear activation function

The perceptrons are connected in parallel and are arranged in consecutive layers to a feedforward MLP network, Figure 9.12. Each of the P inputs affects each perceptron in such a way that in a hidden layer with K perceptrons there exist $(K \cdot P)$ weights w_{kp}. The output neuron is most often a perceptron with a linear activation function, Figure 9.13.

The adaptation of the weights w_i based on measured input and output signals is usually realized by the minimization of the quadratic loss function.

$$J(\mathbf{w}) = \frac{1}{2} \sum_{n=1}^{N-1} e^2(n) \tag{9.142}$$

$$e(n) = y(n) - \hat{y}(n)$$

where e is the model error, y is the measured output signal and $\hat{y}$ is the network output.

As in the case of parameter estimation with the least squares method,

$$\frac{dJ(\mathbf{w})}{d\mathbf{w}} = 0 \tag{9.143}$$

is generated. Due to the nonlinear dependency, a direct solution is not possible. Therefore, e.g. gradient methods for numerical optimization are applied. Because

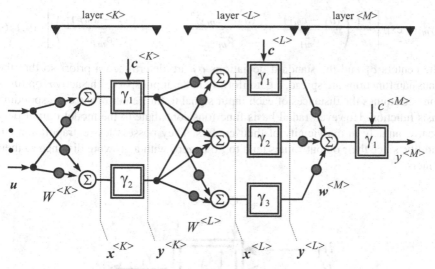

Fig. 9.12. Feedforward multi-layer perceptron network (MLP network). Three layers with (2·3·1) perceptrons. $< K >$ is the first hidden layer

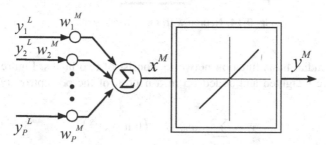

Fig. 9.13. Output neuron as perceptron with linear activation function

of the necessary back-propagation of errors through all hidden layers, the method is called "error back-propagation" or also "delta-rule". The so-called learning rate η has to be chosen (tested) suitably. In principle, gradient methods allow only slow convergence in the case of a large number of unknown parameters.

Radial basis function (RBF) Network
The neurons of RBF networks, Figure 9.14, compute the Euclidean distance in the input operator

$$x = ||\mathbf{u} - \mathbf{c}||^2 \tag{9.144}$$

and feed it to the activation function

$$G_m = \gamma_m \left(||\mathbf{u} - \mathbf{c}||^2 \right) \tag{9.145}$$

The activation function is given by radial basis functions usually in the form of Gaussian functions with

$$\gamma_m = \exp\left[\frac{-1}{2}\left(\frac{(u_1 - c_{m1})^2}{\sigma_{m1}^2} + \frac{(u_2 - c_{m2})^2}{\sigma_{m2}^2} + \ldots + \frac{(u_P - u_{mP})^2}{\sigma_{mP}^2}\right)\right] \quad (9.146)$$

The centers c_j and the standard deviations σ_j are determined a priori so that the Gaussian functions are spread, e.g. uniformly in the input space. The activation function determines the distances of each input signal to the center of the corresponding basis function. However, radial basis functions contribute to the model output only locally, namely in the vicinity of their centers. They possess less extrapolation capability, since their output values tend to go to zero with a growing distance to their centers.

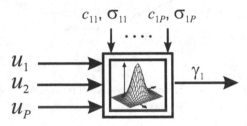

Fig. 9.14. Neuron with radial basis function

Usually, radial basis function networks consist of two layers, Figure 9.15. The outputs γ_i are weighted and added up in a neuron of the perceptron type, Figure 9.13, so that

$$y = \sum_{m=1}^{M} w_m \, G_m \left(\| \mathbf{u} - \mathbf{c} \|^2\right) \quad (9.147)$$

Since the output layer weights are located behind the nonlinear activation functions in the signal flow, the error signal is linear in these parameters and, consequently, the least squares method in its explicit form can be applied. In comparison to MLP networks with gradient methods, a significantly faster convergence can be obtained. However, if the centers and standard deviations have to be optimized too, nonlinear numerical optimization methods are also required.

Local linear model networks
The local linear model network (LOLIMOT) is an extended radial basis function network, [9.42], [9.43]. It is extended by replacing the output layer weights with a linear function of the network inputs (9.148). Furthermore, the RBF network is normalized, such that the sum of all basis functions is one. Thus, each neuron represents a local linear model with its corresponding validity function, see Figure 9.16. The validity functions determine the regions of the input space where each neuron is active. The general architecture of local model networks is extensively discussed in [9.41].

The kind of local model network discussed here utilizes normalized Gaussian validity functions (9.146) and an axis-orthogonal partitioning of the input space.

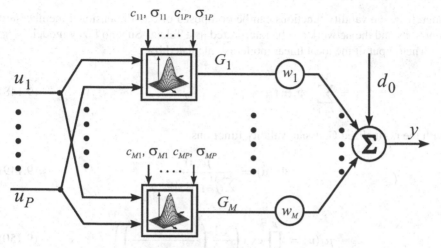

Fig. 9.15. Feedforward radial basis function (RBF) network

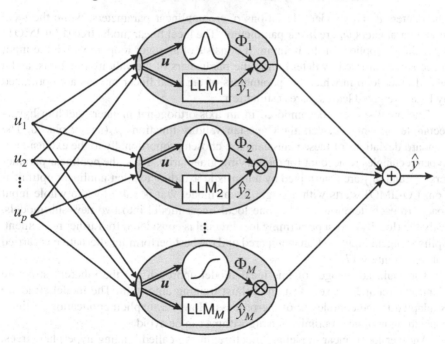

Fig. 9.16. Local linear model network (LOLIMOT)

Therefore, the validity functions can be composed of one-dimensional membership functions and the network can be interpreted as a Takagi-Sugeno fuzzy model.

The output of the local linear model is calculated by

$$\hat{y} = \sum_{i=1}^{M} \Phi_i(\mathbf{u}) \left(w_{i,0} + w_{i,1} u_1 + \ldots + w_{i,p} u_p \right) \tag{9.148}$$

with the normalized Gaussian validity functions

$$\Phi_i(\mathbf{u}) = \frac{\mu(\mathbf{u})}{\sum_{j=1}^{M} \mu_j(\mathbf{u})} \tag{9.149}$$

here

$$\mu_i(\mathbf{u}) = \prod_{j=1}^{p} \exp\left(-\frac{1}{2} \left(\frac{(u_j - c_{i,j})^2}{\sigma_{ij}^2} \right) \right) \tag{9.150}$$

The centers c and standard deviations σ are nonlinear parameters, while the local model parameters w_i are linear parameters. The local linear model tree (LOLIMOT) algorithm is applied for the training. It consists of an outer loop, in which the input space is decomposed by determining the parameters of the validity functions, and a nested inner loop in which the parameters of the local linear models are optimized by local-weighted least squares estimation.

The input space is decomposed in an axis-orthogonal manner, yielding hyper-rectangles in whose centers the Gaussian validity functions $\mu_i(u)$ are placed. The standard deviations of these Gaussians are chosen proportionally to the extension of hyper-rectangles to account for the varying granularity. Thus, the nonlinear parameters $c_{i,j}$ and $\sigma_{i,j}$ are determined by a heuristic-avoiding explicit nonlinear optimization. LOLIMOT starts with a single linear model that is valid for the whole input space. In each iteration, it splits one local linear model into two new sub-models. Only the (locally) worst performing local model is considered for further refinement. Splits along all input axes are compared and the best performing alternative is carried out, see Figure 9.17.

The main advantages of this local model approach are the inherent structure identification and the very fast and robust training algorithm. The model structure is adapted to the complexity of the process. However, explicit application of time-consuming nonlinear optimization algorithms can be avoided.

Another local linear model architecture, the so-called hinging hyperplane trees, is presented in [9.11], [9.56]. These models can be interpreted as an extension of the LOLIMOT networks with respect to the partitioning scheme. While the LOLIMOT algorithm is restricted to axis-orthogonal splits, the hinging hyperplane trees allow an axis-oblique decomposition of the input space. These more complex partitioning strategies lead to an increased effort in model construction. However, this feature is necessary in the case of strong nonlinear model behavior and higher-dimensional input spaces.

The fundamental structures of three artificial neural networks have been described. These models are very well suited to the approximation of measured input/output data of static processes, compare also [9.19], [9.51]. For this, the training data has to be chosen in such a way that the considered input space is as evenly as possible covered with data. After the training procedure, a parametric mathematical model of the static process behavior is available. Consequently, direct computation of the output values $\hat{y}$ for arbitrary input combinations u is possible.

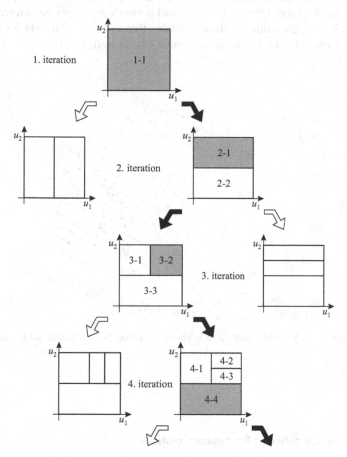

Fig. 9.17. Tree construction of the LOLIMOT algorithm

An advantage of the automatic training procedure is the possibility of using arbitrarily distributed data in the training data set. There is no necessity to know data at exactly defined positions, as in the case of grid-based look-up table models, see Section 9.3.4. This clearly decreases the effort required for measurements in practical applications.

Example 9.1: Artificial neural network for the static behavior of a combustion engine

As an example, the engine characteristics of a six-cylinder SI (spark-ignition) engine is used. Here, the engine torque has to be identified and is dependent on the throttle angle and the engine speed. Figure 9.18 shows the 433 available data points that were measured on an engine test stand.

For the approximation, an MLP network is applied. After the training, an approximation for the measurement data shown in Figure 9.19 is given. For that purpose, 31 parameters are required. Obviously, the neural network possesses good interpolation and extrapolation capabilities. This also means that in areas with only few training data, the process behavior can be approximated quite well, [9.25].

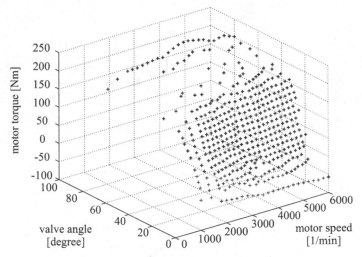

Fig. 9.18. Measured SI engine data (2.5 l, V6 cyl.): unevenly distributed, 433 measurement data points

☐

b) Artificial neural networks for dynamic systems

The memoryless static networks can be extended with dynamic elements to dynamic neural networks. One can distinguish between neural networks with external and internal dynamics, [9.44], [9.32]. ANNs with external dynamics are based on static networks, e.g. MLP or RBF networks. The discrete time input signals $u(k)$ are passed to the network through additional filters $F_i(q-1)$. In the same way, either the measured output signals $y(k)$ or the NN outputs $\hat{y}(k)$ are passed to the network through filters $G_i(q-1)$. The operator q^{-1} denotes a time shift

$$y(k) \cdot q^{-1} = y(k-1) \tag{9.151}$$

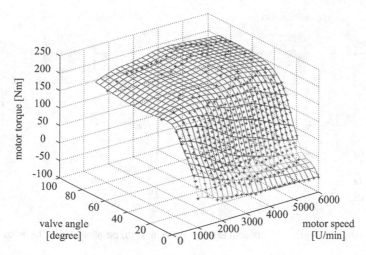

Fig. 9.19. Approximation of measured engine data (+) with an MLP network (2 · 6 · 1): 31 parameters

In the simplest case, the filters are pure time delays, Figure 9.20a

$$\hat{y}(k) = f_{NN}\ [u(k), u(k-1), \dots, \hat{y}(k-1), \hat{y}(k-2), \dots] \tag{9.152}$$

where the time-shifted sampled values are the network input signals. The structure in Figure 9.20a shows a parallel model (equivalent to the output error model for parameter estimation of linear models). In Figure 9.20b, the measured output signal is passed to the network input. Then, the series-parallel model is obtained (equivalent to the equation error model for parameter estimation of linear models). One advantage of the external dynamic approach is the possibility of using the same adaptation methods as in the case of static networks. However, the drawbacks are the increased dimensionality of the input space, possible stability problems and an iterative way of computing the static model behavior, namely through simulation of the model. Then, for example, a step function is used as the input signal and one has to wait until the steady state of the model is reached.

ANNs with internal dynamics realize dynamic elements inside the model structure. According to the kind of included dynamic elements, one can distinguish between recurrent networks, partially recurrent networks and locally recurrent globally feedforward networks (LRGF), [9.44]. The LRGF networks maintain the structure of static networks except that dynamic neurons are utilized, see Figure 9.21. The following can be distinguished: local synapse feedback, local activation feedback and local output feedback. The simplest case is the local activation feedback, [9.2]. Here, each neuron is extended by a linear transfer function, most often of first or second order, see Figure 9.22. The dynamic parameters a_i and b_i are adapted. Static and dynamic behavior can be easily distinguished and stability can be guaranteed.

Usually, MLP networks are used in LRGF structures. However, RBF networks with dynamic elements in the output layer can be applied as well, if a Hammerstein-

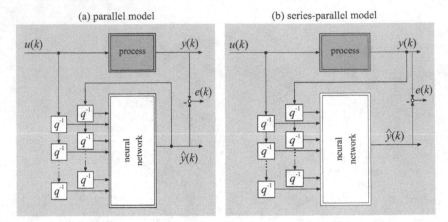

Fig. 9.20. Artificial neural network with external dynamics: (a) parallel model; (b) series-parallel model

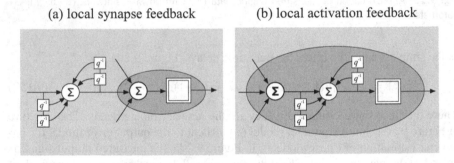

(c) local output feedback

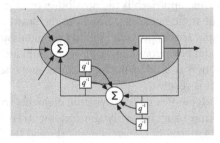

Fig. 9.21. Dynamic neurons for neural networks with internal dynamics: (a) local synapse feedback; (b) local activation feedback; (c) local output feedback

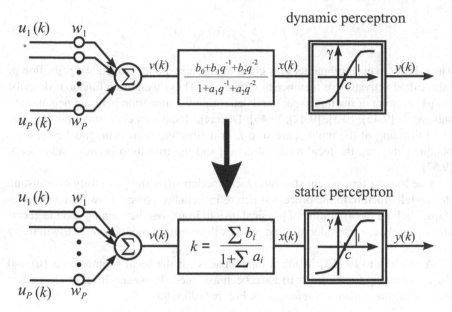

Fig. 9.22. Dynamic perceptron, [9.2]

structure of the process can be assumed, [9.2]. Usually, the adaptation of these dynamic NNs is based on extended gradient methods, [9.44].

Based on the basic structure of ANNs, special structures with particular properties can be built. If, for example, the local linear model network (LOLIMOT) is combined with the external dynamic approach, a model structure with locally valid linear input/output models result.

c) Semi-physical models

Frequently the static or dynamic behavior of processes depends on the operating point, described by the variables $\mathbf{z}$. Then all the inputs have to be separated into manipulated variables $\mathbf{u}$ and operating point variables $\mathbf{z}$. By this separation local linear models can be identified with varying parameters depending on the operating point, also called linear parameter variable models (LPVM), [9.5].

A nonlinear discrete-time dynamic model with p inputs u_i and one output y can be described by

$$y(k) = f(\mathbf{x}(k)) \qquad (9.153)$$

with

$$\mathbf{x}^T(k) = \left[u_1(k-1), \cdots, u_1(k-n_{u1}), \cdots, u_p(k-1), \cdots, u_p(k-n_{um}) \right.$$
$$\left. y(k-1), \cdots, y(k-n_y) \right] \qquad (9.154)$$

For many types of nonlinearities this nonlinear (global) overall model can be represented as a combination of locally active submodels

$$y = \sum_{i=1}^{M} \phi_i(\mathbf{u})\, g_i(\mathbf{u}) \qquad (9.155)$$

The validity of each submodel g_i is given by its corresponding weighting function ϕ_i (also called activation or membership function). These weighting functions describe the partitioning of the input space and determine the transition between neighboring submodel, [9.45], [9.3], [9.41], [9.43]. Different local models result from the way of partitioning of the input space $\mathbf{u}$, e.g. grid structure, axis-orthogonal cuts, axis-oblique cuts, etc., the local model structure and the transition between submodels, [9.57].

Due to their transparent structure, local models offer the possibility of adjusting the model structure to the process structure in terms of physical law based relationships. Such an incorporation of physical insight improves the training and the generalization behavior considerably and reduces the required model complexity in many cases.

According to (9.155) identical input spaces for the local submodels $g_i(\mathbf{u})$ and the membership functions $\Phi(\mathbf{u})$ have been assumed. However, local models allow the realization of *distinct input spaces*, Figure 9.23 with

$$y = \sum_{i=1}^{M} \Phi_i(\mathbf{z})\, g_i(\mathbf{x}) \qquad (9.156)$$

The input vector $\mathbf{z}$ of the weighting functions comprises merely those inputs of the vector $\mathbf{u}$ having significant nonlinear effects which cannot be explained by the local submodels. Only those directions require a subdivision into different parts. The decisive advantage of this procedure is the considerable reduction of the number of inputs in $\mathbf{z}$. Thus, the difficult task of structure identification can be simplified.

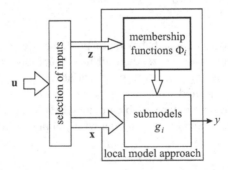

Fig. 9.23. Structure of local model approaches with distinct inputs spaces for local submodels and membership functions

The use of separate input spaces for the local models (vector $\mathbf{x}$) and the membership functions (vector $\mathbf{z}$) becomes more precise by considering another representation of the structure in (9.156). As normally local model approaches are assumed to being linear with reference to their parameters according to

$$g_i(\mathbf{x}) = w_{i0} + w_{i1} x_1 + \cdots + w_{inx} x_{nz} \tag{9.157}$$

(9.156) can be arranged to

$$
\begin{aligned}
y &= w_0(\mathbf{z}) + w_1(\mathbf{z}) x_1 + \cdots + w_p(\mathbf{z})\, x_{nz} \\
&\text{with} \quad w_j(\mathbf{z}) = \textstyle\sum_{i=1}^{M} w_{ij}\, \Phi_i(\mathbf{z})
\end{aligned} \tag{9.158}
$$

Thus, the specified local model approaches can be interpreted as linear-in-the-parameter relationships with operating point dependent parameters $w_j(\mathbf{z})$, whereupon these parameters depend on the input values in vector $\mathbf{z}$. Consequently, the process coefficients $w_j(\mathbf{z})$ still have a physical meaning. Therefore these models are called *semi-physical models*, [9.57].

The choice of *approximate submodel structures* always requires a compromise between submodel complexity and the number of submodels. The most often applied linear submodels have the advantage of being a direct extension of the well known linear models. However, under certain conditions more complex submodels may be reasonable. If the main nonlinear influence of input variables can be described qualitatively by a nonlinear transformation of the input variables (e.g. $f_1(x) = (x_1^2, \; x_1 x_2, \ldots)$), then the incorporation of that knowledge into the submodels leads to a considerable reduction of the required number of submodels. Generally, this approach can be realized by a pre-processing of the input variables $\mathbf{x}$ to the nonlinearly transformed variables, Figure 9.24

$$\mathbf{x}^* = F(\mathbf{x}) = \left[f_1(\mathbf{x}) \; f_2(\mathbf{x}) \cdots f_p(\mathbf{x}) \right]^T \tag{9.159}$$

Besides those heuristically determined model structures, local model approaches also enable the incorporation of fully physically determined models. Furthermore, local models allow the employment of inhomogeneous models. Consequently, different local submodel structures are valid within the different operating regimes.

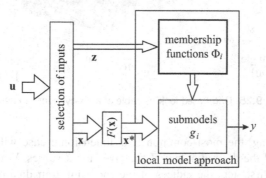

Fig. 9.24. Pre-processing of input variables $\mathbf{x}$ for incorporation of prior knowledge into the submodel structure

9.3.4 Identification with Grid-based Look-up Tables for Static Process

In this section, a further nonlinear model architecture besides the polynomial-based models, neural networks and fuzzy systems is presented. These grid-based look-up tables are the most common type of nonlinear static models used in practice. Especially in the field of nonlinear control, look-up tables are widely accepted as they provide a transparent and flexible representation of nonlinear relationships. Electronic control units of modern automobiles, for example, contain about 100 such grid-based look-up tables, in particular for engine and emission control, [9.8].

In automotive applications, due to cost reasons computational power and storage capacity are strongly restricted. Furthermore, constraints of real-time operation have to be met. Under these conditions, grid-based look-up tables represent a suitable means of storage of nonlinear static mappings. The models consist of a set of data points or nodes positioned on a multidimensional grid. Each node comprises two components. The scalar data point heights are estimates of the approximated nonlinear function at their corresponding data point position. All nodes located on grid lines, as shown in Figure 9.25, are stored, e.g. in the ROM of the control unit. For model generation, usually all data point positions are fixed a priori. The most widely applied method of obtaining the data point heights is to position measurement data points directly on the grid. Then, an optimization can be avoided.

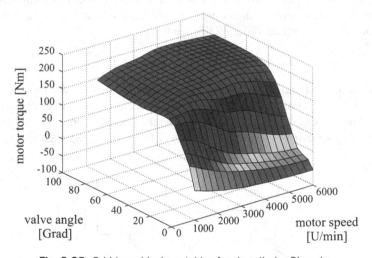

Fig. 9.25. Grid-based look-up table of a six-cylinder SI engine

In the following, the most common two-dimensional case will be considered. The calculation of the desired output Z for given input values X and Y consists of two steps. In the first step, the indices of the enclosing four data points have to be selected. Then, a bilinear area interpolation is performed, [9.54]. For this, four areas have to be calculated, as shown in Figure 9.26, [9.54], [9.56].

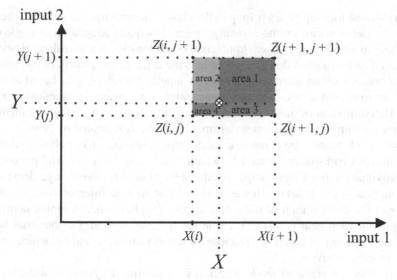

Fig. 9.26. Areas for interpolation within a look-up table

For the calculation of the desired output Z, the four selected data point heights are weighted with the opposite areas and added up. Finally, this result has to be divided by the total area, (9.143).

$$
Z(X, Y) = \left[\; [Z(i, j) \underbrace{(X(i + j) - X) (Y(j + 1) - Y)}_{\text{area 1}}] \right.
$$
$$
+ [Z(i + 1, j) \underbrace{(X - X(i)) (Y(j + 1) - Y)}_{\text{area 2}}]
$$
$$
+ [Z(i, j + 1) \underbrace{(X(i + 1) - X) (Y - Y(j))}_{\text{area 3}}]
$$
$$
\left. + [Z(i + 1, j + 1) \underbrace{(X - X(i)) (Y - Y(j))}_{\text{area 4}}] \right]
$$
$$
\div \underbrace{[(X(i + 1) - X(i)) (Y(j + 1) - Y(j))]}_{\text{overall area}}
$$

(9.160)

Because of the relatively simple computational algorithm, area interpolation rules are widely applied, especially in real-time applications. The accuracy of the method depends on the number of grid positions. For the approximation of "smooth" mappings, a small number of data points is sufficient, while for stronger nonlinear behavior a finer grid has to be chosen.

The area interpolation is based on the assumption that all data point heights are available in the whole range covered by the grid. However, this condition is often not fulfilled.

Grid-based look-up tables belong to the class of nonparametric models. The described model structure has the advantage that a subsequent adaptation of single data point heights due to changing environmental conditions is easy to realize. However, the main disadvantage of this look-up table is the exponential growth of the number of data points with an increasing number of inputs. Therefore, grid-based look-up tables are restricted to one- and two-dimensional input spaces in practical applications. Determination of the heights of the look-up table based on measurements at arbitrary coordinates with parameter estimation methods is treated by [9.40].

Another alternative are parametric model representations, like polynomial models, neural networks or fuzzy models, which clearly require less model parameters to approximate a given input output relationship. Therefore, the storage demand of these models is much lower. However, in contrast to area interpolation, the complexity of the computation of the output is much higher, since complex nonlinear functions for each neuron of a fuzzy rule has to be computed. On the other hand, grid-based look-up tables are not suitable for the identification and modelling of dynamic process behavior.

A detailed overview of model structures for nonlinear system identification is given in [9.43].

9.3.5 Parameter Estimation for Non-continuously Differentiable Nonlinear Processes (Friction and Backlash)

Non-continuously differentiable nonlinear processes appear in mechanical systems, especially in the form of friction and backlash and in electrical systems with magnetization hysteresis.

a) Processes with friction

In many mechanical processes dry and viscous friction appears. Here, dry friction can be modelled as a constant with a velocity-dependent sign. In steady state, a hysteresis curve arises from the dynamic relationship.

For the identification of processes with friction, the hysteresis curve can directly be found pointwise by slow continuous or stepwise changes of the input signal $u(t) = y_1(t)$ and the measurement of $y(t) = y_2(t)$.

If the hysteresis curves are described by

$$y_+(u) = K_{0+} + K_{1+}u$$
$$y_-(u) = K_{0-} + K_{1-}u \qquad (9.161)$$

then the parameters can be estimated from $v = 1, 2, \ldots , N - 1$ measured points with the least squares method

$$\hat{K}_{1\pm} = \frac{N \sum u(v)\, y_\pm (v) - \sum u(v) \sum y_\pm(v)}{N \sum u^2 (v) - \sum u(v) \sum u(v)} \qquad (9.162)$$

$$\hat{K}_{0\pm} = \frac{1}{N} \left[\sum y_{\pm}(v) - \hat{K}_{1\pm} \sum u(v) \right]$$ (9.163)

As the differential equations are linear in the parameters, direct methods of parameter estimation can be applied for processes with dry and viscous friction in motion. For this, both differential and difference equations are well-suited process models. In some cases, it is expedient not only to use velocity-dependent dry friction but also velocity direction-dependent dynamic parameters, e.g. in the form of difference equations

$$y(k) = -\sum_{i=1}^{m} a_{1+} y(k-1) + \sum_{i=1}^{m} b_{i+} u(k-i) + K_{0+}$$ (9.164)

$$y(k) = -\sum_{i=1}^{m} a_{i-} y(k-i) + \sum_{i=1}^{m} b_{1-} u(k-1) + K_{0-}$$ (9.165)

K_{0+} and K_{0-} can be understood as direction-dependent offsets or DC values. Then, the following methods can be applied for the estimation of these offsets:

- implicit estimation of the offset parameters K_{0+} and K_{0-};
- explicit estimation of the offset parameters K_{0+} and K_{0-} with generation of differences $\Delta y(k)$ and $\Delta u(k)$ and parameter estimation for

$$\Delta y(k) = -\sum_{i=1}^{m} \hat{a}_i \, \Delta y(k-i) + \sum_{i=1}^{m} \hat{b}_i \, \Delta u(k-i)$$ (9.166)

with the assumption of velocity-independent dynamic parameters $\hat{a}_i$ and $\hat{b}_i$. Then, for each direction, the parameters $\hat{K}_{0+}$ and $\hat{K}_{0-}$ have to be estimated separately.

For this parameter estimation method with a direction-dependent model, an additional identification requirement has to be considered, which is that the motion takes place in only one direction without reversal. This means it has to satisfy

$$\dot{y}(t) > 0 \quad \text{or} \quad \dot{y}(t) < 0$$ (9.167)

which can be tested by

$$\Delta y(k) > \varepsilon \quad \text{or} \quad \Delta y(k) < \varepsilon$$ (9.168)

for all k with $\epsilon = 0$.

A test signal for proportional acting processes fulfilling this condition was proposed by [9.38], Figure 9.27. The motion in one direction with a certain velocity is generated by a linear ascent. Then, this is followed by a step for the excitation of higher frequencies and a transition to a steady state condition. In the case of a reversal of motion, the parameter estimation has to be stopped (in Figure 9.27 the points 1, 2, 3, ...) and has either to be restarted or continued with values according to the same direction.

The hysteresis curve can be computed from the static behavior of the model (9.164), (9.165)

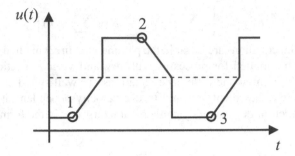

Fig. 9.27. Test signal for parameter estimation of processes with dry friction

$$y_+(u) = \frac{\hat{K}_{0+}}{1 + \sum \hat{a}_{i+}} + \frac{\sum \hat{b}_{i+}}{1 + \sum \hat{a}_{i+}} u \qquad (9.169)$$

$$y_-(u) = \frac{\hat{K}_{0-}}{1 + \sum \hat{a}_{i-}} + \frac{\sum \hat{b}_{i-}}{1 + \sum \hat{a}_{i-}} u \qquad (9.170)$$

For the verification of the parameter estimation based on the dynamic behavior, the computed characteristic curve can be compared with the measured curve resulting directly from the measured static behavior.

For rotary drives, [9.23], [9.22] have developed a special parameter estimation method that correlates the measured torque with the rotational acceleration and estimates the moment of inertia. Following from that, the characteristic curve of the friction torque can be estimated in a nonparametric form.

The methods described above for the identification of processes with friction have been successfully tested in practical applications and applied to digital control with friction compensation, see [9.38], [9.52]. Further treatment is given by [9.1], [9.9].

b) Processes with backlash (dead zone)

As an example again, an oscillator with backlash or dead zone of width $2y_t$ is considered, Figure 9.28. For the oscillator without backlash, it is

$$m \, \ddot{y}_2(t) + d \, \dot{y}_2(t) + c \, y_2(t) = c \, y_3(t) \qquad (9.171)$$

The backlash can be described as follows

$$y_3(t) = \begin{cases} y_1(t) - y_t \text{ for} & y_1(t) > y_t \\ 0 \text{ for } -y_t \le y_1(t) \le y_t \\ y_1(t) + y_t \text{ for} & y_1(t) < y_t \end{cases} \qquad (9.172)$$

This equation leads to the nonlinear characteristic shown in Figure 9.28b. In the case where the backlash is at one restriction ($y_1(t) > y_t$), it is

$$m \, \ddot{y}_2(t) + d \, \dot{y}_2(t) + c \, y_2(t) + c \, y_t = c \, y_1(t) \qquad (9.173)$$

and for the other restriction with $y_1(t) < y_t$

$$m \ \ddot{y}_2(t) + d \ \dot{y}_2(t) + c \ y_2 - c \ y_t = c \ y_1(t) \qquad (9.174)$$

The backlash appears as a constant with a sign depending on the sign of $y_1(t)$. For the range inside the backlash, it is $y_3(t) = 0$ and it holds that the system eigen-behavior

$$m \ \ddot{y}_2(t) + d \ \dot{y}_2(t) + c \ y_2(t) = 0 \qquad (9.175)$$

if point 3 (for instance, because of a friction not considered) is fixed. If point 3 is not fixed and can move arbitrarily inside the backlash, the spring forces do not apply. Then, one has to set $y_2 = y_3$ and in (9.161) $c = 0$.

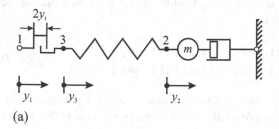

(a)

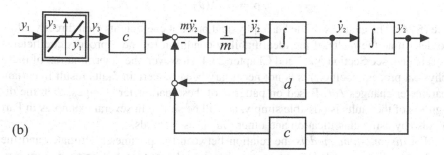

(b)

Fig. 9.28. Mechanical oscillator with backlash (dead zone): (a) schematic set-up; (b) block diagram for the cases $y_1(t) > y_t$ and $y_1(t) < y_t$

One obtains a simplified block diagram for the regions outside the backlash shown in Figure 9.29. The effect of the backlash in these regions can be interpreted as an offset shift of the input signal with changing sign.

9.4 Symptom generation with identification models

a) Parameter estimation

As discussed in Chapter 1, certain process parameters change if they are influenced by faults. This means that either one parameter or several parameters change compared to the normal parameter set θ_0

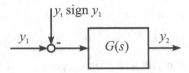

Fig. 9.29. Simplified block diagram for a linear system with backlash for $|y_1(t)| > |y_t|$

$$\Delta\theta = \hat{\theta}(k) - \theta_0 \qquad (9.176)$$

As the parameters usually fluctuate because of signal disturbances and unmodelled effects, thresholds $\pm\Delta\theta_{th}$ have to be defined for normal low and high values, and limit checking has to take place, see Chapter 7. Parameter changes which exceed the thresholds are then called symptoms

$$\left.\begin{array}{l} \Delta\theta_s = \hat{\theta}(k) - (\theta_0 + \Delta\theta_{th}) \ \theta(k) > \theta_0 \\ \Delta\theta_s = \hat{\theta}(k) - (\theta_0 - \Delta\theta_{th}) \ \theta(k) < \theta_0 \end{array}\right\} \qquad (9.177)$$

In the case of parameter estimation with *continuous-time models* the parameter estimates can then be directly related to *physical process coefficients* by using the inverse algebraic relationship

$$\mathbf{p} = f^{-1}(\boldsymbol{\theta})$$

see (5.24). The changes of the basic physical process coefficients then directly allow to determine the faults, if the identifiability condition for these process coefficients is satisfied, see Section 5.2.3 and Chapter 23.4. However, the determination of these physical process coefficients is not necessary because certain faults result in certain parameter changes $\Delta\theta$. Based on patterns of these parameter changes, then the diagnosis of the faults is possible simply, as will be shown in several examples in Part V, also by using classification and other diagnosis methods.

For *discrete-time models* the relation between the parameter estimates and the continuous-time model parameters is rather complicated, as can be seen from a z-transform table. Calculation of the continuous-time parameters is only recommended for first order systems. This means that for processes for orders $m \geq 2$ upwards only classification and other diagnosis methods are applicable. But also in this case simple patterns of parameter changes may be sufficient in practical applications.

b) Neural networks

Because *neural networks* usually contain many parameters as weights, which do not allow a physical interpretation, these parameters seem not to be suitable for fault detection. For nonlinear static models then especially look-up tables allow a direct comparison of output signals, e.g. in form of look-up table differences $\Delta y(u_1, u_2, \dots)$. The identified neural network models are mostly used to generate output signal symptoms.

c) Input excitation

Parameter estimation of dynamic processes requires an excitation through the input signal. For processes which change their inputs in normal operation anyhow, like for servo control systems, actuators, machine tools and robots, vehicles and other mobile systems, identification methods can be applied directly, eventually under the condition of enough rich excitation or persistent excitation, e.g. [9.30], [9.31]. Otherwise special exciting test signals, like step functions, harmonic oscillations or PRBS have to be applied. Depending on the operating conditions, this can in some cases be tolerated from time to time, like for pumps, or be applied without any problems for end-of-line quality control of assembled products.

9.5 Problems

1) The following linear first order difference equation is given $y(k)+a_1 y(k-1) = b_1 u(k-1)$. The measured input and output signals $u(k)$ and $y(k)$ respectively, with $k = 1...N$, are given. What are the estimation equations to calculate the parameters a_1 and b_1 with the least squares method? What has to be changed in the estimation equation if a dead-time of $d = 5$ is added?

2) Derive the estimation equation for the estimation of the weighting function $g(k), k = 0...n$ (impulse response) of a linear process using the least squares method given the measured input and output signals $u(k)$ and $y(k)(k = 1...N)$ respectively?

3) The linear first order system

$$y(k) + a_1 y(k-1) = b_1 u(k-1)$$

is given with the following measurements:

k	0	1	2	3	4	5	6	7	8	9	10
$u(k)$	0	1	-1	1	1	1	1	-1	-1	0	0	0
$y(k$	0	0	0	0	0	0	0	1	0	0	0	0

Calculate the estimates of the parameters a_1 and b_1.

4) State the recursive estimation equations for the least squares method for the linear first order process

$$y(k) + a_1 y(k-1) = b_1 u(k-1)$$

What has to be changed in the estimation equations if a dead-time of $d = 2$ is added?

5) The first order differential equation

$$y(t) + a_1 \dot{y}(t) = b_0 u(t)$$

describes a linear system in continuous-time domain. The input and the output signal $u(t)$ and $y(t)$ and its derivative $\dot{y}(t)$ are measured at discrete time samples $k = 1...N$. Determine the equations to estimate the parameters a_1 and b_0 of the differential equation with the least squares method.

6) What are the conditions for the utilization of direct parameter estimation methods, (e.g. LS method) for the parameters of polynominal nonlinear models?

7) Sketch a neuron of a MLP and a RBF network.

8) The training of MLP networks and the training of the output layer weights of RBF networks should be compared. Which kind of optimization methods are applicable? Characterize the different optimization methods.

9) Consider a two-layer perceptron network with P inputs, K hidden neurons and M output units. Write down an expression for the total number of network weights.

10) Determine the equations for the interpolation of a grid-based look-up table with one input and one output. What kind of interpolation appears in this one-dimensional case?

11) The Coulomb friction of a second order oscillating system increases because of missing lubrications. How can a symptom for fault detection be generated? Use equations.

12) Determine the vectors of (9.120) for parameter estimation with continuous-time signals for the DC motor described in Example 10.3 and design the corresponding state-variable filter.

13) Determine the recursive parameter estimation algorithm for a model of the DC motor in discrete time. Define the required vectors.

10

Fault detection with parity equations

A straightforward way to detect process faults is to compare the process behavior with a process model describing the nominal, i.e. non-faulty behavior. The difference of signals between the process and the model are expressed by *residuals*. Therefore residuals describe discrepancies between the process and the model and check for consistency, [10.8]. The design of the residuals can be made with transfer functions or in state-space formulation. The method of parity equations goes probably back to [10.5] with a formulation for state-space models. Further publications have shown this method for different model structures, like for input-output models by [10.10], [10.11] and enhanced state-space models by [10.23] and others, see Chapter 11.

10.1 Parity equations with transfer functions

Figure 10.1 shows two arrangements for the case of linear processes. To explain the method, first a single-input single-output process is considered. The process is described by the transfer function

$$G_p(s) = \frac{y_p(s)}{u(s)} = \frac{B_p(s)}{A_p(s)} \tag{10.1}$$

and the process model by

$$G_m(s) = \frac{y_m(s)}{u(s)} = \frac{B_m(s)}{A_m(s)} \tag{10.2}$$

The model is assumed to be known and has known, fixed parameters, such that

$$G_p(s) = G_m(s) + \Delta G_m(s) \tag{10.3}$$

where $\Delta G_m(s)$ describes model errors.

The residuals can now be formulated by the output error or the polynomial error, similar to parameter estimation methods.

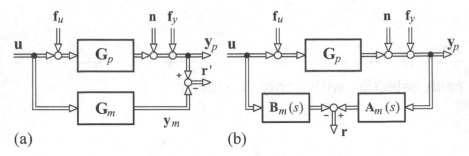

Fig. 10.1. Residual generation with parity equations for a MIMO process with transfer functions:(a) output errors; (b) polynomial errors or equation errors

For the *output error* the residual becomes

$$r'(s) = y_p(s) - y_m(s) = y_p(s) - G_m(s)\,u(s)$$
$$= G_p(s)\,[u(s) + f_u(s)] + n(s) + f_y(s) - G_m(s)\,u(s)$$
$$= \Delta G_m(s)\,u(s) + G_p(s)\,f_u(s) + n(s) + f_y(s) \qquad (10.4)$$

The residual is zero for ideal matching of process and model, no additive faults f_u and f_y and no noise. Usually, it shows deviations depending on the model error ΔG_m and noise n and the exciting input signal u. In this case of additive faults the residual changes are identical with the output fault f_y and filtered by the process G_p for input faults f_u.

The *polynomial error* (or *equation error*) leads to

$$r(s) = A_m(s)\,y_p(s) - B_m(s)\,u(s)$$
$$= A_m(s)\,[G_p(s)\,[u(s) + f_u(s)] + n(s) + f_y(s)] - B_m(s)\,u(s) \quad (10.5)$$

If the process and the model agree, ideally the residual becomes

$$r(s) = A_m(s)\,[f_y(s) + n(s)] + B_m(s)\,f_u(s) \qquad (10.6)$$

Additive input faults f_u are filtered by the model polynomial $B_m(s)$ and additive output faults f_y by the polynomial $A_m(s)$, which both may obtain higher order derivatives. (10.4) and (10.5) are *parity equations*, and r' and r are called *primary residuals*, [10.8].

The residuals of the considered single-input single-output process are in both cases influenced by additive input and output faults, by the noise and by model errors, (e.g. by parameter changes) and a separation is usually not possible. However, the situation improves if more measurements are available, as, for example, for multi-input multi-output (MIMO) processes.

The output residual of a MIMO process with transfer function matrix $\mathbf{G}_p(s)$ is calculated by

$$\mathbf{r}'(s) = \mathbf{y}_p(s) - \mathbf{y}_m(s) = \mathbf{y}_p(s) - \mathbf{G}_m(s)\,\mathbf{u}(s) \qquad (10.7)$$

which is called the *computational form* of the parity equation, [10.8]. By this way the residuals are calculated, using the measured process input and output signals. If the assumed faults are introduced, one obtains

$$\mathbf{r}'(s) = \mathbf{G}_p(s) \left[\mathbf{u}(s) + \mathbf{f}_u(s)\right] + \mathbf{f}_y(s) + \mathbf{n}(s) - \mathbf{G}_m(s)\,\mathbf{u}(s)$$
$$= \Delta\mathbf{G}_m(s)\,\mathbf{u}(s) + \mathbf{G}_p(s)\,\mathbf{f}_u(s) + \mathbf{f}_y(s) + \mathbf{n}(s) \qquad (10.8)$$

which shows the influence of the faults on the residual vector. This is called the *internal form* or *unknown-input-effect form* or *evaluation form*, [10.8], [10.4].

If process and model are identical, the internal form of the residual equation reduces to

$$\mathbf{r}'(s) = \mathbf{G}_p(s)\,\mathbf{f}_u(s) + \mathbf{f}_y(s) + \mathbf{n}(s) \qquad (10.9)$$

Deriving the equation error residuals, one may write the process in the form

$$\mathbf{A}_p(s)\mathbf{y}_p(s) = \mathbf{B}_p(s)\,\mathbf{u}(s) \qquad (10.10)$$

The computational form of the polynomial residual is then

$$\mathbf{r}(s) = \mathbf{A}_m(s)\,\mathbf{y}_p(s) - \mathbf{B}_m(s)\mathbf{u}(s) \qquad (10.11)$$

The internal form becomes

$$\mathbf{r}(s) = \mathbf{A}_m(s)\left[\mathbf{G}_p(s)\,\mathbf{u}(s) + \mathbf{G}_p(s)\,\mathbf{f}_u(s) + \mathbf{f}_y(s) + \mathbf{n}(s)\right] - \mathbf{B}_m(s)\,\mathbf{u}(s) \qquad (10.12)$$

and if process and model are identical

$$\mathbf{r}(s) = \mathbf{A}_m(s)\left[\mathbf{f}_y(s) + \mathbf{n}(s)\right] + \mathbf{B}_m(s)\,\mathbf{f}_u(s) \qquad (10.13)$$

The number of residuals equals the number of output signals.

If only single input or single output faults appear, some elements of the residual vector deviate differently and some do not, which makes a separation or isolation of the additive faults possible as will be shown later.

A comparison of (10.9) and (10.13) shows that the two types of primary residuals are related by

$$\mathbf{r}(s) = \mathbf{A}_m(s)\,\mathbf{r}'(s) \qquad (10.14)$$

Therefore, the polynomial residual includes higher order derivatives of the signals, which may lead to realizability problems and to an amplification of higher frequent noise.

Example 10.1: Comparison of parity equations for a simulated process

a) Simulated process

To show the effects of different faults on the output and equation error, a linear first order process is considered, described by the transfer function

$$G_P(s) = \frac{y(s)}{u(s)} = \frac{Y(s) - Y_0}{U(s) - U_0} = \frac{K}{Ts + 1}$$

where U_0, Y_0 is the operation point in steady-state. This relatively simple process was selected to allow as well simulations as analytical expressions for the different designs of parity relations, [10.19]. First, the system is subject to faults in the process parameters, which are either step- or rampwise. These are the two cases most likely encountered in real systems, mimicking both suddenly appearing faults and slowly drifting faults. Considering faults in both process parameters, the transfer function becomes

$$G(s) = \frac{(K + \Delta K)}{(T + \Delta T)s + 1}$$

where ΔK and ΔT describe the individual *process parameter faults*.

The process input is initially at U_0, the output at Y_0. At the onset of the fault, two different cases will be examined: The system input either remains constant or jumps stepwise from U_0 to U_1. One can derive two parity equations, one based on the *output error*, given as

$$r'(s) = y(s) - \left(\frac{K_M}{T_M s + 1}\right) u(s)$$

where the index M denotes the coefficients of the model. The other scheme for forming a parity equation is the *equation error* or *polynomial error*, yielding

$$r(s) = A_M(s)\, y(s) - B_M(s)\, u(s) = (T_M s + 1)\, y(s) - K_M\, u(s)$$

The corresponding schemes are shown in Figure 10.1.

Besides process parameter faults, the system will also be subject to *additive faults* at the input and output. These faults are also modelled as both a step and a ramp respectively. Furthermore, the system is subject to noise acting at the input or the output of the system respectively. Considering this, the output becomes

$$y(s) = \left(\frac{K}{Ts + 1}\right) (u(s) + f_u(s) + n_u(s)) + f_y(s) + n_y(s)$$

where $f_u(s)$ is the additive fault at the input and $n_u(s)$ the noise injected. $f_y(s)$ denotes an additive fault at the output.

b) Process parameter faults

If the system input remains unchanged and the *gain change* ΔK appears suddenly, stepwise, the response of the residuals will be

$$r'(t) = U_0\, \Delta K \left(1 - e^{-\frac{t}{T}}\right) ; \quad r(t) = U_0\, \Delta K$$

If a step from U_0 to U_1 is applied to the system input at the same time as the fault occurs, the residuals will react as

$$r'(t) = U_1 \, \Delta K \left(1 - e^{-\frac{t}{T}}\right) \; ; \; r(t) = U_1 \, \Delta K$$

If the *time constant* T changes by ΔT then for a constant system input, the residuals are given by

$$r'(t) = 0 \text{ and } r(t) = 0$$

Thus, during steady-state operation, the parity equations are insensitive to faults in T.

Detectability of faults in T is also difficult during dynamic operations. If a step-input is applied to the system, the residuals are only dynamically excited as

$$r'(t) = -K(U_1 - U_0)\left(e^{-\frac{t}{T+\Delta T}} - e^{-\frac{t}{T}}\right) \text{ and } r(t) = -\frac{\Delta T \; K(U_1 - U_0)}{T + \Delta T}e^{-\frac{t}{T+\Delta T}}$$

Figure 10.2 shows the time histories for the residuals for stepwise and also drift-wise changes of the parametric faults.

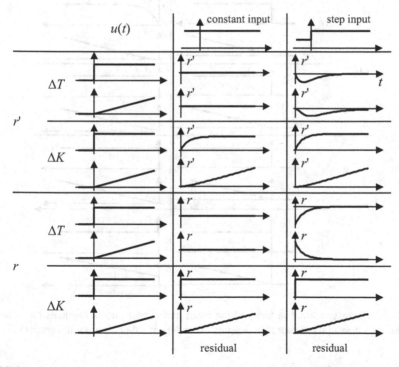

Fig. 10.2. Influence of parameter faults $\Delta K, \Delta T$ on the output residual $r'(t)$ and the equation error residual $r(t)$ for different time history of the faults: step or drift and no input or stepwise input excitation

c) Additive faults at the input/output

If an *additive stepwise fault* is applied at the process input $f_u(t) = \Delta f_u \, \sigma(t)$ with $\sigma(t)$ being the step function, the residuals react as

$$r'(t) = K \, \Delta f_u \left(1 - e^{-\frac{t}{T}}\right) \quad \text{and} \quad r(t) = K \, \Delta f_u$$

If a stepwise additive fault $f_y(t)$ at the output appears independent of the signal that is applied at the process input, the residual react as

$$r'(t) = \Delta f_y \quad \text{and} \quad r(t) = \Delta f_y(1 + T \, \delta(t))$$

Figure 10.3 shows the time histories of the residuals for stepwise and driftwise additive faults.

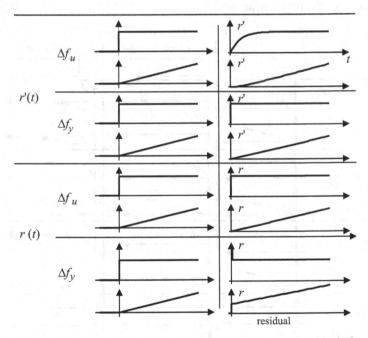

Fig. 10.3. Influence of additive faults at the input and output on the residuals for step- and driftwise fault development over time. Results are independent of input excitation $u(t)$

d) Influence of noise at the input and output

The points of attack of the noise are identical with the points of attack for additive faults. The power density spectra are denoted as $S_{n_u n_u}(\omega)$ and $S_{yy}(\omega)$ respectively.

Noise $n_u(t)$ in the parity equation caused at the input can be calculated as

$$S_{n_{r'}n_{r'}}(\omega) = \frac{K^2}{1 + \omega^2 T^2} S_{n_u n_u}(\omega) \quad \text{and} \quad S_{n_r n_r}(\omega) = K^2 S_{n_u n_u}(\omega)$$

For the output error scheme, input noise appears as a low-pass filtered noise, the constant component ($\omega = 0$) is attenuated by a gain K. For the equation error, all spectral components are attenuated by the same, frequency independent, gain K.

If a noise source $n_y(t)$ at the output is present, the noise $n_y(t)$ is simply added to the output error residual $r'(t)$. For the equation error residual $r(t)$, the noise is high-pass filtered with gains greater one for all frequencies. Thus, high frequencies are attenuated with a large gain, which is undesirable in most applications.

e) Summary

Regarding *threshold-design*, there is no difference between output and equation error for faults in K or simultaneous faults in K and T. The thresholds for both residuals will be equal. Thus, the selection of the residual is dependent on other criteria like noise for instance. With faults in T, the reaction of $r(t)$ for a stepwise fault in the presence of a stepwise excitation is stronger than that of residual $r'(t)$. Thus, one should opt for the equation error, provided there is no output noise present.

Furthermore, the *output error* $r'(t)$ and the *equation error* $r(t)$ react similar for additive faults at the process input and output. The values for $t \to \infty$ are the same, it only takes more time for $r'(t)$ to reach this value. Thus, it depends on the time constant T, if $r(t)$ should be preferred. $r'(t)$ should be preferred if the process has much *noise* $n_u(t)$ with high frequencies at the input. Then, the low-pass behavior of the output error filters this noise out, while $r(t)$ would amplify this noise with gains of K. Small frequencies are amplified with the same gain for $r'(t)$ and $r(t)$. If high frequent noise occurs at the system output, one should prefer the output residual $r'(t)$, because in this case all frequency components remain unchanged. For high frequencies with high amplitudes at the system output, $r(t)$ should not be used either, because this noise is being amplified with $\sqrt{1 + \omega^2 T^2}$. These results are summarized in Table 10.1.

The results given here are an extract of [10.19]. There also the time histories for driftwise faults are calculated, also for simultaneous changes of the gain and the time constant. Furthermore, the sensitivities of the residuals with regard to the faults are given and tabulated.

□

10.2 Parity equations with state-space models

10.2.1 Continuous-time parity approach

The state-space model of a linear multi-input multi-output process according to Figure 10.4 is considered, which is described by

Table 10.1. Properties of output residuals $r'(t)$ and equation residuals $r(t)$

	Input	Fault	Duration of reaction $r(t)$	$r'(t)$	Design of thresholds $r(t)$	$r'(t)$	Reaction due to noise $n_u(t)$ $r(t)$	$r'(t)$	Reaction due to Noise $n_y(t)$ $r(t)$	$r'(t)$
Change in K	⊿	⊿	⏰⏰	⏰⏰	+	+	G	LP	HP	G
	⊿	⊿	⏰⏰	⏰⏰	+	+	G	LP	HP	G
	⊿	⊿	⏰⏰	⏰⏰	+	+	G	LP	HP	G
	⊿	⊿	⏰⏰	⏰⏰	+	+	G	LP	HP	G
Change in $K+T$	⊿	⊿	⏰⏰	⏰⏰	+	+	LP	LP	HP	G
	⊿	⊿	⏰⏰	⏰⏰	+	+	LP	LP	HP	G
	⊿	⊿	⏰⏰	⏰⏰	+	+	LP	LP	HP	G
	⊿	⊿	⏰⏰	⏰⏰	+	+	LP	LP	HP	G
Change in T	⊿	⊿	0	0	0	0	LP	LP	HP	G
	⊿	⊿		⏰	-	-	LP	LP	HP	G
	⊿	⊿	0	0	0	0	LP	LP	HP	G
	⊿	⊿		⏰	-	-	LP	LP	HP	G
f_U	↑?	⊿	⏰⏰	⏰⏰	+	+	G	LP	HP	G
	↑?	⊿	⏰⏰	⏰⏰	+	+	G	LP	HP	G
f_Y	↑?	⊿	⏰⏰	⏰⏰	+	+	G	LP	HP	G
	↑?	⊿	⏰⏰	⏰⏰	+	+	G	LP	HP	G

⏰⏰ long duration   ⏰ short duration   0 no reaction
+ positive   - difficult   0 no reaction
LP low pass   HP high pass   G attenuation by gain

$$\dot{\mathbf{x}}(t) = \mathbf{A}\,\mathbf{x}(t) + \mathbf{B}\,\mathbf{u}(t) + \mathbf{V}\,\mathbf{v}(t) + \mathbf{L}\,\mathbf{f}(t) \tag{10.15}$$

$$\mathbf{y}(t) = \mathbf{C}\,\mathbf{x}(t) + \mathbf{N}\,\mathbf{n}(t) + \mathbf{M}\,\mathbf{f}(t) \tag{10.16}$$

where $\mathbf{n}(t)$ are noise disturbances and $\mathbf{v}(t)$ unmeasurable inputs or disturbances. $\mathbf{f}(t)$ are additive faults which may be composed of additive faults $\mathbf{f}_l(t)$ on the input and $\mathbf{f}_m(t)$ on the output

$$\mathbf{f}^T(t) = \left[\mathbf{f}_m^T(t)\ \mathbf{f}_l^T(t)\right] \tag{10.17}$$

The design of the residuals follows the original approach from [10.5], see also [10.7] and [10.12].

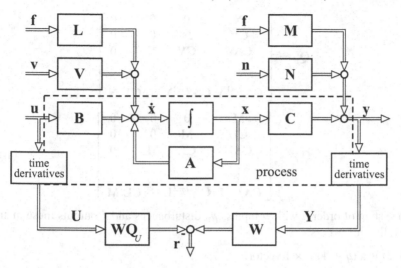

Fig. 10.4. Parity equations based on state-space model for continuous time

Introducing (10.15) in (10.16) yields

$$\dot{\mathbf{y}}(t) = \mathbf{C}\,\dot{\mathbf{x}}(t) + \mathbf{N}\,\dot{\mathbf{n}}(t) + \mathbf{M}\,\dot{\mathbf{f}}(t)$$
$$= \mathbf{C}\,\mathbf{A}\,\mathbf{x}(t) + \mathbf{C}\,\mathbf{B}\,\mathbf{u}(t) + \mathbf{C}\,\mathbf{V}\,\mathbf{v}(t) + \mathbf{C}\,\mathbf{L}\,\mathbf{f}(t) + \mathbf{N}\dot{\mathbf{n}}(t) + \mathbf{M}\dot{\mathbf{f}}(t) \quad (10.18)$$

The second derivative gives

$$\ddot{\mathbf{y}}(t) = \mathbf{C}\,\ddot{\mathbf{x}}(t) + \mathbf{N}\,\ddot{\mathbf{n}}(t) + \mathbf{M}\,\ddot{\mathbf{f}}(t)$$
$$= \mathbf{C}\,\mathbf{A}^2\,\mathbf{x}(t) + \mathbf{C}\,\mathbf{A}\,\mathbf{B}\,\mathbf{u}(t) + \mathbf{C}\,\mathbf{B}\,\dot{\mathbf{u}}(t) + \mathbf{C}\,\mathbf{A}\,\mathbf{V}\,\mathbf{v}(t)$$
$$+ \mathbf{C}\,\mathbf{V}\,\dot{\mathbf{v}}(t) + \mathbf{N}\,\ddot{\mathbf{n}}(t) + \mathbf{C}\,\mathbf{A}\,\mathbf{L}\,\mathbf{f}(t) + \mathbf{C}\,\mathbf{L}\,\dot{\mathbf{f}}(t) + \mathbf{M}\ddot{\mathbf{f}}(t) \quad (10.19)$$

By this way a redundancy in the equations for the same time instant t is generated.
Increasing the number of $q \le n$ derivatives of $y(t)$ leads to the equation system

$$\mathbf{Y}(t) = \mathbf{T}\,\mathbf{x}(t) + \mathbf{Q}_u\mathbf{U}(t) + \mathbf{Q}_v\,\mathbf{V}(t) + \mathbf{Q}_n\,\mathbf{N}(t) + \mathbf{Q}_f\,\mathbf{F}(t) \quad (10.20)$$

with

$$\mathbf{Y}(t) = \begin{bmatrix} \mathbf{y}(t) \\ \dot{\mathbf{y}}(t) \\ \vdots \\ \mathbf{y}^{(q)}(t) \end{bmatrix} \quad \mathbf{U}(t) = \begin{bmatrix} \mathbf{u}(t) \\ \dot{\mathbf{u}}(t) \\ \vdots \\ \mathbf{u}^{(q)}(t) \end{bmatrix} \quad \mathbf{V}(t) = \begin{bmatrix} \mathbf{v}(t) \\ \dot{\mathbf{v}}(t) \\ \vdots \\ \mathbf{v}^{(q)}(t) \end{bmatrix} \quad \mathbf{F}(t) = \begin{bmatrix} \mathbf{f}(t) \\ \dot{\mathbf{f}}(t) \\ \vdots \\ \mathbf{f}^{(q)}(t) \end{bmatrix}$$
$$(10.21)$$

and

$$\mathbf{T} = \begin{bmatrix} \mathbf{C} \\ \mathbf{CA} \\ \mathbf{CA}^2 \\ \vdots \\ \mathbf{CA}^q \end{bmatrix} \quad \mathbf{Q}_U = \begin{bmatrix} 0 & 0 & 0 & \cdots & 0 \\ \mathbf{CB} & 0 & 0 & \cdots & 0 \\ \mathbf{CAB} & \mathbf{CB} & 0 & \cdots & 0 \\ \vdots & \vdots & & \ddots & \vdots \\ \mathbf{CA}^{q-1}\mathbf{B} & \mathbf{CA}^{q-2}\mathbf{B} & \cdots & \mathbf{CB} & 0 \end{bmatrix} \quad (10.22)$$

$$\mathbf{Q}_v = \begin{bmatrix} \mathbf{N} & 0 & 0 & \cdots & 0 \\ \mathbf{CV} & \mathbf{N} & 0 & \cdots & 0 \\ \mathbf{CAV} & \mathbf{CV} & \mathbf{N} & \cdots & 0 \\ \vdots & \vdots & & \ddots & \vdots \\ \mathbf{CA}^{q-1}\mathbf{V} & \mathbf{CA}^{q-2}\mathbf{V} & \cdots & \mathbf{CV} & \mathbf{N} \end{bmatrix}$$

$$\mathbf{Q}_f = \begin{bmatrix} \mathbf{M} & 0 & 0 & \cdots & 0 \\ \mathbf{CL} & \mathbf{M} & 0 & \cdots & 0 \\ \mathbf{CAL} & \mathbf{CL} & \mathbf{M} & \cdots & 0 \\ \vdots & \vdots & & \ddots & \vdots \\ \mathbf{CA}^{q-1}\mathbf{L} & \mathbf{CA}^{q-2}\mathbf{L} & \cdots & \mathbf{CL} & \mathbf{M} \end{bmatrix} \tag{10.23}$$

For a system of order n with p inputs, p_v disturbances and r outputs these matrices have following orders:

- $\mathbf{Y}(t)$ is a $(q+1)\, r \times 1$ vector;
- $\mathbf{U}(t)$ is a $(q+1)\, p \times 1$ vector;
- $\mathbf{T}$ is a $(q+1)\, r \times n$ matrix;
- $\mathbf{Q}_U$ is a $(q+1)\, r \times (q+1)p$ matrix;
- $\mathbf{Q}_V$ is a $(q+1)\, r \times (q+1)p_v$ matrix;

As the state vector $\mathbf{x}(t)$ and the disturbance $\mathbf{v}(t)$ are unknown (10.20) is multiplied by a vector $\mathbf{w}^T$

$$\mathbf{w}^T \mathbf{Y}^T = \mathbf{w}^T \mathbf{T}\,\mathbf{x}(t) + \mathbf{w}^T \mathbf{Q}_u \mathbf{U}(t) + \mathbf{w}^T \mathbf{Q}_v \mathbf{V}(t) + \mathbf{w}^T \mathbf{Q}_n \mathbf{N}(t) + \mathbf{w}^T \mathbf{Q}_f \mathbf{F}(t) \tag{10.24}$$

By selecting $\mathbf{w}^T$ (dimension $1 \times (q+1)\, r$) such that

$$\mathbf{w}^T \mathbf{T} = 0 \quad \text{and} \quad \mathbf{w}^T \mathbf{Q}_v = 0 \tag{10.25}$$

a residual vector in the computational form is obtained

$$\mathbf{r}(t) = \mathbf{w}^T \mathbf{Y}(t) - \mathbf{w}^T \mathbf{Q}_u \mathbf{U}(t) \tag{10.26}$$

Through (10.25) a part of the elements of $\mathbf{w}^T$ is determined according to order of $\mathbf{T}$ and $\mathbf{Q}_v$. The remaining elements can be used to design different parity equations. Inserting (10.24) and (10.26) yields the internal form of the parity equation

$$r(t) = \mathbf{w}^T \mathbf{Q}_f \mathbf{F}(t) + \mathbf{w}^T \mathbf{Q}_n \mathbf{N}(t) \tag{10.27}$$

which shows how the residual is affected by the faults $\mathbf{F}(t)$ and noise $\mathbf{N}(t)$. If (10.25) is satisfied, the residual is independent of the unknown input $\mathbf{v}(t)$ and the states $\mathbf{x}(t)$. More residuals are obtained by selecting several different vectors $\mathbf{w}^T$, thus forming a matrix $\mathbf{W}$ and the residual vector becomes

$$\mathbf{r}(t) = \mathbf{W}\,\mathbf{Y}(t) - \mathbf{W}\,\mathbf{Q}_u\,\mathbf{U}(t) \tag{10.28}$$

compare Figure 10.4.

The order of $\mathbf{W}$ determines the number of parity equations. For Laplace transformation of (10.28) one needs

$$\mathbf{Y}(s) = \begin{bmatrix} \mathbf{y}(s) \\ s\,\mathbf{y}(s) \\ \vdots \\ s^q\,\mathbf{y}(s) \end{bmatrix} = \begin{bmatrix} \mathbf{I} \\ s\,\mathbf{I} \\ \vdots \\ s^q\,\mathbf{I} \end{bmatrix} \mathbf{y}(s) = \mathbf{L}_y(s)\,\mathbf{y}(s)$$

$$\mathbf{U}(s) = \begin{bmatrix} \mathbf{u}(s) \\ s\,\mathbf{u}(s) \\ \vdots \\ s^q\,\mathbf{u}(s) \end{bmatrix} = \begin{bmatrix} \mathbf{I} \\ s\,\mathbf{I} \\ \vdots \\ s^q\,\mathbf{I} \end{bmatrix} \mathbf{u}(s) = \mathbf{L}_u(s)\,\mathbf{u}(s)$$

and obtains

$$\mathbf{r}(s) = \mathbf{W}\,\mathbf{L}_y(s)\,\mathbf{y}(s) - \mathbf{W}\,\mathbf{Q}_u\,\mathbf{L}_u(s)\,\mathbf{u}(s) \tag{10.29}$$

which shows similarities to the equation error residual for parity equations with transfer functions, (10.5).

The discussed approach of parity equations with state-space models needs higher order derivatives of order q for the measured inputs and outputs. They can be obtained by state-variable filters, e.g. [10.17] and Chapter 23.2. As they use the information of the signal in one time instant relatively noisy results can be expected in practice. Therefore discrete-time versions should be preferred as treated in the next section.

10.2.2 Discrete-time parity approach

The parity equations with state-space models are now considered for discrete-time models. They will be simpler to derive and to implement in this form than for continuous time. According to Figure 10.5 the basic process equations are

$$\mathbf{x}(k+1) = \mathbf{A}\,\mathbf{x}(k) + \mathbf{B}\,\mathbf{u}(k) + \mathbf{V}\,\mathbf{v}(k) + \mathbf{L}\,\mathbf{f}(k) \tag{10.30}$$

$$\mathbf{y}(k) = \mathbf{C}\,\mathbf{x}(k) + \mathbf{N}\,\mathbf{n}(k) + \mathbf{M}\,\mathbf{f}(k) \tag{10.31}$$

where $\mathbf{v}(k)$ and $\mathbf{n}(k)$ are unmeasurable disturbance signals. $\mathbf{f}(k)$ are additive faults which may be composed of additive faults $\mathbf{f}_l(t)$ on the input and $\mathbf{f}_m(t)$ on the output.

The design of the residuals follows the original approach from [10.5], see also [10.7] and [10.12]. To simplify the notations, the state-space model without faults and disturbances is used

$$\mathbf{x}(k+1) = \mathbf{A}\,\mathbf{x}(k) + \mathbf{B}\,\mathbf{u}(k) \tag{10.32}$$

$$\mathbf{y}(k) = \mathbf{C}\,\mathbf{x}(k) \tag{10.33}$$

Introducing (10.32) in (10.33) yields

$$\mathbf{y}(k+1) = \mathbf{C}\,\mathbf{A}\,\mathbf{x}(k) + \mathbf{C}\,\mathbf{B}\,\mathbf{u}(k) \tag{10.34}$$

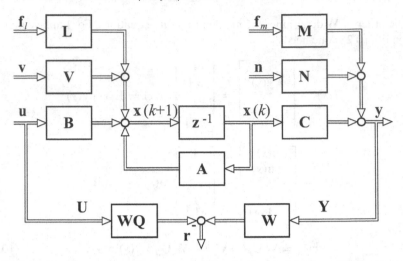

Fig. 10.5. Residual generation with parity equations in discrete time for a MIMO process with a state-space model

u	process input vector $(1 \times p)$		n	noise vector $(1 \times r)$
x	process state vector $(m \times m)$		y	input noise vector $(1 \times m)$
y	process output vector $(1 \times r)$		f	fault vector

For the next sampled output it holds

$$y(k + 2) = C\,x(k + 2)$$
$$= C\,A\,x(k + 1) + C\,B\,u(k + 1)$$
$$= C\,A^2\,x(k) + C\,A\,B\,u(k) + C\,B\,u(k + 1) \qquad (10.35)$$

and for the qth sample $(q \leq m)$

$$y(k + q) = C\,A^q\,x(k) + C\,A^{q-1}\,B\,u(k) + \ldots + C\,B\,u(k + q - 1) \qquad (10.36)$$

Here redundant equations are generated for different time instants. Now, the equations for a *time window* of length $q + 1$ are summarized and lead to

$$Y(k + q) = T\,x(k) + Q\,U(k + q) \qquad (10.37)$$

or time shifted by q backwards

$$Y(k) = T\,x(k - q) + Q\,U(k) \qquad (10.38)$$

with the vectors

$$Y(k) = \begin{bmatrix} y(k - q) \\ y(k - q + 1) \\ \vdots \\ y(k) \end{bmatrix} \qquad U(k) = \begin{bmatrix} u(k - q) \\ u(k - q + 1) \\ \vdots \\ u(k) \end{bmatrix} \qquad (10.39)$$

and the matrices

$$
\mathbf{T} = \begin{bmatrix} \mathbf{C} \\ \mathbf{C\,A} \\ \mathbf{C\,A}^2 \\ \vdots \\ \mathbf{C\,A}^q \end{bmatrix} \quad \mathbf{Q} = \begin{bmatrix} \mathbf{0} & \mathbf{0} & \cdots & \mathbf{0} \\ \mathbf{C\,B} & \mathbf{0} & \cdots & \\ \mathbf{C\,A\,B} & \mathbf{C\,B} & & \\ \vdots & \vdots & \ddots & \vdots \\ \mathbf{C\,A}^{q-1}\mathbf{B} & \mathbf{C\,A}^{q-2}\mathbf{B} & \cdots & \mathbf{C\,B\,0} \end{bmatrix} \tag{10.40}
$$

Hence (10.38) describes the $(q + 1)$ input and output signals and the initial state vector $\mathbf{x}(k - q)$ over a time interval of length $(q + 1)$, thus forming a *temporal redundancy*, [10.23].

As the state vector $\mathbf{x}(k - q)$ is unknown, (10.38) is multiplied by a vector $\mathbf{w}^T$

$$
\mathbf{w}^T\,\mathbf{Y}(k) = \mathbf{w}^T\,\mathbf{T}\,\mathbf{x}(k - q) + \mathbf{w}^T\,\mathbf{Q}\,\mathbf{U}(k) \tag{10.41}
$$

By selecting $\mathbf{w}^T$ such that

$$
\mathbf{w}^T\,\mathbf{T} = \mathbf{0} \tag{10.42}
$$

an input-output relation results and residual can be defined (computational form)

$$
\mathbf{r}(k) = \mathbf{w}^T\,\mathbf{Y}(k) - \mathbf{w}^T\,\mathbf{Q}\,\mathbf{U}(k) \tag{10.43}
$$

The dimension of $\mathbf{w}^T$ is $1 \times (q + 1)r$ where r is the number of outputs. If the order of $\mathbf{A}$ is m, the matrix $\mathbf{T}$ has order $m \times (q + 1)r$. Through (10.42) m elements of $\mathbf{w}^T$ are determined. However, the remaining $(q + 1)r - m$ elements of $\mathbf{w}^T$ can be chosen freely.

More residuals can be determined by multiplying (10.38) with a matrix $\mathbf{W}$ and the condition

$$
\mathbf{W\,T} = \mathbf{0} \tag{10.44}
$$

The residual vector then results as, see also Figure 10.5,

$$
\mathbf{r}(k) = \mathbf{W}\,[\mathbf{Y}(k) - \mathbf{Q}\,\mathbf{U}(k)] \tag{10.45}
$$

$\mathbf{W}$ contains the vectors $\mathbf{w}^T$, where some elements are determined by (10.42). The remaining elements can be chosen such that special properties of the residual, e.g. structured residuals are obtained as will be shown in Section 10.3. An example is given in [10.16], see also [10.14]. The length $q + 1$ of the time window is free. Usually $q = n$ is a proper choice, see [10.5].

If unknown inputs $\mathbf{v}(k)$ act on the process, the residuals can be made independent of these inputs if in addition to (10.44)

$$
\mathbf{W\,Q}_v = \mathbf{0} \tag{10.46}
$$

is satisfied, where $\mathbf{Q}_v$ corresponds to $\mathbf{Q}$ but with $\mathbf{V}$ instead of $\mathbf{B}$, see [10.14].

After introducing the noise and fault terms from (10.30) and (10.31), the internal form of the parity equation is obtained from (10.45)

$$\mathbf{r}(t) = \mathbf{w}^T \, \mathbf{Q}_F \, \mathbf{f}(t) + \mathbf{w}^T \, \mathbf{Q}_n \, \mathbf{n}(t) \qquad (10.47)$$

where $\mathbf{Q}_F$ and $\mathbf{Q}_n$ are matrices like $\mathbf{Q}$, see [10.14]. The residuals then only depend on the additive faults and the noise.

Practical experience shows that the residuals result with relative large variance due to noise and deviations between the process and its model. Therefore the residuals must be low-pass filtered. In comparison to the transfer function approach the state-space approach gives more freedom in selecting the residual generating vector $\mathbf{W}$. However it uses less time history of the signals and is therefore more sensitive to noise, is more computationally involved and not so straightforward as the transfer function version.

10.3 Properties of residuals

In the ideal case the residuals should only be influenced by the faults to be detected. However, because of the existence of modelling errors, unknown input signals, stationary and instationary disturbances at the outputs, the residuals will vary continuously. This means that wider thresholds for the residuals have to be used. But too large thresholds do not allow to detect small faults. There exist following ways to improve this conflict situation:

- enhanced residuals for particular faults;
- filtering of high frequency effects like noise from low frequent changing faults by low-pass filters;
- maximizing of the fault sensitivity of residuals;
- robustness against modelling errors;
- adaptive thresholds, e.g. depending on the input excitation.

All the measures have the goal to make the residuals sensitive to faults and robust against disturbing effects.

10.3.1 Generation of enhanced residuals

The primary residuals $\mathbf{r}$ arising from the output model or the polynomial model do not necessarily allow to separate the faults by observing their deviations. Furtheron the state-space approach leaves some freedom in the design of the weighting matrix $\mathbf{W}$. The idea of *enhanced residuals* is to give the residual vector special properties in order to diagnose or at least to isolate the faults from each other. Two known approaches are the generation of *structured* and *directional residuals*, [10.8].

a) Structured residuals

The goal of the design of structured residuals is that the faults influence some residuals and some not. Then, in dependence on the faults, certain patterns appear in the vector (or table). Generally speaking, this means that the residuals only influence certain subspaces in the vector-space, compare Figure 10.6a. Therefore structured residual vectors are at least independent (or decoupled) from one of the faults. The resulting *residual patterns* are also called *fault signatures*.

b) Directional residuals

The design of directional residuals intends to reach a certain vector in the residual space for each fault, such that the direction is fixed, but the length of the vector depends on fault size, see Figure 10.6b. Also in dynamic transients the direction of the residual vector should be maintained.

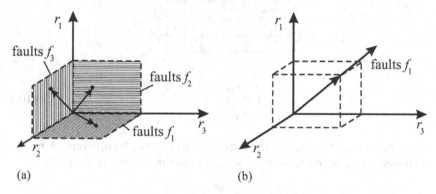

(a)   (b)

Fig. 10.6. Enhanced residuals: (a) structured residuals; (b) directional residuals

The residuals are usually checked against a threshold r_{th}. It is now assumed that the threshold is the same for positive and negative deflections and that the exceeding of the maximal or minimal threshold is triggered. This limit checking yields binary outputs

$$r_i^* = \begin{cases} 0 \text{ if } |r_i(t)| < r_{thi} \\ 1 \text{ if } |r_i(t)| > r_{thi} \end{cases} \tag{10.48}$$

$r_i^* = 1$ means that one of the thresholds was exceeded. Then different isolating properties according to Table 10.2 can be distinguished. These structural matrices are called *strongly isolating* if by an error of one residual no other fault is isolated. *Weak isolation* means that by one error another (wrong) fault is isolated. If the patterns are indistinguishable, the residuals are not isolable. Some more, special cases are discussed in [10.8].

Table 10.3 shows the possible patterns for two residuals. Hence, 3 faults can be (weakly) isolated or generally

$$\kappa = 2^n - 1$$

faults, if n is the number of residuals.

More possibilities for the isolation and diagnosis of faults results if the signs and also the size of the residuals are taken into account. Including the sign leads to

Table 10.2. Structured residuals; 1= residual is sensitive; 0 = residual is insensitive against the respective fault

	strongly isolating				strongly isolating				weakly isolating				not isolating			
	f_1	f_2	f_3	f_4	f_1	f_2	f_3	f_4	f_1	f_2	f_3	f_4	f_1	f_2	f_3	f_4
r_1^*	1	0	0	0	1	1	1	0	0	1	1	0	0	0	1	0
r_2^*	0	1	0	0	1	1	0	1	1	1	1	1	1	1	1	0
r_3^*	0	0	1	0	1	0	1	1	1	1	0	1	1	1	0	1
r_4^*	0	0	0	1	0	1	1	1	0	0	1	1	1	1	0	1

Table 10.3. Structured residuals without considering the sign of r_i

residuals	no fault	f_1	f_2	f_3
r_1^*	0	1	0	1
r_2^*	0	0	1	1

$$r_i^* = \begin{cases} 0 \text{ if } r_{thi}^- \le r_i(t) \le r_{thi}^+ \\ 1 \text{ if } r_i(t) > r_{thi}^+ \\ -1 \text{ if } r_i(t) < r_{thi}^- \end{cases} \tag{10.49}$$

Here the positive and negative thresholds r_{thi}^+ and r_{thi}^- can be different. A maximal set of patterns for two residuals in Table 10.4 shows that now 7 or generally

$$\kappa = 3^n - 1$$

faults are isolable. For $n = 3$ residuals the number of (weakly) distinguishable faults is 7 without sign and 26 with sign. Therefore the consideration of the sign of the residuals increases the number of detectable faults considerably.

Table 10.4. Structured residuals including the sign of r_i

residuals	no fault	f_1	f_2	f_3	f_4	f_5	f_6	f_7
r_1^*	0	1	-1	0	0	1	-1	1
r_2^*	0	0	0	1	-1	1	1	-1

A further improvement for diagnosis is obtained by using the size of the residual deflections. If the residuals are marked by

$$r_i^* = \begin{cases} 0 & \text{if } r_{thi}^- \le r_i(t) \le r_{thi}^+ \\ + & \text{if } r_i(t) \text{ is positive small} \\ ++ & \text{if } r_i(t) \text{ is positive large} \\ - & \text{if } r_i(t) \text{ is negative small} \\ -- & \text{if } r_i(t) \text{ is negative large} \end{cases} \tag{10.50}$$

then theoretically 24 combinations are possible for 2 residuals, see Table 10.5 or generally

$$\kappa = 5^n - 1$$

In practical cases, of course, not all combinations will appear. But this consideration shows that the residuals become better isolable and allow a more detailed fault diagnosis if the sign and the size of the residual deflection is evaluated. This improves also the detection of multiple faults.

Table 10.5. Structured residuals including the sign and size of r_i

residuals	no fault	f_1	f_2	f_3	f_4	f_5	f_6	f_7	f_8
r_1^*	0	+	+	+	+	+	++	++	++
r_2^*	0	0	+	++	–	––	0	+	++

	f_9	f_{10}	f_{11}	f_{12}	f_{13}	f_{14}	f_{15}	f_{16}	f_{17}
r_1^*	++	++	–	–	–	–	–	––	––
r_2^*	–	––	0	+	++	–	––	0	+

	f_{18}	f_{19}	f_{20}	f_{21}	f_{22}	f_{23}	f_{24}
r_1^*	––	––	––	0	0	0	0
r_2^*	++	–	––	+	++	–	––

10.3.2 Generation of structured residuals

The goal in designing structured residuals is to generate good isolating patterns of the residual vector. This means that the residual should be at least independent (decoupled) from the faults to be detected. As this is directly possible with additive faults $\mathbf{f}_u$ on the inputs and $\mathbf{f}_y$ on the outputs mainly this case will be considered.

The polynomial error (or equation error) of a MIMO process with p inputs and r outputs is according to (10.4)

$$\mathbf{r}(s) = \mathbf{A}_m(s)\,\mathbf{y}_p(s) - \mathbf{B}_m(s)\,\mathbf{u}_p(s) \tag{10.51}$$

To generate structured residuals this equation is multiplied by a *residual generating matrix* $\mathbf{W}$

$$\mathbf{r}^*(s) = \mathbf{W}(s)\,\left[\mathbf{A}_m(s)\,\mathbf{y}_p(s) - \mathbf{B}_m\,\mathbf{u}_p(s)\right] \tag{10.52}$$

or

$$\mathbf{r}^*(s) = \mathbf{W}(s)\left\{ \begin{bmatrix} A_1(s) & 0 & \cdots & 0 \\ 0 & A_2(s) & \ddots & \\ 0 & 0 & \cdots & A_r(s) \end{bmatrix} \begin{bmatrix} y_1(s) \\ y_2(s) \\ \vdots \\ y_r(s) \end{bmatrix} - \right.$$

$$\left. \begin{bmatrix} B_1(s) & 0 & \cdots & 0 \\ 0 & B_2(s) & \cdots & 0 \\ & & \ddots & \\ 0 & 0 & \cdots & B_p(s) \end{bmatrix} \begin{bmatrix} u_1(s) \\ u_2(s) \\ \vdots \\ u_p(s) \end{bmatrix} \right\} \tag{10.53}$$

To obtain zeros in $\mathbf{r}^*$ the matrix $\mathbf{W}$ has to be selected such that the elements of $\mathbf{r}^*$ become independent on one measurement each (elements of $\mathbf{y}_p$ and $\mathbf{u}_p$).

Therefore

$$\mathbf{W}_y(s)\,\mathbf{A}_m(s) = \mathbf{0} \quad \text{and} \quad \mathbf{W}_u(s)\,\mathbf{B}_m(s) = \mathbf{0} \tag{10.54}$$

with $\mathbf{W}^T = \begin{bmatrix} \mathbf{W}_y & \mathbf{W}_u \end{bmatrix}$

Changing the notation of (10.52) leads to

$$\begin{bmatrix} r_1^* \\ r_2^* \\ \vdots \\ r_r^* \end{bmatrix} = \begin{bmatrix} \mathbf{w}_{y1}^T(s) \\ \mathbf{w}_{y2}^T(s) \\ \vdots \\ \mathbf{w}_{u1}^T(s) \\ \mathbf{w}_{u2}^T(s) \\ \vdots \end{bmatrix} \tag{10.55}$$

$$[\mathbf{A}_1(s)y_1(s) + \mathbf{A}_2(s)y_2(s) + \ldots + \mathbf{B}_1(s)u_1(s) + \mathbf{B}_2(s)u_2(s) + \ldots]$$

and with

$$\mathbf{w}_{y1}^T(s)\,\mathbf{A}_1(s) = 0 \text{ independence on } y_1(s)$$
$$\mathbf{w}_{y2}^T(s)\,\mathbf{A}_2(s) = 0 \text{ independence on } y_2(s)$$

$$\vdots$$

$$\mathbf{w}_{u1}^T(s)\,\mathbf{B}_1(s) = 0 \text{ independence on } u_1(s)$$
$$\mathbf{w}_{u2}^T(s)\,\mathbf{B}_2(s) = 0 \text{ independence on } u_2(s)$$

$$\vdots$$

is reached. By this way the (dynamic) polynomials $\mathbf{w}_i^T(s)$ are designed which give the residuals $\mathbf{r}^*(s)$ according to (10.52) the required independence on specific input and output signals.

This procedure will be demonstrated by two examples.

Example 10.2: Structured residuals for a single-input two-output process

Two parallel first order processes with one input u and two outputs y_1 and y_2 are considered, Figure 10.7,

$$G_1(s) = \frac{y_1(s)}{u(s)} = \frac{K_1}{1 + T_1 s} \; ; \; G_2(s) = \frac{y_2(s)}{u(s)} = \frac{K_2}{1 + T_2 s}$$

(The input could be the steering angle δ of a vehicle and the outputs the yaw rate $\dot{\psi}$ and the lateral acceleration $\ddot{y}$ and the goal is to detect offset faults in all 3 sensors).

The rearranged process equations are

$$0 = y_1(s) + T_1 s\, y_1(s) - K_1 u(s)$$

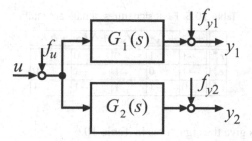

Fig. 10.7. Single-input two-output process with additive faults

$$0 = y_2(s) + T_2 s\, y_2(s) - K_2 u(s)$$

or in vector notation

$$\begin{bmatrix} 0 \\ 0 \end{bmatrix} = \begin{bmatrix} 1 + T_1 s \\ 0 \end{bmatrix} y_1(s) + \begin{bmatrix} 0 \\ 1 + T_2 s \end{bmatrix} y_2(s) + \begin{bmatrix} -K_1 \\ -K_2 \end{bmatrix} u(s)$$

To obtain an independence of the residuals from 3 signals, this equation is multiplied with the residual generating matrix

$$\mathbf{W}(s) = \begin{bmatrix} \mathbf{w}_{y1}^T(s) \\ \mathbf{w}_{y2}^T(s) \\ \mathbf{w}_{u1}^T(s) \end{bmatrix}$$

Independence on y_1, y_2 and u is reached by

$$\mathbf{w}_{y1}^T(s) \begin{bmatrix} 1 + T_1 s \\ 0 \end{bmatrix} = 0 \rightarrow \mathbf{w}_{y1}^T(s) = [0\ 1]$$

$$\mathbf{w}_{y2}^T(s) \begin{bmatrix} 0 \\ 1 + T_2 s \end{bmatrix} = 0 \rightarrow \mathbf{w}_{y2}^T(s) = [1\ 0]$$

$$\mathbf{w}_{u1}^T(s) \begin{bmatrix} -K_1 \\ -K_2 \end{bmatrix} = 0 \rightarrow \mathbf{w}_{u1}^T(s) = [K_2\ -K_1]$$

After multiplying the vector notation equation by $\mathbf{W}(s)$ three residuals result

$$r_1^*(s) = (1 + T_2 s)\, y_2(s) - K_2 u(s)$$
$$r_2^*(s) = (1 + T_1 s)\, y_1(s) - K_1 u(s)$$
$$r_3^*(s) = K_2(1 + T_1 s)\, y_1(s) - K_1(1 + T_2 s)\, y_2(s)$$

It is interesting to see that these residuals also follow directly from the transfer functions

$$\frac{y_1(s)}{u(s)} = G_1(s)\ ;\quad \frac{y_2(s)}{u(s)} = G_2(s)\ ;\quad \frac{y_1(s)}{y_2(s)} = \frac{G_1(s)}{G_2(s)}$$

if after cross-multiplication all terms brought on one side and 0 is replaced by r_i^*.

Table 10.6. Fault signatures of different faults

	$+f_u$	$+f_{y2}$	$+f_{y1}$	$+\Delta K_1$	$+\Delta K_2$
r_1^*	-1	1	0	0	-1
r_2^*	-1	0	1	-1	0
r_3^*	0	-1	1	-1	1

These residuals give the signatures in Table 10.6.

Hence, the residuals are strongly isolating for offset faults f_u, f_{y1} and f_{y2}, also if their signs are different. Gain changes ΔK_1 and ΔK_2 are only distinguishable from f_{y1} and f_{y2} if their deviations have the same sign. As this is usually not known, they are practically undistinguishable. These results hold as well for changing as for constant inputs. Changes of the time constants lead to residual deviations if the input $u(t)$ changes. But they result only in passing small residuals and are practically not detectable.

The residuals contain derivatives of the output signals which may be a problem if higher frequent noise is present. Then low-pass filters have to be applied, also damping the residual responses.

If the same procedure is applied for the output error (10.6), following residuals result:

$$r_1'^*(s) = y_2(s) - \frac{K_2}{1+T_2 s}u(s)$$
$$r_2'^*(s) = y_1(s) - \frac{K_1}{1+T_1 s}u(s)$$
$$r_3'^*(s) = \frac{K_2}{1+T_2 s}\, y_1(s) - \frac{K_1}{1+T_1 s}\, y_2(s)$$

This set of equations does not need extra low-pass filters because no realizability problems exist.

Summarizing, additive faults of all 3 signal measurements can be isolated by structured parity equations.

$\square$

Example 10.3: Structured residuals for a DC motor

As a further example a DC motor with brush commutation is considered. Modelling, resulting equations and a signal-flow diagram are shown in Chapter 20. By neglecting minor effects, a classical DC motor can be described by linear differential equations and the signal flow diagram in Figure 10.8. For the electrical part, the armature circuit, it holds

$$U_A(t) = R_A\, I_A(t) + L_A\, \dot{I}_A(t) + \Psi\, \omega(t)$$

and for the mechanical part

$$\Psi\, I_A(t) = J\, \dot{\omega}(t) + M_F\, \omega(t) + M_L(t)$$

(The symbols are explained in Chapter 20).

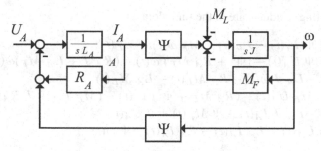

Fig. 10.8. Signal flow diagram of a DC motor. Parameters $R_A = 1.53\Omega$; $L_A = 6.8\ 10^{-3}\Omega s$; $\Psi = 0.34Vs$; $M_F = 0.37\ 10^{-3}Nms$; $J = 1.9\ 10^{-3}kgm^2$

Laplace transformation leads to

$$0 = U_A(s) - R_A\ I_A(s) - L_A s\ I_A(s) - \Psi\ \omega(s)$$
$$0 = \Psi\ I_A(s) - J\ s\ \omega(s) - M_F\ \omega(s) - M_L(s)$$

or in vector notation

$$\mathbf{0} = \begin{bmatrix} 1 \\ 0 \end{bmatrix} U_A(s) - \begin{bmatrix} R_A + L_A s \\ -\Psi \end{bmatrix} I_A(s) - \begin{bmatrix} \Psi \\ M_F + J\,s \end{bmatrix} \omega(s) + \begin{bmatrix} 0 \\ 1 \end{bmatrix} M_L(s)$$

To obtain structured residuals, multiplication with the residual generation matrix yields

$$\mathbf{r}(s) = \mathbf{W}(s)\left[\begin{bmatrix} 1 \\ 0 \end{bmatrix} U_A(s) - \begin{bmatrix} R_A + L_A s \\ -\Psi \end{bmatrix} I_A(s) - \begin{bmatrix} \Psi \\ M_F + Js \end{bmatrix} \omega(s) \right.$$
$$\left. + \begin{bmatrix} 0 \\ 1 \end{bmatrix} M_L(s)\right]$$

Independence of the residuals on the measured signals and the unknown load torque M_L is obtained by

$$\mathbf{w}_1^T(s)\begin{bmatrix} 1 \\ 0 \end{bmatrix} \qquad = 0 \rightarrow \mathbf{w}_1^T(s) = [0\ 1] \qquad \text{(independent on } U_A)$$

$$\mathbf{w}_2^T(s)\begin{bmatrix} R_A + L_A s \\ -\Psi \end{bmatrix} = 0 \rightarrow \mathbf{w}_2^T(s) = [\Psi\ \ R_A + L_A s] \quad \text{(independent on } I_A)$$

$$\mathbf{w}_3^T(s)\begin{bmatrix} \Psi \\ M_F + Js \end{bmatrix} = 0 \rightarrow \mathbf{w}_3^T(s) = [M_F + J\,s\ \ -\Psi] \text{ (independent on } \omega)$$

$$\mathbf{w}_4^T(s)\begin{bmatrix} 0 \\ 1 \end{bmatrix} \qquad = 0 \rightarrow \mathbf{w}_4^T(s) = [1\ 0] \qquad \text{(independent on } M_L)$$

Hence, the residual generation matrix becomes

$$\mathbf{W} = \begin{bmatrix} 0 & 1 \\ \Psi & R_A + L_A\ s \\ M_F + J\ s & -\Psi \\ 1 & 0 \end{bmatrix}$$

and the resulting residuals are in the time domain

$$r_1(t) = \Psi \ I_A(t) - J \ \dot{\omega}(t) - M_F \ \omega(t) - M_L(t)$$
$$r_2(t) = \Psi \ U_A(t) - (\Psi^2 + R_A \ M_F) \ \omega(t) - (R_A \ J + L_A \ M_F)\dot{\omega}(t)$$
$$\qquad -L_A \ J \ \ddot{\omega}(t) - R_A \ M_L(t) - L_A \ \dot{M}_L(t)$$
$$r_3(t) = M_F \ U_A(t) - (R_A \ M_F + \Psi^2) \ I_A(t) - (M_F \ L_A + J \ R_A)\dot{I}_A(t)$$
$$\qquad -L_A \ J \ \ddot{I}_A(t) + \Psi \ M_L(t) + J \ \dot{U}_A(t)$$
$$r_4(t) = U_A(t) - L_A \ \dot{I}_A(t) - R_A \ I_A(t) - \Psi \ \omega(t)$$

These residuals require the 1st order derivatives of $U_A(t), I_A(t)$ and $\omega(t)$ and the 2nd order derivatives of $I_A(t)$ and $\omega(t)$. This is a practical problem and can only be solved by low-pass filtering of the measured signals and especially by using state variable filters, see [10.14] and Chapter 23.2. A drawback is the required knowledge of the load torque M_L for all residuals except r_4.

As already mentioned, the residuals can also be obtained if the Laplace transformed state variables $\omega(s)$ and $I_A(s)$ are calculated based on the signal flow diagram of Figure 10.8 or the basic two equations as functions on the following measurable variables and input signals

$$\omega(s) = f[I_A(s), M_L(s)] \quad \text{independent on } U_A(s)$$
$$\omega(s) = f[U_A(s), M_L(s)] \quad \text{independent on } I_A(s)$$
$$I_A(s) = f[U_A(s), M_L(s)] \quad \text{independent on } \omega(s)$$
$$I_A(s) = f[U_A(s), \omega(s)] \quad \text{independent on } M_L(s)$$

For example, the first equation yields

$$0 = \Psi \ I_A(s) - J \ s \ \omega(s) - M_F \ \omega(s) - M_L(s)$$

If then the zero is replaced by $r_1(s)$, the corresponding equation results. Hence, the residuals form special transfer functions of the process, where each one measurement is eliminated.

The resulting fault signatures of the residuals for additive and parametric faults are given in the Table 10.7. Hence, the residuals are strongly isolating for the additive offset faults of the sensors of U_A, I_A and ω, and for load changes $\Delta \ M_L$. Parameter changes are practically not distinguishable for R_A and L_A and for J and M_F and are only weakly isolated. Change of Ψ is also only weakly isolated from the others. Under steady state conditions, changes $\Delta \ R_A$ are not distinguishable from ΔU_A, and changes ΔM_F not from ΔM_L. Summarizing, additive sensor faults can be detected and isolated under the condition that the process parameters remain constant. The considered parity equations are not suitable to detect and isolate parameter faults. An application of parity equations to real DC motors is shown in Chapter 20.

This example of the DC motor can also be used to study the behavior of the residuals in the case of *multiple faults*. If two additive sensor faults arise simultaneously and only the sign of the residuals is considered and not their size, Table 10.8 shows the result. The effect of the single faults on the residuals partially compensate and partially add to each other. However, in this case all fault signatures are different and

Table 10.7. Fault signatures of a DC motor for different single faults; ∗ means, that deviations are only obtained by dynamic excitation of the process

	Additive sensor faults				Parameter faults				
	$+f_{UA}$	$+f_{IA}$	$+f_\omega$	$+f_{ML}$	$+f_{RA}$	$+f_{LA}$	$+f_\Psi$	$+f_J$	$+f_{MF}$
r_1	0	+1	−1	−1	0	0	+1	−1*	−1
r_2	+1	0	−1	−1	−1	−1*	±1	−1*	−1
r_3	+1	−1	0	+1	−1	−1*	±1	±1*	±1
r_4	+1	−1	−1	0	−1	−1*	−1	0	0

Table 10.8. Fault signatures of a DC motor for double additive sensor faults

	Additive sensor faults				
	f_{UA} AND f_{IA}	f_{UA} AND Δf_ω	f_{UA} AND f_{ML}	f_{IA} AND f_ω	f_{IA} AND f_{ML}
r_1	+1	−1	−1	0	0
r_2	+1	0	0	−1	−1
r_3	0	+1	+1	−1	0
r_4	0	0	+1	−1	−1

mostly even strongly isolating. Hence, by taking the signatures of Table 10.8 into account, it is possible to detect and isolate also double sensor faults.

□

10.3.3 Sensitivity of parity equations

The sensitivity of structured residuals with regard to additive and parametric faults is considered in the sequel, following [10.15], see also [10.9].

Writing (10.26) or (10.51) for one residual leads to

$$r(t) = [\mathbf{w}^T - \mathbf{w}^T \ \mathbf{Q}] \begin{bmatrix} \mathbf{Y}(t) \\ \mathbf{U}(t) \end{bmatrix} = \boldsymbol{\theta}^T(\mathbf{p}) \ \boldsymbol{\psi}(t) \tag{10.56}$$

where $\boldsymbol{\theta}^T(\mathbf{p})$ contains the process parameters in dependence on their physical parameters $\mathbf{p}$ and $\boldsymbol{\psi}(t)$ the measured input and output signals and their derivatives.

Small *additive faults* $\Delta\boldsymbol{\psi}(t)$ then yield, with $\boldsymbol{\theta}_0^T(\mathbf{p})$ and $\boldsymbol{\psi}_0(t)$ for the fault free case,

$$r(t) = \boldsymbol{\theta}_0^T(\mathbf{p}) [\boldsymbol{\psi}_0(t) + \Delta\boldsymbol{\psi}(t)] = \boldsymbol{\theta}_0^T(\mathbf{p}) \Delta\boldsymbol{\psi}(t) \tag{10.57}$$

as

$$\boldsymbol{\theta}_0^T(\mathbf{p})\boldsymbol{\psi}_0(t) = 0 \tag{10.58}$$

if there is no fault. If the process parameters remain constant the sensitivity becomes

$$\frac{\partial r(t)}{\partial \boldsymbol{\psi}(t)} = \boldsymbol{\theta}_0^T(\mathbf{p}) \tag{10.59}$$

The residual is therefore proportional to the additive faults $\Delta\boldsymbol{\psi}(t)$ with the sensitivity coefficient (gain) $\boldsymbol{\theta}_0^T(\mathbf{p})$ and does not depend on the input excitation $\mathbf{U}(t)$.

In the case of *parametric faults* $\Delta \mathbf{p}(t)$ the residuals are calculated according to (10.56)

$$r(t) = \boldsymbol{\theta}_0^T(\mathbf{p}) \boldsymbol{\psi}(t) \qquad (10.60)$$

where $\boldsymbol{\theta}_0(\mathbf{p})$ are the fixed parameters in the model. In the fault free case, it is $r(t) = 0$. After a small fault in the parameters the process follows a new differential equation

$$
\begin{aligned}
0 &= \boldsymbol{\theta}_0^T(\mathbf{p}) \boldsymbol{\psi}_0(t) + \tfrac{\partial}{\partial \mathbf{p}}[\boldsymbol{\theta}^T(\mathbf{p}) \, \boldsymbol{\psi}(t)] \, \Delta \mathbf{p} \\
&= \boldsymbol{\theta}_0^T(\mathbf{p}) \boldsymbol{\psi}_0(t) + \tfrac{\partial \boldsymbol{\theta}^T(\mathbf{p})}{\partial \mathbf{p}} \boldsymbol{\psi}(t) \, \Delta \mathbf{p} + \boldsymbol{\theta}_0^T(\mathbf{p}) \tfrac{\partial \boldsymbol{\psi}(t)}{\partial \mathbf{p}} \Delta \mathbf{p} \qquad (10.61) \\
&= \boldsymbol{\theta}_0^T(\mathbf{p})[\boldsymbol{\psi}_0(t) + \Delta \boldsymbol{\psi}(t)] + \tfrac{\partial \boldsymbol{\theta}^T(\mathbf{p})}{\partial \mathbf{p}} \boldsymbol{\psi}(t) \, \Delta \mathbf{p}
\end{aligned}
$$

where

$$\Delta \boldsymbol{\psi}(t) = \frac{\partial \, \boldsymbol{\psi}(t)}{\partial \, \mathbf{p}} \Delta(\mathbf{p}) \qquad (10.62)$$

describes the changes of the signals due to the parameter changes $\Delta \mathbf{p}$.

The calculated residual with the fixed parameters according to (10.66) but the changed signals then becomes

$$r(t) = \boldsymbol{\theta}_0^T(\mathbf{p})[\boldsymbol{\psi}_0(t) + \Delta \boldsymbol{\psi}(t)] \qquad (10.63)$$

and from (10.61) is follows

$$r(t) = \frac{\partial \boldsymbol{\theta}^T(\mathbf{p})}{\partial \mathbf{p}} \, \boldsymbol{\psi}(t) \, \Delta \mathbf{p} \qquad (10.64)$$

The residual sensitivity to parametric faults is therefore

$$-\frac{\partial r(t)}{\partial \mathbf{p}} = -\frac{\partial \boldsymbol{\theta}^T(\mathbf{p})}{\partial \mathbf{p}} \boldsymbol{\psi}(t) \qquad (10.65)$$

and depends on the parameter sensitivity of the process model with regard to physical parameter changes $\Delta \mathbf{p}$ and the measured signals $\boldsymbol{\psi}(t)$. Therefore the sensitivity of parity equations to parametric faults depends on the input excitation $U(t)$ of the process. For $U(t) = 0$, no parametric faults can be detected.

10.4 Parity equations for nonlinear processes

10.4.1 Parity equations for special nonlinear processes

Because of the large variety of nonlinear process models it is not possible to describe general applicable ways to generate parity equations. If, however, the nonlinear model can be expressed by

$$y_m(t) = f_{NL}[\dot{y}(t), \ddot{y}(t), \ldots; \, u(t), \dot{u}(t), \ddot{u}(t) \ldots] \qquad (10.66)$$

output residuals

$$r(t) = y_p(t) - y_m(t) \tag{10.67}$$

can be directly generated. This holds also for discrete time polynomial models like the Hammerstein model or the parametric Volterra model, described in Section 9.3.2. The output residual for the Hammerstein model is, for example,

$$r'(k) = y_p(k) - A_m(q^{-1}) y_m(k) + B_{1m}(q^{-1}) u(k) + B_{2m}(q^{-1}) u^2(k) + \dots \tag{10.68}$$

For these nonlinear polynomial models even equation error residuals are possible, like

$$r(k) = \left(1 - A_m(q^{-1})\right) y_p(k) - B_{1m}(q^{-1}) u(k) - B_{2m}(q^{-1}) u^2(k) - \dots \tag{10.69}$$

In Section 5.2.4 a bilinear state variable model is considered, where the input signal $U(k)$ is multiplied with the state variables $\mathbf{x}(k)$. [10.14] has shown for the case of continuous time how parity equations can be generated by decoupling them from the bilinear term $\mathbf{U}_i \mathbf{X}$ similar as in Section 10.2.1. Artificial neural networks, as described in Section 9.3.3, can directly be applied to generate output residuals, see, e.g. some case studies in [10.18].

The structure of the nonlinear process models depends strongly on the considered process.

10.4.2 Parity equation for nonlinear, local linear models[1]

The considered approach is based on a Takagi-Sugeno (TS) *fuzzy model* of the nominal process, [10.25] and follows [10.2]. The model can be built from heuristic knowledge and/or by means of identification algorithms from measurement data. The transparent inner structure of the model is used for the generation of symptoms that indicate the occurrence and the locations of *sensor faults*. This enables a continuous fault detection for nonlinear processes in all ranges of operation. The model is run in parallel (multiple-step prediction) and in series-parallel (one-step prediction) to the process, which leads to symptoms with different sensitivity to the faults. The suitability of the two models is discussed and in the final fault-detection scheme both approaches are combined in order to exploit the advantages of each. The applicability of the proposed method is illustrated for an industrial-scale heat-exchanger pilot plant, see [10.2], [10.18].

a) Fuzzy-neuro process models

In the discussion that follows, MISO (multi-input single-output) processes are considered. The description is made in the discrete time domain, but can directly be transferred to the continuous-time domain.

A MISO process with m inputs $u(z)$ and one output $y(z)$ can generally be described in the z-domain by the following nonlinear dynamic function

---

[1] follows Peter Ballè, [10.3]

$$y(z) = f_{NL}(\mathbf{u}(z)); \quad \mathbf{u}(z) = [u_1(z), u_2(z), \ldots, u_m(z)]^T \tag{10.70}$$

The function $f_{NL}(\cdot)$ is nonlinear and dynamic and therefore contains terms of z^{-i} with $i = 1, \cdots, n$. The function $f_{NL}(\cdot)$ is now modelled by a Takagi-Sugeno fuzzy model with linear functions in the rule consequent part. The rulebase of the TS model consists of M rules in the form of

$$R_i : \text{IF} \ < u_{p1} \text{ is } P_1 \ > \text{ AND } \cdots \text{ AND } \ < u_{pp} \text{ is } P_p \ > \text{ THEN}$$
$$< y(z) = y_i(z) >$$

The rule consequent part is a linear dynamic function

$$y_i(z) = B_{i1} u_{c1}(z) + \cdots + B_{ic} u_{cc}(z) - A_i(z) y_i(z)$$
$$\text{with } B_{ik}(z) = b_{0ik} + b_{1ik} z^{-1} + \cdots + b_{nik} z^{-n} \tag{10.71}$$
$$A_i(z) = a_{1i} z^{-1} + \cdots + a_{ni} z^{-n}$$

Here $b_{0ik} \cdots b_{nik}$ are the parameters of the linear regression model of the ith rule with respect to input $u_{ck}(z)$, $B_{ij}(z)$ and $A_i(z)$ are polynomials in z (MA filters). The operator z is now omitted for the sake of simplicity. The inputs $\mathbf{u}_c = [u_{c1}, u_{c2} \cdots u_{cc}]$ of the rule consequent part are a subset of all inputs $\mathbf{u} : \mathbf{u}_c \subseteq \mathbf{u}$. The inputs of the rule premise parts are defined by $\mathbf{u}_p = [u_{p1}, u_{p1} \cdots u_{pp}]$ and are also a subset of the model inputs $\mathbf{u}; \mathbf{u}_p \subseteq \mathbf{u}$. The inputs $\mathbf{u}_c$ define the linear local dynamic behavior of the process, while $\mathbf{u}_p$ should contain all the variables that influence its nonlinear (operating point depending) characteristic (corresponds to $\mathbf{z}$ in Section 9.3.3c)). In many applications, it is useful or even necessary to choose different input spaces $\mathbf{u}_p$ and $\mathbf{u}_c$ for rule premise part and rule consequent part. The fuzzy sets P_i are defined over the universe of discourse of $\mathbf{u}_p$. In this approach, normalized Gaussian functions are applied. With the product operator as t-norm, the model output is calculated as the weighted sum over all M rule consequents:

$$y(z) = \sum_{i=1}^{M} (B_{i1}(z)u_1(z) + \cdots + B_{im}(z)u_m(z) - A_i(z)y_i(z))\phi_i(\mathbf{u}_p, \mathbf{c}_i, \sigma_i) \tag{10.72}$$

where $\phi_i(\mathbf{x}, \mathbf{c}_i, \sigma_i)$ is the normalized Gaussian weighting function for the ith model with center c_i and standard deviation σ_i.

$$\phi_i(\mathbf{x}, \mathbf{c}_j, \sigma_i) = \frac{\mu_i(\mathbf{u}_p, \mathbf{c}_j, \sigma_i)}{\sum_{j=i}^{M} \mu_j(\mathbf{u}_p, \mathbf{c}_j, \sigma_j)} \tag{10.73}$$

This calculation of the model output leads to a nonlinear interpolation between the rule consequents. Figure 10.9 shows a simple numerical example for a static system with two inputs (x_1, x_2), three local linear models and three rules of the form

IF $< x_1$ is large $>$ AND $< x_2$ is small $>$ THEN $< y = y_1 >$
IF $< x_2$ is large $>$ THEN $< y = y_2 >$
IF $< x_1$ is small $>$ AND $< x_2$ is small $>$ THEN $< y = y_3 >$

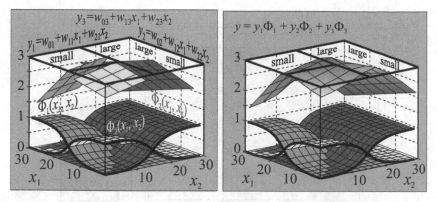

Fig. 10.9. Example of a TS fuzzy model with three membership functions ϕ_i and three rules with linear regression models in the consequent parts. On the left the membership functions and the linear regression models are displayed; on the right the overall model output is shown.

The advantage of the model is the interpretability of the rule consequences. They describe the process in the operating point by the rule premise.

For construction of the model, different algorithms exist. Here, the LOcal LInear MOdel Tree (LOLIMOT) algorithm proposed in [10.20] is applied, see Section 9.3.3. The parameters c_i and σ_i are determined by a tree-like construction algorithm. After the structure has been determined in an outer loop, the parameters of the rule consequents can be estimated by a weighted linear least-squares algorithm. For alternative construction approaches refer, e.g. to [10.24].

Below, ϕ_i is used instead of $\phi_i(\mathbf{u}_p, \mathbf{c}_j, \boldsymbol{\sigma}_i)$ for the sake of simplicity. (10.72) can also be written in the form

$$
\begin{aligned}
y(z) &= \left(\sum_{i=1}^{M} B_{i1}(z)\,\phi_i\right) u_1(z) + \cdots + \left(\sum_{i=1}^{M} B_{im}(z)\,\phi_i\right) u_m(z) - \\
&\quad \left(\sum_{i=1}^{M} A_i(z)\,\phi_i\right) y(z) \\
&= B_1(z, \mathbf{u}_p)\, u_1(z) + \cdots + B_m(z, \mathbf{u}_p) u_m(z) \\
&\quad - A(z, \mathbf{u}_p)\, y(z)
\end{aligned}
\tag{10.74}
$$

Here, each parameter of the polynomial B_i and A is a nonlinear interpolation between the parameters of the rule consequent parts. This is similar to a linear function with time-variant parameters, Figure 10.10. The superimposed parameters describe the dynamic behavior of the process at the actual operating point. Therefore, [10.6] refers to this as a *dynamic linearization* of the fuzzy model. The dynamic linearization has some appealing advantages, compared to normal linearization. Applying the latter to Takagi-Sugeno fuzzy models results in "bubble" effects, [10.1] which are caused by the rule premise part. Dynamic linearization overcomes this disadvantages, [10.6], [10.21]. The described approach can also be seen as a neural radial basis function weighting of a local linear model, as used for the LOLIMOT approach, [10.21].

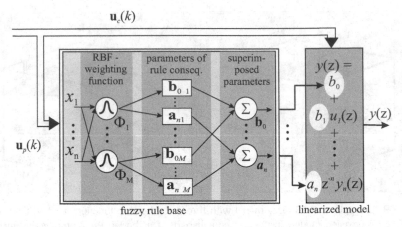

$\mathbf{u}_c(k)$

Fig. 10.10. Scheme for dynamic linearization of the TS fuzzy model or local linear neuronal model. $\mathbf{u}_c$: inputs for local linear models, $\mathbf{u}_p$: operating point dependent behavior.

b) Fault-detection scheme for sensor faults

Generation of structured parity equations with TS models
As shown above, a linearized model with varying parameters can be derived from the TS fuzzy model using (10.73). These linearized models are now used for the generation of structured residuals. Assume a nonlinear MIMO system with r outputs $y_{1...r}$, and m inputs $u_{1...m}$ of order n

$$
\begin{aligned}
\mathbf{y}(k) &= f_{NL}(\mathbf{u}(k), \mathbf{y}(k)) \\
\mathbf{y}(k) &= [y_1(k), y_2(k), \dots, y_r(k)]^T \\
\mathbf{u}(k) &= [u_1(k), u_2(k), \dots, u_m(k)]
\end{aligned}
\tag{10.75}
$$

For each output $y_i(k)$ of $f_{NL}(\cdot)$ a local linear model in the form of (10.71) can be identified, which generally depends on all the inputs and outputs of the process. The model can then be described in the form

$$
\begin{aligned}
A_{11}(z)\, y_1(z) &= B_{11}(z)\, u_1(z) + \cdots + A_{1r}(z)\, y_r(z) + c_0^{(1)} \\
A_{rr}(z)\, y_r(z) &= B_{r1}(z)\, u_1(z) + \cdots + A_{r(r-1)}(z)\, y_{r-1}(z) + c_0^{(r)}
\end{aligned}
\tag{10.76}
$$

$A_{ij}(z)$ and $B_{ji}(z)$ are moving average (MA) filters with parameters $a_{ij}^{(l)}$, $b_{ij}^{(l)}$ and $c_o^{(l)}$ which depend on $\mathbf{u}_p$ similarly to (10.74). (10.76) can also be written in a vector form

$$
\begin{aligned}
\mathbf{r}(z) &= \mathbf{A}_1(z)\, y_1(z) + \cdots + \mathbf{A}_r\, y_r(z) \\
&\quad - \mathbf{B}_1(z)\, u_1(z) - \cdots - \mathbf{B}_m(z) u_m(z) - \mathbf{c}_0 \approx \mathbf{0}
\end{aligned}
\tag{10.77}
$$

with the residuals vector $\mathbf{r}(z)$ and the vectors

$$
\begin{aligned}
\mathbf{A}_i(z) &= [A_{1i}(z) \cdots A_{ri}(z)]^T, \quad \mathbf{B}_i(z) = [B_{1i}(z) \cdots B_{ri}(z)]^T, \\
\mathbf{r}(z) &= [r_1 \cdots r_r]^T
\end{aligned}
\tag{10.78}
$$

The residuals of (10.77) depend on all the inputs and outputs of the system. In order to generate *structured residuals*, (10.77) has to be transformed to decouple the residuals from the different input and output measurements. Therefore, both sides of (10.77) are multiplied with a residual generator $\mathbf{w}(z)$:

$$r_i^*(z) = \mathbf{w}^T(z)\, \mathbf{r}(z) = \mathbf{w}^T(z)\, (\mathbf{A}_1(z)\, y_1(z) \\ + \cdots - \mathbf{B}_m(z)\, u_m(z) - \mathbf{c}_0) \tag{10.79}$$

$\mathbf{w}(z)$ is a vector of length r with MA filters $w_{ij}(z)$ as elements. As can be seen from (10.77), choosing $\mathbf{w}(z)$ to satisfy the condition

$$\mathbf{w}^T(z)\, \mathbf{B}_i(z) = \mathbf{0} \quad \text{or} \quad \mathbf{w}^T(z)\, \mathbf{A}_i(z) = \mathbf{0} \tag{10.80}$$

leads to residuals which are decoupled from input u_i or from output y_i, respectively. Note that it is always possible to find a $\mathbf{w}^T(z)$ that satisfies (10.80), but sometimes the residual generators for different inputs and outputs are equal, and no isolation between these faults is possible. The single residual generator vectors build the rows of a residual generator matrix $\mathbf{W}(z)$, and (10.79) becomes

$$\mathbf{r}^*(z) = \mathbf{W}(z)\, (\mathbf{A}_i(z)\, y_1(z) + \cdots - \mathbf{B}_m(z)\, u_m(z) - \mathbf{c}_0$$
$$\mathbf{W}(z) = \begin{bmatrix} W_{11}(z) & \cdots & W_{1r}(z) \\ \vdots & \ddots & \vdots \\ W_{1r}(z) & \cdots & W_{rr}(z) \end{bmatrix} \tag{10.81}$$

As mentioned above, the vectors $\mathbf{A}_i$ and $\mathbf{B}_i$ depend on the rule premise inputs $\mathbf{u}_p$. As the same holds for the elements of the residual generator, the residual generator has to be calculated online for each sample interval. The residual generation is shown in Figure 10.11 for both structures.

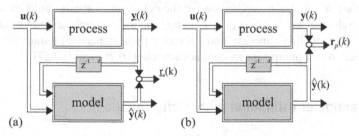

Fig. 10.11. Parallel and series parallel model, and output residuals: (a) series parallel model; (b) parallel model

Parallel and series parallel model
There are two principal ways of modelling dynamic systems with an auto-regressive part. The first is the series-parallel model (also called the equation error model) which uses previous process outputs $\mathbf{y}(k - d)$ for the prediction of the next process

output. The second is the parallel model (or output error model), where previous model outputs are fedback, Figure 10.11. Due to the different structure, the derived residuals show different behavior, [10.13]. In the series-parallel residual, measured noisy output signals are used instead of noise-free estimated ones, and hence, their signal to noise ratio is smaller. Therefore, series-parallel residuals have often to be low-pass filtered to separate the residual deviation caused by a fault from that caused by noise. On the other hand, parallel models are likely to drift, due to model uncertainties and the auto-regressive part. Thus, the performance of the parallel model is worse and the model can also become unstable, [10.22]. The pros and cons are as follows:

Parallel-model parity equations

+   even smaller faults can lead to higher residual deflections;
+   signal-to-noise is relatively high;
+   no filter is needed;
−   model is difficult to build, and its performance is worse;
−   models are likely to drift, due to unprecise modeling;
−   generated parity equations have an auto-regressive part.

Series-parallel-model parity equations

+   lead to high peaks, when sudden changes occur;
+   very fast response, but only small deflection;
+   no auto-regressive part in the parity equation;
+   no drift effects → output error is smaller ;
−   possess only a small signal-to-noise ratio;
−   need for low-pass filtering.

In order to improve the performance of the overall fault-detection scheme, both approaches can be combined. The sensitivity is computed, and the most sensitivity residuals are selected. An application for the detection of sensor faults of a nonlinear heat exchanger is described in [10.2] and [10.18] The application of neuro-fuzzy models with local linear models and operating point dependent parameters for the fault detection of a pump-pipe system is shown in [10.26] and [10.27].

10.5 Parameter estimation with parity equations

The consideration of the sensitivity of the residuals of parity equations in Section 10.3.3 has shown that the residuals in the case of a small change of a parameter p_j yield

$$r(t) = -\frac{\partial \boldsymbol{\theta}^T(\mathbf{p})}{\partial p_j} \boldsymbol{\psi}(t) \, \Delta p_j = -\boldsymbol{\beta}_j(t) \, \Delta p_j \qquad (10.82)$$

with the scalar (residual sensitivity)

$$\boldsymbol{\beta}_j(t) = \frac{\partial \boldsymbol{\theta}^T(\mathbf{p})}{\partial p_j} \, \boldsymbol{\psi}(t) \tag{10.83}$$

which follows from the process model. The parameter changes Δp_j can be determined by a least squares estimation. Therefore (10.82) is written as

$$r(k) = -\boldsymbol{\beta}_j(k)\Delta p_j + e(k)$$

where $e(k)$ is an equation error and $k = t/T_0 = 0, 1, 2, \ldots$ this discrete sampling time. By using N measured samples

$$\mathbf{r}(k) = [r(k-1) \, r(k-2) \, \cdots \, r(k-N)] \tag{10.84}$$

$$\boldsymbol{\beta}_j(k) = [\beta_j(k-1) \, \beta_j(k-2) \, \cdots \, \beta_j(k-N)] \tag{10.85}$$

the nonrecursive least squares estimate is according to Section 9.2.1

$$\Delta \hat{p}_j(k) = -[\boldsymbol{\beta}_j^T(k)\boldsymbol{\beta}_j(k)]^{-1}\boldsymbol{\beta}_j^T(k)\mathbf{r}(k) = -\frac{\sum_{v=1}^{N} \beta_j(k-v)r(k-v)}{\sum_{v=1}^{N} \beta_j^2(k-v)} \tag{10.86}$$

This estimation performs a correlation of the residual sensitivity β_j to the parameter p_j with the residual r.

A corresponding recursive parameter estimation with exponential forgetting follows according to (9.98), (9.99) and (9.100)

$$\Delta p_j(k+1) = \Delta p_j(k) + \gamma(k)[r(k+1) + \beta_j(k+1)\Delta p_j(k)] \tag{10.87}$$

$$\gamma(k) = \frac{1}{P(k)\beta_j^2(k+1) + \lambda} P(k)\beta_j(k+1) \tag{10.88}$$

$$P(k+1) = [1 - \gamma(k)\beta_j(k+1)]P(k)\frac{1}{\lambda} \tag{10.89}$$

with $\lambda < 1.0$ and starting value $P(0) = \alpha$ $(\alpha \geq 1000)$.

For each parity equation with one residual $r_j(k)$ therefore one parameter deviation $\Delta p_j(k)$ can be estimated. This may be used for adapting one parameter of the fault-detection procedure with parity equations if one knows that these single parameters are time varying due to normal operating conditions. However, if faults are expected which change parameters, then direct parameter estimation of all process parameters should be preferred.

Example 10.4: Adaptive parity equations for a DC motor

The residual for the electrical part of the DC motor is, see Example 10.3,

$$r_4(t) = \boldsymbol{\theta}^T(p)\boldsymbol{\psi}(t) = U_A(t) - L_A \, \dot{I}_A(t) - R_A \, I_A(t) - \Psi \, \omega(t)$$

with

$$\theta^T(p) = [1 \ L_A \ R_A \ \Psi]$$

$$\psi^T(t) = [U_A(t) \ - \ \dot{I}_A(t) \ - \ I_A(t) \ - \ \omega(t)]$$

As the armature current depends strongly on the temperature, this parameter is time varying, depending on the load and cooling of the motor. Therefore R_A will be estimated. (10.83) leads to

$$\beta_j(t) = \frac{\partial \, \theta^T(p)}{\partial \, \hat{R}_A} \psi(t) = -I_A(t)$$

The estimation equation then follows (10.86)

$$\Delta\hat{R}_A(k) = \frac{\sum_{\nu=1}^{N} I_A(k - \nu)r_4(k - \nu)}{\sum_{\nu=1}^{N} I_A^2(k - \nu)}$$

Correspondingly one obtains from residual r_1 for the friction parameter

$$\Delta\hat{M}_F = \frac{\sum_{\nu=1}^{N} \omega(k - \nu)r_1(k - \nu)}{\sum_{\nu=1}^{N} \omega^2(k - \nu)}$$

see [10.14].

Figure 10.12 shows a simulation run for a DC motor with $U_A = 25$ V $=$ const, $M_{load} = 2$Nm $=$ const and with an increase of the armature resistance from $R_A = 1.53 \ \Omega$ to $R_A = 1.72 \ \Omega$ due to an increase of temperature during the operation within the first 50 min run-time. Without adapting the resistance parameter the residual exceeds a threshold after about 10 min. With parameter adaptation of R_A in the residual equation all 5 min the residual is set back to zero and a wrong alarm is avoided, see also the experimental investigations in Chapter 20.

□

10.6 Problems

1) Derive the equations for the primary residuals of parity equations with output error and equation error in the form of differential equations for a linear second order mass-spring-damper system of Example 5.5. Which measures have to be made to come to realizable solutions?
2) Solve the same tasks as in 1), but with discrete-time signals, z-transfer functions and difference equations. Compare realizability with 1).
3) What are the advantages of structured residuals compared to primary residuals?
4) What is the difference between the evaluation form and computational form of parity equations? Write the corresponding equations for transfer functions.

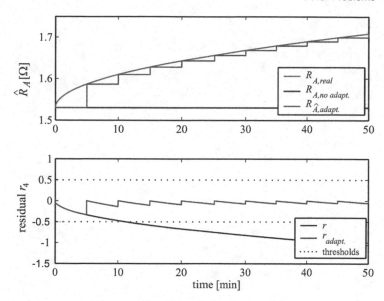

Fig. 10.12. Simulated residual $r_4(t)$ of the DC motor for increasing armature resistance $R_A(t)$ due to warming-up without and with adaptation (according to I. Unger)

5) An electrical drive system for electrical steering of vehicles with a gear in series connection is given with simplified transfer functions:

$$I(t) \longrightarrow \boxed{\dfrac{K_m}{Js + F_f}} \xrightarrow{\;\omega(t)\;} \boxed{\dfrac{K_g}{s}} \xrightarrow{\;y(t)\;}$$

The DC current $I(t)$, the speed $\omega(t)$ and the position $y(t)$ of a rack can be measured. The task consists in the fault detection of sensor faults of all three signals and faults in bearings and windings of the DC motor.

a) Derive equations for structured residuals.
b) Show the effect of offset sensor faults in a fault-symptom table, if no process faults happen.
c) Are the sensor faults strongly or weakly isolable?
d) How do the residuals for the process react, i.e. a change of F_f and K_g? Add the fault-symptom table. Which faults can be diagnosed?
e) Is a slowly drifting fault of the position sensor detectable?

Fault detection with state observers and state estimation

As state observers use an output error between a measured process output and an adjustable model output, they are a further alternative for model-based fault detection. It is assumed, as in the case of parity equation approaches, that the structure and the parameters of the model are precisely known. State observers adjust the state variables according to initial conditions and to the time course of the measured input and output signals.

Several approaches have been proposed for fault detection which are based on the classical Luenberger state observer, Kalman filter and the so-called output observer. Some of the basic methods are treated in this chapter.

11.1 State observers

A linear-time invariant process is considered which can be described by the state-space model

$$\dot{\mathbf{x}}(t) = \mathbf{A}\,\mathbf{x}(t) + \mathbf{B}\,\mathbf{u}(t) \tag{11.1}$$

$$\mathbf{y}(t) = \mathbf{C}\,\mathbf{x}(t) \tag{11.2}$$

Here p input signals $u(t)$ and r output signals $\mathbf{y}(t)$ are assumed because the fault-detection methods are especially suitable for multi-variable processes. With the assumption that the structure and the parameters of the model are known, a state observer is used to reconstruct the unmeasurable state variable based on measured inputs and outputs

$$\dot{\hat{\mathbf{x}}}(t) = \mathbf{A}\,\hat{\mathbf{x}}(t) + \mathbf{B}\,\mathbf{u}(t) + \mathbf{H}\,\mathbf{e}(t) \tag{11.3}$$

$$\mathbf{e}(t) = \mathbf{y}(t) - \mathbf{C}\,\hat{\mathbf{x}}(t) \tag{11.4}$$

compare Figure 11.1. $\mathbf{e}(t)$ is an output error which acts through the observer matrix $\mathbf{H}$ on the reconstructed state derivatives $\dot{\hat{\mathbf{x}}}(t)$. Inserting (11.4) in (11.3) yields the implementation form of the state observer

$$\dot{\hat{\mathbf{x}}}(t) = [\mathbf{A} - \mathbf{H}\,\mathbf{C}]\,\hat{\mathbf{x}}(t) + \mathbf{B}\,\mathbf{u}(t) + \mathbf{H}\,\mathbf{y}(t) \tag{11.5}$$

where it is assumed that the system is observable.

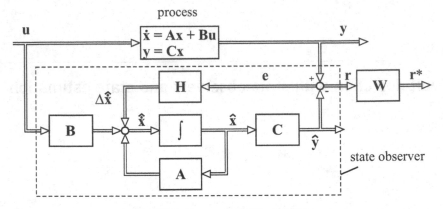

Fig. 11.1. Process and state observer

The state error

$$\tilde{\mathbf{x}}(t) = \dot{\mathbf{x}}(t) - \hat{\dot{\mathbf{x}}}(t) \tag{11.6}$$

between the real process states and the observed states becomes under the assumption that process and model parameters are identical and by introducing (11.1) and (11.5)

$$\dot{\tilde{\mathbf{x}}}(t) = [\mathbf{A} - \mathbf{H}\,\mathbf{C}]\,\tilde{\mathbf{x}}(t) \tag{11.7}$$

Hence, the state error vanishes asymptotically

$$\lim_{t\to\infty} \tilde{\mathbf{x}}(t) = \mathbf{0}$$

for any initial state deviation $[\mathbf{x}(0) - \hat{\mathbf{x}}(0)]$ if the observer is stable, which can be reached by proper design of the observer feedback matrix $\mathbf{H}$, e.g. by pole placement.

11.1.1 Additive faults

The process is now influenced by unmeasurable disturbances $\mathbf{v}(t)$ and $\mathbf{n}(t)$ and additive faults $\mathbf{f}_l(t)$ and $\mathbf{f}_m(t)$ as follows, compare Figure 11.2,

$$\dot{\mathbf{x}}(t) = \mathbf{A}\,\mathbf{x}(t) + \mathbf{B}\,\mathbf{u}(t) + \mathbf{V}\mathbf{v}(t) + \mathbf{L}\,\mathbf{f}_l(t) \tag{11.8}$$

$$\mathbf{y}(t) = \mathbf{C}\,\mathbf{x}(t) + \mathbf{N}\,\mathbf{n}(t) + \mathbf{M}\,\mathbf{f}_m(t) \tag{11.9}$$

$\mathbf{L}$ and $\mathbf{M}$ are fault entry matrices. Introducing this process equations into the observer equation according to (11.5) leads to the state error

$$\dot{\tilde{\mathbf{x}}}(t) = [\mathbf{A} - \mathbf{H}\,\mathbf{C}]\,\tilde{\mathbf{x}}(t) + \mathbf{V}\mathbf{v}(t) + \mathbf{L}\,\mathbf{f}_l(t) - \mathbf{H}\,\mathbf{N}\,\mathbf{n}(t) - \mathbf{H}\,\mathbf{M}\,\mathbf{f}_m(t) \tag{11.10}$$

and the output error becomes

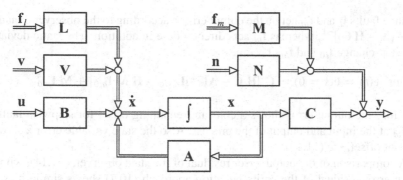

Fig. 11.2. Multi-variable process with disturbances **v**, **n** and fault signals $\mathbf{f}_l$, $\mathbf{f}_m$

u	process input vector $[p \times 1]$	**v**	input disturbance vector $[m \times 1]$
x	process state vector $[m \times m]$	$\mathbf{f}_l$	additive input fault vector $[m \times 1]$
y	process output vector $[r \times 1]$	$\mathbf{f}_m$	additive output fault vector $[r \times 1]$
n	output disturbance vector $[1 \times r]$		

$$\mathbf{e}(t) = \mathbf{y}(t) - \mathbf{C}\,\hat{\mathbf{x}}(t) = \mathbf{C}\,\tilde{\mathbf{x}}(t) + \mathbf{N}\,\mathbf{n}(t) + \mathbf{M}\,\mathbf{f}_m(t) \qquad (11.11)$$

After initial state deviations $[\mathbf{x}(0) - \hat{\mathbf{x}}(0)]$ are asymptotically vanished, the state error $\tilde{\mathbf{x}}$ and the output error $\mathbf{e}(t)$ depend on the disturbances $\mathbf{v}(t)$ and $\mathbf{n}(t)$ and the faults $\mathbf{f}_l(t)$ and $\mathbf{f}_m(t)$. $\tilde{\mathbf{x}}$ can be used as residuals if primary faults $\mathbf{f}_l(t)$ on the states (as for leak detection) are of interest. In general, however, the output error $\mathbf{e}(t) = \mathbf{r}(t)$ will be used as residual. The residual is zero, if no disturbances and faults are present and it deviates from zero, if faults $\mathbf{f}_l(t)$ or $\mathbf{f}_m(t)$ appear (and also if $\mathbf{n}(t) \neq 0$ and $\mathbf{v}(t) \neq 0$). It is interesting to see that the residuals do not depend on the input signal $\mathbf{u}(t)$.

To obtain the input-output relation of the state observer (11.5) is Laplace transformed

$$[s\,\mathbf{I} - \mathbf{A} + \mathbf{H}\,\mathbf{C}]\,\hat{\mathbf{x}}(s) = \mathbf{B}\,\mathbf{u}(s) + \mathbf{H}\,\mathbf{y}(s) \qquad (11.12)$$

which leads to

$$\hat{\mathbf{x}}(s) = [s\,\mathbf{I} - \mathbf{A} + \mathbf{H}\,\mathbf{C}]^{-1}\,[\mathbf{B}\,\mathbf{u}(s) + \mathbf{H}\,\mathbf{y}(s)] \qquad (11.13)$$

Introduction into the output error residual (11.4) yields

$$\mathbf{r}(s) = \mathbf{e}(s) = -\mathbf{C}\,[s\,\mathbf{I} - \mathbf{A} + \mathbf{H}\,\mathbf{C}]^{-1}\,\mathbf{B}\,\mathbf{u}(s) + \left[\mathbf{I} - \mathbf{C}[s\mathbf{I} - \mathbf{A} + \mathbf{H}\,\mathbf{C}]^{-1}\,\mathbf{H}\right]\mathbf{y}(s) \qquad (11.14)$$

This is the Laplace transformed computational form of the residuals for state observers. As (11.10) and (11.11) show the influence of the faults in the residuals $\tilde{\mathbf{x}}(t)$ or $\mathbf{e}(t)$, they correspond to the internal form of parity equations, compare (10.8). Applying the Laplace transform to the state observer equations with additive faults (11.10) and (11.11) and omitting the disturbance terms leads to

$$\mathbf{r}(s) = \mathbf{e}(s) = \mathbf{C}\,[s\,\mathbf{I} - (\mathbf{A} - \mathbf{H}\,\mathbf{C})]^{-1}\,[\mathbf{L}\,\mathbf{f}_l\,(s) - \mathbf{H}\,\mathbf{M}\,\mathbf{f}_m(s)] + \mathbf{M}\,\mathbf{f}_m\,(s) \quad (11.15)$$

Additive faults $\mathbf{f}_l$ and $\mathbf{f}_m$ act on the output error $\mathbf{e}$ according to the observer dynamics $[s\,\mathbf{I} - (\mathbf{A} - \mathbf{H}\,\mathbf{C})]^{-1}$, whereas $\mathbf{f}_m$ acts directly on $\mathbf{e}$ in addition. The static deviation for a step-change $\mathbf{f}_{l0}$ and $\mathbf{f}_{m0}$ becomes

$$\lim_{t \to \infty} \mathbf{e}(t) = \mathbf{e}(s = 0) = \mathbf{C}\,[\mathbf{H}\,\mathbf{C} - \mathbf{A}]^{-1}\,[\mathbf{L}\,\mathbf{f}_{l0} - \mathbf{H}\,\mathbf{M}\,\mathbf{f}_{m0}] + \mathbf{M}\,\mathbf{f}_{m0} \quad (11.16)$$

Therefore the output error shows a constant remaining offset for stepwise faults $\mathbf{f}_l$ and $\mathbf{f}_m$ at the input and output of the process. Also the state variable error $\tilde{\mathbf{x}}$ shows a constant offset, see (11.10).

A comparison of the output error residual of the state observer, (11.14), with the output error residual of the parity equation approach (10.7) shows similarities. Instead of the process transfer function $\mathbf{G}_p(s)$ the observer dynamics $[s\mathbf{I} - (\mathbf{A} - \mathbf{H}\,\mathbf{C})]^{-1}$ act between additive input faults and the output error.

11.1.2 Multiplicative faults

If faults appear as changes in the parameter matrices $\Delta\mathbf{A}$, $\Delta\mathbf{B}$ or $\Delta\mathbf{C}$, the process behavior becomes (after the transients have been settled)

$$\dot{\mathbf{x}}(t) = [\mathbf{A} + \Delta\,\mathbf{A}]\mathbf{x}(t) + [\mathbf{B} + \Delta\,\mathbf{B}]\mathbf{u}(t) \quad (11.17)$$

$$\mathbf{y}(t) = [\mathbf{C} + \Delta\,\mathbf{C}]\mathbf{x}(t) \quad (11.18)$$

and the state and output error without disturbances

$$\dot{\tilde{\mathbf{x}}}(t) = [\mathbf{A} - \mathbf{H}\,\mathbf{C}]\,\tilde{\mathbf{x}}(t) + [\Delta\mathbf{A} - \mathbf{H}\,\Delta\,\mathbf{C}]\mathbf{x}(t) + \Delta\,\mathbf{B}\,\mathbf{u}(t) \quad (11.19)$$

$$\mathbf{e}(t) = \mathbf{C}\,\tilde{\mathbf{x}}(t) + \Delta\,\mathbf{C}\,\mathbf{x}(t) \quad (11.20)$$

Hence, the state and output error depend on the parameter changes multiplied with the input signal $\mathbf{u}(t)$ and the state variables $\mathbf{x}(t)$. Therefore the analysis of the behavior is not as straightforward as for additive faults.

11.1.3 Fault isolation with state observers

If the output error $\mathbf{e}(t) = \mathbf{r}(t)$ is used as *primary residual*, its elements will deflect depending on the faults as shown, e.g. by (11.15). In order to determine the single faults, *enhanced residuals* have to be generated, as in the case of parity equations, Section 10.3. This can be reached by *fault-detection filters* or by *dedicated observers*, mostly in form of a bank of observers. Note that these approaches require *multiple process outputs*.

Fault-detection filters for multi-output processes (fault-sensitive observers)

There is some freedom in the design of the observer feedback matrix **H**. Usually the poles s_i of of the characteristic equation of the observer

$$\det \left[s\, \mathbf{I} - \mathbf{A} + \mathbf{H}\, \mathbf{C} \right] = (s - s_1)(s - s_2)\dots(s - s_m) = 0 \qquad (11.21)$$

are selected such, that they lie in the left half s-plane in order to guarantee a stable behavior and that they are faster and better damped than the process poles. This means, they will be shifted to the left as compared to the process. However, if the gains become too large the output noise $n(t)$ will be amplified too much. Therefore a proper compromise has to be found.

Now the observer feedback matrix **H** can also be used to give the residuals **r** a proper structure for fault isolation, as proposed by [11.3] and [11.16] as a fault sensitive observer. The considered state equation is

$$\dot{\mathbf{x}} = \mathbf{A}\,\mathbf{x} + \mathbf{B}\,\mathbf{u} + \mathbf{l}_i\; f_{li} \qquad (11.22)$$

For each fault f_{li} a fault influence vector $\mathbf{l}_i$ is determined, such that m linear independent vectors $\mathbf{l}_i$ result. As residual the output error

$$\mathbf{r} = \mathbf{e} = \mathbf{y} - \hat{\mathbf{y}} \qquad (11.23)$$

is used. It is now assumed that the system is completely measurable. Therefore from the output equation

$$\mathbf{y} = \mathbf{C}\,\mathbf{x} \qquad \text{Rank}\ \mathbf{C} = m \qquad (11.24)$$

it follows

$$\mathbf{x} = \mathbf{C}^{-1}\,\mathbf{y} \qquad (11.25)$$

(which means that all states are measurable). Introducing (11.25) in (11.22) leads to

$$\dot{\mathbf{y}} = \mathbf{C}\,\mathbf{A}\,\mathbf{C}^{-1}\,\mathbf{y} + \mathbf{C}\,\mathbf{B}\,\mathbf{u} + \mathbf{C}\,\mathbf{l}_i\; f_{li} \qquad (11.26)$$

For the observer holds

$$\dot{\hat{\mathbf{x}}} = \mathbf{A}\,\hat{\mathbf{x}} + \mathbf{B}\,\mathbf{u} + \mathbf{H}\,\mathbf{r} \qquad (11.27)$$

and with (11.25)

$$\dot{\hat{\mathbf{y}}} = \mathbf{C}\,\mathbf{A}\,\mathbf{C}^{-1}\,\hat{\mathbf{y}} + \mathbf{C}\,\mathbf{B}\,\mathbf{u} + \mathbf{C}\,\mathbf{H}\,\mathbf{r} \qquad (11.28)$$

where the states are now eliminated.

Furtheron the residual (11.23) becomes with (11.26), (11.28)

$$\dot{\mathbf{r}} = \left[\mathbf{C}\,\mathbf{A}\,\mathbf{C}^{-1} - \mathbf{C}\,\mathbf{H} \right]\mathbf{r} + \mathbf{C}\,\mathbf{l}_i\; f_{li} \qquad (11.29)$$

To decouple the residuals from each other a diagonal matrix with fast equal eigenvalues λ is introduced

$$\left[\mathbf{C}\,\mathbf{A}\,\mathbf{C}^{-1} - \mathbf{C}\,\mathbf{H} \right] = \lambda\,\mathbf{I} \qquad (11.30)$$

Laplace transformation of (11.29) leads to

$$\mathbf{r}(s) = [s\,\mathbf{I} - \lambda\,\mathbf{I}]^{-1}\,\mathbf{C}\,\mathbf{l}_i\;f_{li} = \mathbf{C}\,\mathbf{l}_i\;\frac{f_{li}}{s - \lambda} \tag{11.31}$$

The searched observer feedback matrix $\mathbf{H}$ now follows from (11.30)

$$\mathbf{H} = [\mathbf{A}\;-\lambda\,\mathbf{I}]\;\mathbf{C}^{-1} \tag{11.32}$$

If the states are not measurable some transformations of the model have to be performed, [11.3], [11.16]. An example is given in [11.12].

A further procedure is given by [11.21] and [11.30].

Bank of observers (dedicated observers):

Another possibility to distinguish between different faults, is to use special observers with different inputs and outputs. The observer with the missing inputs or outputs then do not show changes for corresponding input and output faults. The following schemes are described for additive sensor faults

$$\mathbf{y} = \mathbf{C}\,\mathbf{x} + \mathbf{M}\,\mathbf{f}_m \tag{11.33}$$

with

$$\mathbf{f}_m^T = [f_{m1}\;\;f_{m2}\cdots f_{mr}] \tag{11.34}$$

A further observer bank is obtained by using for all observers all inputs $\mathbf{u}$ but only one output, compare Figure 11.3a, [11.7]

$$y_i = \mathbf{c}_i^T\,\mathbf{x} + f_{mi} \tag{11.35}$$

and the observer equation (11.5) results in

$$\hat{\mathbf{x}} = \left[\mathbf{A} - \mathbf{h}_i\,\mathbf{c}_i^T\right]\hat{\mathbf{x}} + \mathbf{B}\,\mathbf{u} + \mathbf{h}_i\;y_i \tag{11.36}$$

leading to a residual as output error

$$r_i = y_i - \mathbf{c}^T\,\hat{\mathbf{x}} \tag{11.37}$$

Therefore, only the k-th output error or residual is affected by a fault f_{mk} and all other residuals r_i, $i = 1\ldots r$; $i \neq k$ should be zero. Thus by analyzing the patterns of the residuals the additive sensor fault can be isolated.

Another possibility is to use for the observers all outputs $\mathbf{y}$ but only one input u_i, [11.31], [11.26], Figure 11.3b. This scheme is designed to detect input faults f_{li}.

The bank of observers scheme can be expanded to all inputs and all outputs, except one input or one output for which the fault is modelled. Then the observability or controllability is improved. This is called the *generalized observer scheme*, [11.9].

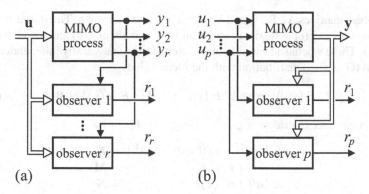

Fig. 11.3. Bank of observers (dedicated observers): (a) all inputs, one output; (b) one input, all outputs

11.2 State estimation (Kalman filter)

For the linear multi-input multi-output process with discrete time signals and without stochastic disturbances

$$\mathbf{x}(k + 1) = \mathbf{A}\,\mathbf{x}(k) + \mathbf{B}\,\mathbf{u}(k) \tag{11.38}$$

$$\mathbf{y}(k) = \mathbf{C}\,\mathbf{x}(k) \tag{11.39}$$

the state observer equation becomes, corresponding to (11.3), (11.4),

$$\hat{\mathbf{x}}(k + 1) = \mathbf{A}\,\hat{\mathbf{x}}(k) + \mathbf{B}\,\mathbf{u}(k) + \mathbf{H}\,[\mathbf{y}(k) - \mathbf{C}\,\hat{\mathbf{x}}(k)] \tag{11.40}$$

The equation error as an output error then is

$$\mathbf{e}(k) = \mathbf{y}(k) - \mathbf{C}\,\hat{\mathbf{x}}(k) \tag{11.41}$$

and the state error equation according to (11.7) becomes

$$\tilde{\mathbf{x}}(k + 1) = [\mathbf{A} - \mathbf{H}\,\mathbf{C}]\,\tilde{\mathbf{x}}(k) \tag{11.42}$$

If no disturbances act on the process, the observer converges to the true state variables if the eigenvalues $\mathbf{A} - \mathbf{H}\,\mathbf{C}$ are asymptotically stable. The speed of convergence can be made fast by a large influence of the observer gain $\mathbf{H}$. However, under the influence of stochastic disturbances the state reconstruction with these observers is not optimal. The state reconstruction must then simultaneously follow the true state variables and reject the noise effects which then leads to an estimation problem.

The process is now supplemented by stochastic noise $\mathbf{v}(k)$ at the input and $\mathbf{n}(k)$ at the output

$$\mathbf{x}(k + 1) = \mathbf{A}\,\mathbf{x}(k) + \mathbf{B}\,\mathbf{u}(k) + \mathbf{V}\,\mathbf{v}(k) \tag{11.43}$$

$$\mathbf{y}(k) = \mathbf{C}\,\mathbf{x}(k) + \mathbf{n}(k) \tag{11.44}$$

The process matrices $\mathbf{A}$, $\mathbf{B}$, $\mathbf{C}$ and $\mathbf{V}$ are assumed to be known. The initial state $\mathbf{x}(0)$ is not known, but probabilistic information is known about $\mathbf{x}(0)$ and also about $\mathbf{v}(k)$ and $\mathbf{n}(k)$. These stochastic variables are assumed to be statistically independent, have a normal (Gaussian) distribution with the mean values

$$E\{\mathbf{x}(0)\} = \mathbf{x}_0 \; ; \quad E\{\mathbf{v}(k)\} = \mathbf{0} \; ; \quad E\{\mathbf{n}(k)\} = \mathbf{0} \tag{11.45}$$

and the covariance matrices

$$\begin{aligned} E\left\{(\mathbf{x}(0) - \mathbf{x}_0)(\mathbf{x}(0) - \mathbf{x}_0)^T\right\} &= \mathbf{X}_0 \\ E\left\{\mathbf{v}(k)\,\mathbf{v}^T(k)\right\} &= \mathbf{M} \\ E\left\{\mathbf{n}(k)\,\mathbf{n}^T(k)\right\} &= \mathbf{N} \end{aligned} \tag{11.46}$$

Furtheron, it is assumed that $\mathbf{M}$ and $\mathbf{N}$ are known in order to have a measure about the size of the noises.

As the state estimation error cannot converge to zero, a best estimate has to be found for the state vector $\mathbf{x}(k)$ based on the measured input variables $\mathbf{u}(k)$ and output variables $\mathbf{y}(k)$. A least squares estimation then requires

$$\min \; \|\mathbf{x}(k) - \hat{\mathbf{x}}(k|j)\|^2 \tag{11.47}$$

Two different time instances are used here. k means the present time and j the used time instant of the measurements. The state estimation can then be given different names, [11.23];

$$k > j \text{ prediction problem;}$$
$$k = j \text{ filtering problem;}$$
$$k < j \text{ smoothing problem.}$$

The filtering problem and one-step ahead prediction is further considered. The used measurements of the output are

$$\mathbf{Y}_j = \{\mathbf{y}(0), \; \mathbf{y}(1), \ldots, \mathbf{y}(j)\} \tag{11.48}$$

Following notations are used

Optimal estimates:

$$\hat{\mathbf{x}}(k|j) = E\left\{\mathbf{x}(k)|\mathbf{Y}_j\right\} \tag{11.49}$$

Estimation error:

$$\tilde{\mathbf{x}}(k|j) = \mathbf{x}(k) - \hat{\mathbf{x}}(k|j) \tag{11.50}$$

Covariance matrices of the estimation error:

$$\mathbf{P}^-(k+1) = E\left\{\tilde{\mathbf{x}}(k+1|k)\,\tilde{\mathbf{x}}^T(k+1|k)\right\} \tag{11.51}$$

$$\mathbf{P}(k+1) = E\left\{\tilde{\mathbf{x}}(k+1|k+1)\,\tilde{\mathbf{x}}^T(k+1|k+1)\right\} \tag{11.52}$$

For time instant $k+1$ the state variable $\mathbf{x}(k+1)$ can be predicted by using the state model (11.43) with the information at time k

$$\hat{\mathbf{x}}(k + 1|k) = \mathbf{A}\,\hat{\mathbf{x}}(k|k) + \mathbf{B}\,\mathbf{u}(k) + \mathbf{V}\,\bar{\mathbf{v}} \tag{11.53}$$

as the exact $\mathbf{v}(k)$ is unknown.

With the assumption $E\,\{\mathbf{v}(k)\} = \bar{\mathbf{v}} = \mathbf{0}$ it yields

$$\hat{\mathbf{x}}(k + 1|k) = \mathbf{A}\,\hat{\mathbf{x}}(k|k) + \mathbf{B}\,\mathbf{u}(k) \tag{11.54}$$

At time $k + 1$ also the measurement of the output $\mathbf{y}(k + 1)$ is available. It holds

$$\mathbf{y}(k + 1) = \mathbf{C}\,\mathbf{x}(k + 1) + \mathbf{n}(k + 1) \tag{11.55}$$

However, $\mathbf{x}(k + 1)$ is unknown. The prediction $\hat{\mathbf{x}}(k + 1|k)$ is disturbed by the noise $\mathbf{v}(k)$ and the measurable output $\mathbf{y}(k + 1)$ for $\mathbf{x}(k + 1)$ by $\mathbf{n}(k + 1)$. It is assumed the $\mathbf{v}(k)$ and $\mathbf{n}(k)$ are statistically independent.

If both, $\hat{\mathbf{x}}(k + 1|k)$ and $\mathbf{x}(k + 1)$ would be known, one could calculate a weighted mean as an estimate.

$$\begin{aligned}\hat{\mathbf{x}}(k + 1|k + 1) &= (\mathbf{I} - \mathbf{K}')\,\hat{\mathbf{x}}(k + 1|k) + \mathbf{K}'\,\mathbf{x}(k + 1) \\ &= \hat{\mathbf{x}}(k + 1|k) + \mathbf{K}'[\mathbf{x}(k + 1) - \hat{\mathbf{x}}(k + 1|k)]\end{aligned} \tag{11.56}$$

where $\mathbf{K}'$ is a $(m{\times}m)$ weighting matrix which is to be chosen such that the covariance of the estimation error $\mathbf{P}(k - 1)$ becomes a minimum. Now, instead of $\mathbf{x}(k + 1)$ the measurable output vector $\mathbf{y}(k + 1)$ is used. Then with $\mathbf{K}' = \mathbf{K}\,\mathbf{C}$ holds

$$\begin{aligned}\hat{\mathbf{x}}(k + 1|k + 1) &= \hat{\mathbf{x}}(k + 1|k) + \mathbf{KC}[\mathbf{x}(k + 1) - \hat{\mathbf{x}}(k + 1|k)] \\ &= \hat{\mathbf{x}}(k + 1|k) + \mathbf{K}[\mathbf{y}(k + 1) - \mathbf{C}\hat{\mathbf{x}}(k + 1|k)] \\ &= [\mathbf{I} - \mathbf{KC}]\,\hat{\mathbf{x}}(k + 1|k) + \mathbf{K}\,\mathbf{y}(k + 1)\end{aligned} \tag{11.57}$$

This equation contains:

$\hat{\mathbf{x}}(k + 1|k)$: the model prediction of $\mathbf{x}(k + 1)$ based on the last estimate $\hat{\mathbf{x}}(k|k)$, (11.54),

$\mathbf{y}(k + 1)$: the new measurement

A recursive estimation algorithm then follows from (11.57)

$$\hat{\mathbf{x}}(k + 1|k + 1) = \hat{\mathbf{x}}(k + 1|k) + \mathbf{K}(k + 1)[\mathbf{y}(k + 1) - \mathbf{C}\,\hat{\mathbf{x}}(k + 1|k)] \tag{11.58}$$

where the correction matrix $\mathbf{K}(k + 1)$ has to be chosen as to minimize the covariance matrix of the estimation error. As this variance changes with time, also $\mathbf{K}(k + 1)$ must be time-variant.

The error of the prediction is

$$\tilde{\mathbf{x}}(k + 1|k) = \hat{\mathbf{x}}(k + 1|k) - E\,\{\hat{\mathbf{x}}(k + 1|k)\} \tag{11.59}$$

and the error in the measurement, see (11.44)

$$\tilde{\mathbf{y}}(k + 1) = \mathbf{y}(k + 1) - E\,\{\mathbf{y}(k + 1)\} = \mathbf{n}(k) \tag{11.60}$$

The corresponding covariance matrices are

$$\mathbf{P}^-(k+1) = E\left\{\tilde{\mathbf{x}}(k+1|k)\,\tilde{\mathbf{x}}^T(k+1|k)\right\} \quad (11.61)$$

$$\mathbf{Y} = E\left\{\tilde{\mathbf{y}}(k+1)\,\tilde{\mathbf{y}}^T(k+1)\right\} = E\left\{\mathbf{n}(k)\,\mathbf{n}^T(k)\right\} = \mathbf{N} \qquad (11.62)$$

The covariance matrix of the recursive estimate $\hat{\mathbf{x}}(k+1|k+1)$ then becomes with (11.57)

$$
\begin{aligned}
\mathbf{P}(k+1) &= E\left\{\tilde{\mathbf{x}}(k+1|k+1)\,\tilde{\mathbf{x}}^T(k+1|k+1)\right\}\\
&= E\{[(\mathbf{I}-\mathbf{K}(k+1)\,\mathbf{C})(\hat{\mathbf{x}}(k+1|k) - E\{\hat{\mathbf{x}}(k+1|k)\}) + \mathbf{K}(k+1)(\mathbf{y}(k+1)\\
&\quad -E\{\mathbf{y}(k+1)\}]\,[(\mathbf{I}-\mathbf{K}(k+1)\,\mathbf{C})(\hat{\mathbf{x}}(k+1|k)) \qquad (11.63)\\
&\quad -E\{\hat{\mathbf{x}}(k+1|k)\} + \mathbf{K}(k+1)(\mathbf{y}(k+1)) - E\{\mathbf{y}(k+1)\}]^T\}\\
&= [\mathbf{I}-\mathbf{K}(k+1)\mathbf{C}]\,\mathbf{P}^-(k+1)\,[\mathbf{I}-\mathbf{K}(k+1)\mathbf{C}]^T + \mathbf{K}(k+1)\mathbf{N}\,\mathbf{K}^T(k+1)
\end{aligned}
$$

Now a value of $\mathbf{K}(k+1)$ is sought which minimizes the variance of the covariance of the estimation error. To find this minimum without differentiation, (11.63) is modified. As shown in [11.1] and [11.14] it can be formed into a complete square in $\mathbf{K}(k+1)$ which after several matrix calculations leads to

$$\mathbf{K}(k+1) = \mathbf{P}^-(k+1)\mathbf{C}^T[\mathbf{C}\,\mathbf{P}^-(k+1)\mathbf{C}^T + \mathbf{N}]^{-1} \qquad (11.64)$$

and

$$\mathbf{P}(k+1) = \mathbf{P}^-(k+1) - \mathbf{K}(k+1)\mathbf{C}\,\mathbf{P}^-(k+1) \qquad (11.65)$$

Herewith $\mathbf{P}^-$ follows from (11.51) and (11.59) with

$$
\begin{aligned}
E\{\hat{\mathbf{x}}(k+1|k)\} &= E\{\mathbf{x}(k+1)\}\\
\mathbf{P}^-(k+1) &= E\{(\hat{\mathbf{x}}(k+1|k) - E\{\mathbf{x}(k+1)\})(\hat{\mathbf{x}}(k+1|k)\\
&\quad -E\{\mathbf{x}(k+1)\})^T\} \qquad (11.66)\\
&= \mathbf{A}\,\mathbf{P}(k)\,\mathbf{A}^T + \mathbf{V}\,\mathbf{M}\,\mathbf{V}^T
\end{aligned}
$$

with covariance matrix $\mathbf{P}(k)$ of the estimation error $\tilde{\mathbf{x}}(k|k)$, according to (11.52). Hence, the sequence of calculations is

Prediction: (from (11.53) and (11.66))

$$\hat{\mathbf{x}}(k+1|k) = \mathbf{A}\,\hat{\mathbf{x}}(k|k) + \mathbf{B}\,\mathbf{u}(k) \qquad (11.67)$$

$$\mathbf{P}^-(k+1) = \mathbf{A}\,\mathbf{P}(k)\,\mathbf{A}^T + \mathbf{V}\,\mathbf{M}\,\mathbf{V}^T \qquad (11.68)$$

Correction: (from (11.64), (11.58) and (11.65))

$$\mathbf{K}(k+1) = \mathbf{P}^-(k+1)\,\mathbf{C}^T[\mathbf{C}\,\mathbf{P}^-(k+1)\,\mathbf{C}^T + \mathbf{N}]^{-1} \qquad (11.69)$$

$$
\begin{aligned}
\hat{\mathbf{x}}(k+1|k+1) &= \hat{\mathbf{x}}(k+1|k) + \mathbf{K}(k+1)[\mathbf{y}(k+1)\\
&\quad -\mathbf{C}\,\hat{\mathbf{x}}(k+1|k)] \qquad (11.70)
\end{aligned}
$$

$$\mathbf{P}(k+1) = [\mathbf{I}-\mathbf{K}(k+1)\mathbf{C}]\mathbf{P}^-(k+1) \qquad (11.71)$$

If the prediction (11.67) is inserted in (11.70) it follows

$$\hat{\mathbf{x}}(k+1|k+1) = \underset{\text{new estimate}}{\mathbf{A}\,\hat{\mathbf{x}}(k|k)} + \underset{\text{old estimate}}{\mathbf{B}\,\mathbf{u}(k)}$$

$$+ \underset{\substack{\text{correction} \\ \text{matrix}}}{\mathbf{K}(k+1)} \;\; \underset{\substack{\text{new measure-} \\ \text{ment}}}{[\mathbf{y}(k+1)} \;\; - \underset{\substack{\text{predicted measurement} \\ \text{based on old estimate}}}{\mathbf{C}(\mathbf{A}\,\hat{\mathbf{x}}(k|k) + \mathbf{B}\,\mathbf{u}(k))]} \qquad (11.72)$$

These filtering equations are called the *Kalman filter*.

The state estimation is a recursive estimation of the state $\hat{\mathbf{x}}(k+1|k+1)$ based on a predicted state $\hat{\mathbf{x}}(k+1|k)$ by the process model and a correction based on the new measurement $\mathbf{y}(k+1)$. A comparison with the state observer (11.40) shows that the observer only uses past information $\hat{\mathbf{x}}(k)$ and $\mathbf{y}(k)$ and not predicted values $\hat{\mathbf{x}}(k+1|k)$ and the new measurement $\mathbf{y}(k+1)$.

The correction matrix or gain $\mathbf{K}(k+1)$ depends on the covariance matrices $\mathbf{M}$ of the state noise $\mathbf{v}(k)$ and $\mathbf{N}$ of the output noise. It can be computed in advance, as it does not depend on measured signals.

If the process matrices $\mathbf{A}$, $\mathbf{B}$ and $\mathbf{C}$ and the noise covariance matrices do not depend on time, the Kalman filter gain $\mathbf{K}(k+1)$ approaches asymptotically a steady state value $\bar{\mathbf{K}}$. The steady state estimation error covariance matrix $\mathbf{P}^-$ follows from the elimination of $\mathbf{P}(k)$ from (11.68) using (11.71) and (11.69) leading to

$$\mathbf{P}^-(k+1) = \mathbf{A}\,\mathbf{P}^-(k)\,\mathbf{A}^T - \mathbf{A}\,\mathbf{P}^-(k)\mathbf{C}^T[\mathbf{C}\,\mathbf{P}^-(k)\,\mathbf{C}^T + \mathbf{N}]^{-1}\,\mathbf{C}\mathbf{P}^-(k)\mathbf{A}^T + \mathbf{V}\,\mathbf{M}\,\mathbf{V}^T \qquad (11.73)$$

which is a Riccati equation. (One recognizes the duality to the optimal state control). Its asymptotic solution results in the steady state matrix $\mathbf{P}^-$. Then the steady state Kalman filter gain becomes

$$\bar{\mathbf{K}} = \mathbf{P}^-\,\mathbf{C}^T[\mathbf{C}\,\mathbf{P}^-\,\mathbf{C}^T + \mathbf{N}]^{-1} \qquad (11.74)$$

The calculations then reduce to

prediction:
$$\hat{\mathbf{x}}(k+1|k) = \mathbf{A}\,\hat{\mathbf{x}}(k|k) + \mathbf{B}\,\mathbf{u}(k) \qquad (11.75)$$

correction:
$$\hat{\mathbf{x}}(k+1|k+1) = \hat{\mathbf{x}}(k+1|k) + \bar{\mathbf{K}}[\mathbf{y}(k+1) - \mathbf{C}\,\hat{\mathbf{x}}(k+1|k)] \qquad (11.76)$$

To obtain a comparison to a state observer the previous correction (11.76)

$$\hat{\mathbf{x}}(k|k) = \hat{\mathbf{x}}(k|k-1) + \bar{\mathbf{K}}[\mathbf{y}(k) - \mathbf{C}\,\hat{\mathbf{x}}(k|k-1)] \qquad (11.77)$$

is inserted in the prediction (11.75), leading to

$$\hat{\mathbf{x}}(k+1|k) = \mathbf{A}\,\hat{\mathbf{x}}(k|k-1) + \mathbf{B}\,\mathbf{u}(k) + \mathbf{A}\,\bar{\mathbf{K}}[\mathbf{y}(k) - \mathbf{C}\,\hat{\mathbf{x}}(k|k-1)] \qquad (11.78)$$

A comparison with the observer (11.40) shows that if the observer gain is chosen as

$$\mathbf{H} = \mathbf{A}\,\bar{\mathbf{K}} \tag{11.79}$$

the observer equation corresponds to the Kalman filter.

Figure 11.4 shows a signal flow diagram of the Kalman filter by using (11.77) and (11.75). (This scheme, as well as (11.77) and (11.78), is depicted for the case that at time instant k the input signal $u(k)$ and the output signal $y(k)$ is available, as for an observer. If, however, the next output sample $y(k + 1)$ is used for estimating $\hat{\mathbf{x}}(k + 1|k + 1)$ the correction follows (11.76)).

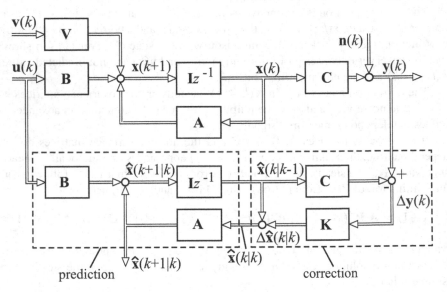

Fig. 11.4. Signal flow diagram of a Kalman filter with the prediction according to (11.75) and the correction according (11.77)

In the original work of [11.17] the recursive state estimator was derived by applying the orthogonality condition between the estimation errors and the measurements

$$E\left\{\tilde{\mathbf{x}}(i)\,\mathbf{Y}^T\,(j)\right\} = \mathbf{0} \quad \text{for} \quad j < i \tag{11.80}$$

An alternative derivation of the Kalman filter follows using the conditional expectation of a least squares estimate

$$E\left\{\mathbf{x}(k)|\mathbf{Y}_j\right\} \tag{11.81}$$

see (11.47) and (11.48). Hence, the estimate is the vertical projection of $\mathbf{x}(k)$ on $\mathbf{Y}_j$

$$\hat{\mathbf{x}}(k|j) = E\left\{\mathbf{x}(k)|\mathbf{Y}_j\right\} \tag{11.82}$$

with the estimation error

$$\tilde{\mathbf{x}}(k|j) = \mathbf{x}(k) - \hat{\mathbf{x}}(k|j)$$

see, e.g. [11.23]. Other references for the Kalman filter are [11.4], [11.10], [11.15], [11.2], [11.19], [11.28], [11.29].

An early application of a Kalman filter for fault detection, excited by the outputs of a multi-output process is shown by [11.22]. The residuum (innovation) changes the character of zero mean white noise with known covariance, if a fault appears. See also [11.31]. In principle the application of the Kalman filter is similar to that of state observers. But it is especially suitable for processes with relative large state variable and output noise.

11.3 Output observers

The classical state observer or Kalman filter primarily reconstruct or estimate the state variables. However, if faults in the state variables are not of interest, so called *output observers* or *functional observers* can be applied for fault detection. The goal is to reconstruct the output by a state space model which only generates residuals for faults $\mathbf{f}(t)$, but not for the measurable input $\mathbf{u}(t)$ and nonmeasurable (unknown) inputs $\mathbf{v}(t)$. There exists a rich literature on unknown output observers, see, e.g. [11.27], [11.6], [11.5]. The following derivation follows [11.8] and [11.13].

The process is described by

$$\dot{\mathbf{x}}(t) = \mathbf{A}\,\mathbf{x}(t) + \mathbf{B}\,\mathbf{u}(t) + \mathbf{V}\,\mathbf{v}(t) + \mathbf{L}\,\mathbf{f}(t) \tag{11.83}$$

$$\mathbf{y}(t) = \mathbf{C}\,\mathbf{x}(t) + \mathbf{M}\,\mathbf{f}(t) \tag{11.84}$$

The goal is now to generate residuals which are independent of the unknown inputs $\mathbf{v}(t)$. Therefore a linear transformation

$$\boldsymbol{\xi}(t) = \mathbf{T}_1\,\mathbf{x}(t) \tag{11.85}$$

is applied to build an observer with new state variables $\boldsymbol{\xi}(t)$. The output of the observer be $\boldsymbol{\eta}(t)$. Therefore no direct output error with the process output $\mathbf{y}(t)$ results, but an output error with a transformation

$$\boldsymbol{\eta}(t) = \mathbf{T}_2\,\mathbf{y}(t) \tag{11.86}$$

The transformed process model without fault and disturbance influences then becomes

$$\dot{\boldsymbol{\xi}}(t) = \mathbf{A}_\xi\,\boldsymbol{\xi}(t) + \mathbf{B}_\xi\,\mathbf{u}(t) \tag{11.87}$$

$$\boldsymbol{\eta}(t) = \mathbf{C}_\xi\,\boldsymbol{\xi}(t) \tag{11.88}$$

and the corresponding state observer, compare Figure 11.5

$$\dot{\hat{\boldsymbol{\xi}}}(t) = \mathbf{A}_\xi\,\hat{\boldsymbol{\xi}}(t) + \mathbf{B}_\xi\,\mathbf{u}(t) + \mathbf{H}_\xi\,\mathbf{y}(t) \tag{11.89}$$

$$\hat{\eta}(t) = \mathbf{C}_\xi \, \hat{\xi}(t) \tag{11.90}$$

This observer then does not feed back an error signal of the outputs, but has the character of a parallel process model. The error of the states is

$$\tilde{\xi}(t) = \hat{\xi}(t) - \mathbf{T}_1 \, \mathbf{x}(t) \tag{11.91}$$

Inserting (11.83) and (11.85) leads to

$$\begin{aligned}
\dot{\tilde{\xi}} &= \dot{\hat{\xi}}(t) - \mathbf{T}_1 \, \dot{\mathbf{x}}(t) \\
&= \mathbf{A}_\xi \, \hat{\xi}(t) + \left(\mathbf{A}_\xi \, \mathbf{T}_1 + \mathbf{H}_\xi \, \mathbf{C} - \mathbf{T}_1 \, \mathbf{A}\right) \mathbf{x}(t) + \left(\mathbf{B}_\xi - \mathbf{T}_1 \, \mathbf{B}\right) \mathbf{u}(t) \\
&\quad - \mathbf{T}_1 \, \mathbf{V} \, \mathbf{v}(t) + \left(\mathbf{H}_\xi \, \mathbf{M} - \mathbf{T}_1 \, \mathbf{L}\right) \mathbf{f}(t)
\end{aligned} \tag{11.92}$$

and the residual becomes with (11.84)

$$\mathbf{r}(t) = \hat{\eta}(t) - \eta(t) = \mathbf{C}_\xi \, \hat{\xi}(t) - \mathbf{T}_2 \, \mathbf{y}(t) \tag{11.93}$$

$$= \mathbf{C}_\xi \, \tilde{\xi}(t) + \left(\mathbf{C}_\xi \, \mathbf{T}_1 - \mathbf{T}_2 \, \mathbf{C}\right) \mathbf{x}(t) + \mathbf{T}_2 \, \mathbf{M} \, \mathbf{f}(t) \tag{11.94}$$

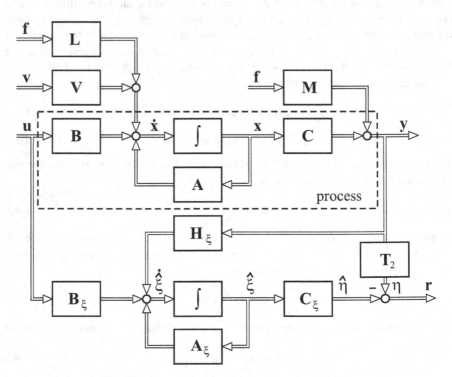

Fig. 11.5. Output observer for the detection of faults

To decouple now the state error $\tilde{\xi}(t)$ and the residual $\mathbf{r}(t)$ from the unknown state $\mathbf{x}(t)$ and the unknown disturbance $\mathbf{v}(t)$ and from $\mathbf{u}(t)$, following equations must be satisfied

$$\mathbf{T}_1 \, \mathbf{A} - \mathbf{A}_\xi \, \mathbf{T}_1 = \mathbf{H}_\xi \, \mathbf{C} \qquad (11.95)$$

$$\mathbf{B}_\xi = \mathbf{T}_1 \, \mathbf{B} \qquad (11.96)$$

$$\mathbf{T}_1 \, \mathbf{V} = \mathbf{0} \qquad (11.97)$$

$$\mathbf{C}_\xi \, \mathbf{T}_1 - \mathbf{T}_2 \, \mathbf{C} = \mathbf{0} \qquad (11.98)$$

In addition, the observer matrix $\mathbf{A}_\xi$ is selected as a diagonal matrix with stable poles. The set of equations (11.95) to (11.98) can be solved by transforming the state equations into Kronecker canonical form, [11.9], by using an eigen-structure assignment for the observer, [11.5], or by an iterative procedure based on singular value decomposition, [11.18]. [11.24] gave a relative simple solution, see also [11.13].

The state error finally becomes

$$\dot{\tilde{\xi}}(t) = \mathbf{A}_\xi \, \tilde{\xi}(t) + \left(\mathbf{H}_\xi \, \mathbf{M} - \mathbf{T}_1 \, \mathbf{L} \right) \mathbf{f}(t) \qquad (11.99)$$

and the residual

$$\mathbf{r} = \mathbf{C}_\xi \, \tilde{\xi}(t) + \mathbf{T}_2 \, \mathbf{M} \, \mathbf{f}(t) \qquad (11.100)$$

If the state errors asymptotically approach zero, the residual only depends on the faults $\mathbf{f}(t)$ and not on the unknown input $\mathbf{v}(t)$.

Laplace transformation of (11.89) and (11.93) yields

$$\mathbf{r}(s) = \mathbf{C}_\xi \, \hat{\tilde{\xi}}(s) - \mathbf{T}_2 \, \mathbf{y}(s) = \left[\mathbf{C}_\xi \, (s \, \mathbf{I} - \mathbf{A})^{-1} \, \mathbf{H}_\xi - \mathbf{T}_2 \right]$$
$$\mathbf{y}(s) + \mathbf{C}_\xi \, [s \, \mathbf{I} - \mathbf{A}]^{-1} \, \mathbf{B}_\xi \, \mathbf{u}(s) \qquad (11.101)$$

This is, as for parity equations, the computational form of the residuals by measuring $\mathbf{u}(t)$ and $\mathbf{y}(t)$. After inserting (11.99) in (11.100) one obtains the internal form

$$\mathbf{r}(s) = \mathbf{C}_\xi \, \left(s \, \mathbf{I} - \mathbf{A}_\xi \right)^{-1} \, \left(\mathbf{H}_\xi \, \mathbf{M} - \mathbf{T}_1 \, \mathbf{L} \right) \mathbf{f}(s) + \mathbf{T}_2 \, \mathbf{M} \, \mathbf{f}(s) \qquad (11.102)$$

Here the dynamic response of faults on the residual can be seen. This shows that in addition to the above requirements

$$\mathbf{C}_\xi \, \left(\mathbf{H}_\xi \, \mathbf{M} - \mathbf{T}_1 \, \mathbf{L} \right) + \mathbf{T}_2 \, \mathbf{M} \neq 0 \qquad (11.103)$$

also must be satisfied.

A comparison of (11.102) with the output error residual for parity equations (10.8) indicates similarities. The faults $\mathbf{f}(s)$ are just filtered by another transfer function, here however, independent on input disturbances $\mathbf{v}(t)$.

The design of this output observer requires that the unknown disturbance entry matrix $\mathbf{V}$ is known precisely because it determines the transformation matrix $\mathbf{T}_1$. The other conditions (11.95), (11.96) and (11.99) let also suppose that the process model $(\mathbf{A}, \mathbf{B}, \mathbf{C})$ must be accurately known. The output observer approach gives relative much freedom for the design, but on cost of design effort and transparency.

11.4 Comparison of the parity- and observer-based approaches

It was already mentioned that similarities exist between the different approaches for fault detection with parity equations and observers. They all use the same measurable input signals $\mathbf{u}(t)$ and output signals $\mathbf{y}(t)$, assuming that the structure and parameters of the process are exactly known and do not change (fixed model) and if unknown inputs $\mathbf{v}(t)$ have to be compensated, that the disturbance entry matrix $\mathbf{V}$ is known.

11.4.1 Comparison of residual equations

The resulting computational form of the residual equations for continuous time are in the Laplace domain without noise terms:

1) Parity equations with transfer functions.
   - output error, (10.7):

$$\mathbf{r}'(s) = \mathbf{y}(s) - \mathbf{G}_m(s)\,\mathbf{u}(s)$$

   - equation error, (10.11):

$$\mathbf{r}(s) = \mathbf{A}_m\,\mathbf{y}(s) - \mathbf{B}_m(s)\,\mathbf{u}(s)$$

   with structured residuals, (10.52):

$$\mathbf{r}^*(s) = \mathbf{W}(s)\,[\mathbf{A}_m(s)\,\mathbf{y}(s) - \mathbf{B}_m(s)\,\mathbf{u}(s)]$$

2) Parity equations with state space models, (10.29):

$$\mathbf{r}(s) = \mathbf{W}\left[\mathbf{L}_y(s)\,\mathbf{y}(s) - \mathbf{Q}_u\,\mathbf{L}_u(s)\,\mathbf{u}(s)\right]$$

3) State observer, (11.14):

$$\mathbf{r}(s) = \left[\mathbf{I} - \mathbf{C}\,[s\,\mathbf{I} - \mathbf{A} + \mathbf{H}\,\mathbf{C}]^{-1}\,\mathbf{H}\right]\,\mathbf{y}(s) - \mathbf{C}\,[s\,\mathbf{I} - \mathbf{A} + \mathbf{H}\,\mathbf{C}]^{-1}\,\mathbf{B}\,\mathbf{u}(s)$$

4) Output observer, (11.101):

$$\mathbf{r}(s) = \left[\mathbf{T}_2 - \mathbf{C}_\xi\,[s\,\mathbf{I} - \mathbf{A}]^{-1}\,\mathbf{H}_\xi\right]\,\mathbf{y}(s) - \mathbf{C}_\xi\left[s\,\mathbf{I} - \mathbf{A}^{-1}\right]\,\mathbf{B}_\xi\,\mathbf{u}(s)$$

Hence, the structure of the residual equations is very similar. They differ in the way the input and output measurements are filtered.

Several authors have compared the various methods for residual generation. In the case of discrete-time realization, it was shown in [11.9] that the unknown input observers and parity equation approach are equivalent if the observer is designed with a deadbeat behavior, see also [11.5]. [11.20] summarize some results and show equivalence between Luenberger state observers and parity equations, see also [11.11]. [11.13] has compared a state observer and a reduced order observer with unknown input and parity equations based on a state space model for a DC motor and shown theoretically and by experiments that all three approaches lead to the same residual deflection for a fault in the armature resistance, see also the next section.

11.4.2 Comparison by simulations[1]

To compare the resulting residuals of different fault-detection methods based on parity equations and observers by simulations a linear model of a DC motor is considered, see Figure 11.6. As described in Chapter 20 the (simplified) dynamic model results from the armature circuit equation

$$L_A \, \dot{I}_A(t) + R_A \, I_A(t) + \Psi \, \omega(t) = U_A(t) \qquad (11.104)$$

and the equation for the mechanical part

$$J \, \dot{\omega}(t) = \Psi \, I_A(t) - M_F \, \omega(t) - M_L(t) \qquad (11.105)$$

where

U_A armature voltage (input)
I_A armature current
ω angular speed (output)
M_L load torque (disturbance)
M_F viscous friction coefficient
L_A armature inductance
R_A armature resistance
Ψ flux linkage

The state space formulation becomes

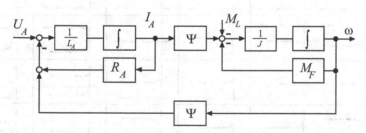

Fig. 11.6. Signal flow diagram of a permanently excited DC motor

$$\begin{bmatrix} \dot{I}_A(t) \\ \dot{\omega}(t) \end{bmatrix} = \begin{bmatrix} -\frac{R_A}{L_A} & -\frac{\Psi}{L_A} \\ \frac{\Psi}{J} & -\frac{M_F}{J} \end{bmatrix} \begin{bmatrix} I_A(t) \\ \omega(t) \end{bmatrix} + \begin{bmatrix} \frac{1}{L_A} & 0 \\ 0 & -\frac{1}{J} \end{bmatrix} \begin{bmatrix} U_A(t) \\ M_L(t) \end{bmatrix}$$

$$\mathbf{y}(t) = \begin{bmatrix} I_A(t) \\ \omega(t) \end{bmatrix} \qquad (11.106)$$

For the generation of *parity equations* the transfer function

$$G_M(s) = \frac{\omega(s)}{U_A(s)} = \frac{\Psi}{(L_A s + R_A)(M_F + Js) + \Psi^2} \qquad (11.107)$$

---

[1] compiled by Iris Unger, [11.25]

is required, assuming $M_L(s) = 0$. As $L_A \ll R_A$, the model can be simplified to

$$G_M(s) = \frac{\Psi}{J R_A s + \Psi^2 + R_A M_F} = \frac{K_M}{T_M s + 1}$$
$$K_M = \frac{\Psi}{\Psi^2 + R_A M_F}; \quad T_M = \frac{J R_A}{\Psi^2 + R_A M_F} \tag{11.108}$$

The residual equations with the output error then become

$$r'(s) = \omega(s) - G_M(s) U_A(s) \tag{11.109}$$

and with the equation error

$$r(s) = K_M U_A(s) - (T_M s + 1)\omega(s) \tag{11.110}$$

The reaction of these residuals is shown in Figures 11.7 to 11.9. Expected deflections result for the additive faults. Usable deflections of the residuals for parametric faults are only obtained for $\Delta\Psi$ as well as for both, no input and input change $U_A(t)$. If the output is noisy, the output error should better be used than the equation error.

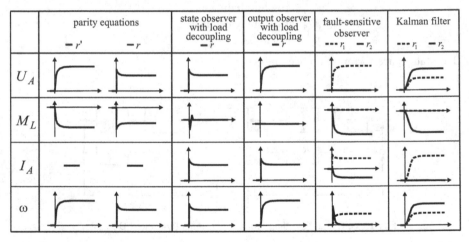

Fig. 11.7. Comparison of different residuals for a DC motor with *additive faults*, without input excitation

Observer-based residuals are generated by a state observer with one measured output, a state observer with two measured outputs, an output observer (unknown input observer) and a fault sensitive observer. The *observer with one measured input and output* and output residual leads to approximately the same results as the parity equations with output error. If two outputs $I_A(t)$ and $\omega(t)$ are measured, (11.61) can be used to design a *Luenberger state observer*. The load torque $M_L(t)$ can then be modelled as disturbance with

$$\mathbf{V} = \begin{bmatrix} 0 \\ -\frac{1}{J} \end{bmatrix}$$

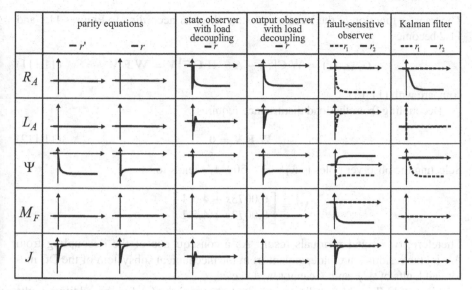

Fig. 11.8. Comparison of different residuals for a DC motor with *parametric faults*, without input excitation

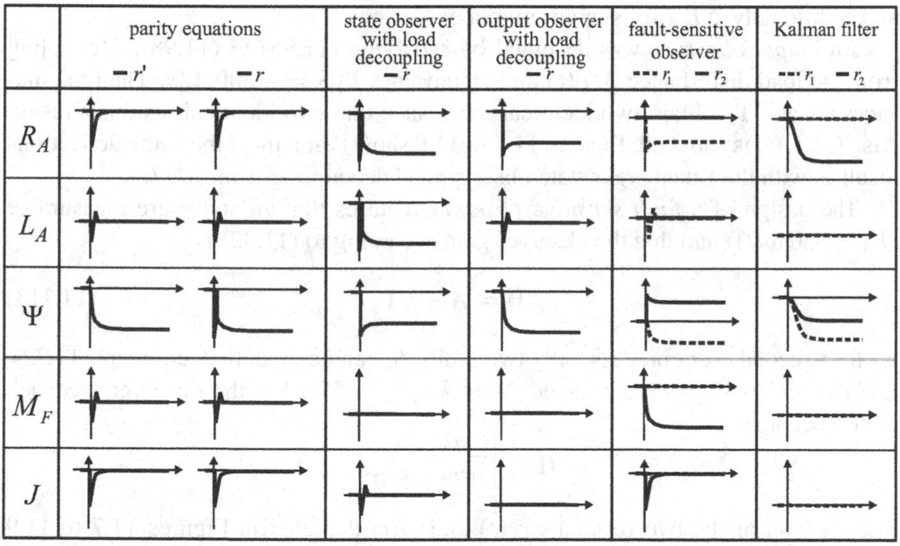

Fig. 11.9. Comparison of different residuals for a DC motor with *parametric faults* and stepwise input excitation $U_A(t)$

The transfer function of the Luenberger state observer according to Figures 11.1 and 11.2 becomes

$$G_{rM_L}(s) = \mathbf{W} \ \mathbf{C}[s \ \mathbf{I} - \mathbf{A} + \mathbf{H} \ \mathbf{C}]^{-1}\mathbf{V} = \mathbf{W} \ \mathbf{F} \ \mathbf{V} \qquad (11.111)$$

compare with (11.15).

Decoupling from the load disturbance requires

$$\mathbf{W} \ \mathbf{F} \ \mathbf{V} = \mathbf{0} \qquad (11.112)$$

Selecting the observer poles to $\lambda_{1,2} = -500 \ 1/s$ leads to

$$\mathbf{W} = \begin{bmatrix} 0.0073s + 4.5 \ 1 \\ 0.0073s + 4.5 \ 1 \end{bmatrix}$$

Therefore, two equal residuals result. As a consequence of the decoupling from $M_L(t)$ the residuals are independent from the mechanical subsystem of the DC motor and faults of M_F and J cannot be detected.

Figures 11.7 to 11.9 indicate expected changes of $r(t)$ for the additive faults $\Delta U_A(t)$, $\Delta I_A(t)$ and $\Delta \omega(t)$. However, significant residual changes are only obtained for the parametric faults ΔR_A and $\Delta \Psi$ in the case of no input excitation, and additionally ΔL_A for stepwise input excitation.

An *output observer* was designed by satisfying (11.95) to (11.98), decoupling from the load disturbance $M_L(t)$ and eigenvalues $\lambda_{1,2} = -500 \ 1/s$. Then the matrices $\mathbf{C}_\xi$ and $\mathbf{T}_2$ obtain two identical rows, such that two identical residuals result. Also (11.103) is satisfied. Figures 11.7 to 11.9 show that almost the same deflections result as with the Luenberger state observer and decoupling from $M_L(t)$.

The design of a *fault sensitive observer* requires that all states are measurable ($I_A(t)$ and $\omega(t)$) and that the observer gain according to (11.32) is

$$\mathbf{H} = \mathbf{A} - \lambda \ \mathbf{I} \qquad (11.113)$$

As for a second order process, only two faults f_{li} can be modelled, compare (11.22), $\Delta U_A(t)$ and $\Delta M_L(t)$ are selected. With $\lambda_{1,2} = -500 \ 1/s$ the observer feedback matrix becomes

$$\mathbf{H} = \begin{bmatrix} 275 & -50 \\ 179 & 500 \end{bmatrix}$$

The reaction of the two residuals $r_1(t)$ and $r_2(t)$ depicted in Figures 11.7 to 11.9 shows deflections for all additive faults and for the parametric faults ΔR_A, $\Delta \Psi$ and ΔM_F without and with input excitation $\Delta U_A(t)$. Table 11.1 shows the sign of the deflections. Here, short impulse like deflections are ignored, because they cannot be distinguished from disturbances in practical cases. Among the additive faults, ΔI_A and ΔM_L can be isolated but ΔU_A and $\Delta \omega$ cannot be distinguished. For the parametric faults only ΔR_A, $\Delta \Psi$ and ΔM_F show significant changes. However, ΔM_F and ΔM_L cannot be distinguished and ΔL_A and ΔJ are not detectable. If the sign of the fault is unknown following faults cannot be distinguished from each

Table 11.1. Residual deflections for the fault-sensitive observer and positive faults

	Additive faults				Parametric faults				
	ΔU_A	ΔI_A	$\Delta\omega$	ΔM_L	ΔR_A	ΔL_A	$\Delta\Psi$	ΔM_F	ΔJ
r_1	+	+	+	0	−	0	−	0	0
r_2	0	−	0	−	0	0	+	−	+

Table 11.2. Residual deflections for the Kalman filter and positive faults

	Additive faults				Parametric faults				
	ΔU_A	ΔI_A	$\Delta\omega$	ΔM_L	ΔR_A	ΔL_A	$\Delta\Psi$	ΔM_F	ΔJ
r_1	+	+	+	0	0	0	−	0	0
r_2	+	0	+	−	−	0	0(−)	0	0

other: ΔU_A, $\Delta\omega$, ΔR_A; ΔI_A, $\Delta\Psi$; ΔM_L, ΔM_F. Hence, from 9 considered faults, 7 faults can be detected. But no unique isolation is possible, only groups of possible faults can be indicated.

A *Kalman filter* was realized as discrete time model with sampling time $T_0 = 0.001\ s$. The state estimates are $\hat{I}_A(k)$ and $\hat{\omega}(k)$. For the design of the Kalman gain the noise variances were chosen as $\sigma_I^2 = 100\ A^2$ and $\sigma_\omega^2 = 100\ 1/s^2$. The measured signals and the state estimates were averaged over a time window of 500 samples and the residuals

$$r_1(k) = \bar{I}_{A\,\text{meas}}(k) - \bar{\hat{I}}_{A\,\text{est}}(k)$$
$$r_2(k) = \bar{\omega}_\text{meas}(k) - \bar{\hat{\omega}}_\text{est}(k)$$

over a time window of 10 samples.

Figures 11.7 to 11.9 show the results for a process with small noise ($\sigma_\omega^2 = 1\ 1/s^2; \sigma_I^2 = 1\ A^2$), to allow a direct comparison with the other residuals. The additive faults give deflections of r_1 and/or r_2 for all cases. For parametric faults r_1 and/or r_2 show changes for ΔR_A and $\Delta\Psi$, but no reactions for ΔL_A, ΔM_F and ΔJ, for both, without or with stepwise input excitation $\Delta U_A(k)$. A comparison of the sign of the residual changes in Table 11.2 indicates that the additive faults ΔI_A and ΔM_L can be distinguished from ΔU_A and $\Delta\omega$, which both have the same deflections. The parametric faults ΔR_A and $\Delta\Psi$ can be detected. But ΔR_A and ΔM_L are not distinguishable. If the sign of the fault is unknown, only groups of faults can be indicated: ΔU_A, $\Delta\omega$; ΔI_A, $\Delta\Psi$; ΔR_A, ΔM_L. Therefore also with the Kalman filter no unique isolation of faults is possible.

Summing up, these simulation results for a DC motor lead to following conclusions (note that the parity equations were designed with a first order model):

1) *Additive faults*
   - The considered parity equations and state observers lead to about the same residual deflections;
   - The considered state observer and the output observer, both designed for decoupling the main disturbance, the load torque, show almost identical residuals;

- The Kalman filter shows large similarities to the observer-based methods and parity equations without disturbance decoupling.

2) *Parametric faults*
  - Out of 5 process parameters only 2 (R_A and Ψ) are detectable as changes by the observer-based methods, but cannot be distinguished from some additive faults;
  - Input excitation has in most cases no effect on the residuals, except in the case of ΔL_A for the state observer with load decoupling;
  - The parametric faults ΔL_A and ΔJ cannot be detected by all methods (except ΔL_A for the state observer).

A comparison of parity equations and recursive parameter estimation with measurements for real DC motors is presented in Chapter 20.

11.5 Problems

1) State the differences between fault detection with state observers and parity equations for processes with one input - one output and two inputs and two outputs. Compare the a priori knowledge, design effort, computational effort, noise sensitivity, detection of additive and multiplicative faults.
2) What are the differences between state observers and a Kalman filter with regard to fault detection?
3) What are the advantages of output observers compared to state observers and parity equations?
4) Design a state observer for Problem 5) in Chapter 10 (electrical steering system).

Fault detection of control loops

The main goals for using automatic control loops are precise following of reference variables (set points), a faster response than in open loop, compensation of all kind of external disturbances on the controlled variable, stabilization of unstable processes, reduction of the influence of process parameter changes with regard to the static and dynamic behavior, partial compensation of actuator and process nonlinearities, and, of course, replacement of manual control by humans. The performance of a SISO control loop with regard to the control error (deviation)

$$e(k) = w(k) - y(k) \tag{12.1}$$

i.e. the deviation of the controlled variable $y(k)$ from the reference variable $w(k)$ depends on many facts, compare Figure 12.1, like:

- external disturbance $w(k), u_v(k), v_i(k)$;
- structure and parameters of the controller G_c and controller faults f_c;
- changes of the structure and parameters of the process G_p and process faults f_p;
- changes and faults of actuator G_a and f_a;
- faults f_s in the sensor G_s and measurement noise n_s.

Hence, many changes and faults influence the performance of closed loops. Usually, only the control deviation e and the control variable y are monitored.

12.1 Effects of faults on the closed loop performance

Small faults in the actuator and process, be they additive or multiplicative, will usually be compensated by the feedback controller (with integral action) and they will not be detectable by considering $e(k)$ and $y(k)$ only, as long as the control deviation turns back to approximately zero. Also small sensor offset faults will not be detected. The controller will just make the wrong sensor signal equal to the reference variable. Only by a redundant sensor or other redundant information for the controlled variable, the offset fault can usually be detected.

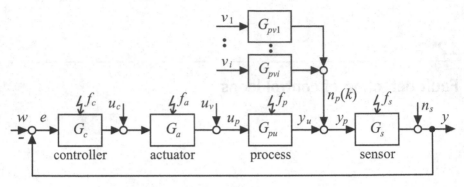

Fig. 12.1. Control loop with variables and fault influences

y	controlled variable	w	reference variable
u_p	manipulated variable	e	control deviation
v_i	process disturbances	n_s	measurement noise
u_v	process input disturbances	$f_{c,a,p,s}$	faults of the controller,
n_p	sum of process disturbances		actuator, process and
y_p	process output to be controlled		sensor

Table 12.1 shows the effect of *larger faults* on the closed-loop behavior. The different faults have a similar effect on the considered changes of closed loop behavior. In addition, some of the behavior is also observed after external disturbances under normal operation, Table 12.2. Therefore it is not easily possible to diagnose the various faults by observing the listed properties. On the other side, at least for larger plants with hundreds of control loops, it would be very practical to have an automatic fault detection for the control loops. For example, [12.4] reports that up to 30 % of all loops oscillate in pulp and paper processes. In many cases they are then put to manual mode or wrongly detuned, [12.5]. As reason frequently stip-slick effects of valves is mentioned (which is less the case for valves with position controller).

12.2 Signal-based methods for closed-loop supervision

The problem of control performance monitoring is treated in several publications. A survey is given, e.g. by [12.8] and [12.3]. First contributions assume a *stochastic behavior* of $y(k)$ and a process with dead time, determine the variance σ_y^2 and relate it to the output variance σ_{MV}^2 of an optimal minimum variance controller. This leads to a performance index $I_p = \sigma_y^2/\sigma_{MV}^2 \geq 1$, [12.7]. The only knowledge on the process is the dead time of the discrete time process model. Modifications of this idea were followed by [12.18], [12.17]. [12.16] considered constraints on the control structure. [12.5] proposed methods to detect oscillations and sluggish control. [12.19] suggested a performance index by relating the closed-loop settling time after a *deterministic disturbance* to the process time-delay. [12.8] modified the performance index by Harris and changed the MV-controller with one pole not in the origin and proposed methods for the detection of oscillations and the diagnosis of

Table 12.1. Possible effects of large faults on closed loop behavior

fault	observation			
	sluggish behavior	oscillatory behavior	large control error	actuator at restriction
change of process structure & parameters	✓	✓	✓	✓
actuator				
• friction		✓	✓	✓
• backlash		✓	✓	✓
• stuck	✓		✓	
sensor				
• offset				✓
• gain	✓	✓		✓
• variance		✓	✓	
controller				
• parameters	✓	✓	✓	✓
• noise			✓	
• wrong tuning	✓	✓	✓	✓

Table 12.2. Effects of external disturbances on closed loop behavior (well-tuned controller)

disturbance	observation			
	sluggish behavior	oscillatory behavior	large control error	actuator at restriction
large, aperiodic				
• low frequent			✓	✓
• medium frequent			✓	
• high frequent			✓	
periodic				
• low frequent	✓			
• medium frequent		✓	✓	✓
• high frequent		✓	✓	✓

valve stiction, see Section 12.3. However, all these methods do not solve all diagnosis problems in closed loop, especially with normal operating data, which are not always stochastic, see also [12.3].

An advantage of using minimum-variance controllers (MV) as reference is that they provide the best possible variance of the control variable for colored stochastic disturbances at the output. In this case the controlled variable becomes for a process without dead-time

$$G_p(z) = \frac{B(z)}{A(z)} \tag{12.2}$$

a white-noise process

$$E\left\{y^2(k)\right\} = \sigma_y^2 = \lambda^2 \, \sigma_v^2 \tag{12.3}$$

where

$$n(z) = \frac{D(z)}{C(z)} \lambda \, v(z) \tag{12.4}$$

is the noise-filter. For processes with dead time it holds

$$E\left\{y^2(k)\right\} = \left[1 + f_1^2 + \ldots + f_d^2\right]\lambda^2 \tag{12.5}$$

see [12.12], where the f_i depend on the noise model. However, the MV-controllers then require a high manipulation effort and more practical controllers are the generalized minimum variance controller where a weighting factor r for $u(k)$ is used. In addition MV-controllers exhibit a poor control behavior for deterministic disturbances and show a remaining offset for lasting disturbances. Therefore, they seem not to be good reference controllers for practical purposes.

A further investigation for signal-based methods was performed by [12.2]. Figure 12.2 shows the time responses to a step change of the reference variable and different faults with a size of 10 %. Following signal-based performance criteria were used to evaluate the effects for the different faults:

- overshoot:

$$\Delta y_{0s} = y_{max} - y(\infty) \tag{12.6}$$

- settling time:

$$T_s = N_s \, T_0 \quad \text{until} \quad |e(k) < \varepsilon| \tag{12.7}$$

- root of mean-squared control deviation:

$$rms_e = S_e = \sqrt{\frac{1}{N} \sum_{k=0}^{N-1} e^2(k)} \tag{12.8}$$

- root of mean-squared, manipulation effort:

$$rms_u = S_u = \sqrt{\frac{1}{N} \sum_{k=0}^{N-1} -[u(k) - u(\infty)]^2} \tag{12.9}$$

- number of zero crossings of control variable for:

$$0 \leq t < T_s : \kappa_e$$

- change of the steady state value of manipulated variable: $\Delta u(\infty)$

The faults F1 and F5 have no effect on the controlled variable $y(t)$ and the manipulated variable $u(t)$. F4 does not influence $y(t)$ but leads to a drastic reduction of $u(k)$. An increased process gain F2 or an increased sensor gain F6 force the loop to strong oscillations and lower steady state values of $u(k)$. In addition an increase of Coulomb friction finally leads to oscillations with much lower frequency and much lower settling time.

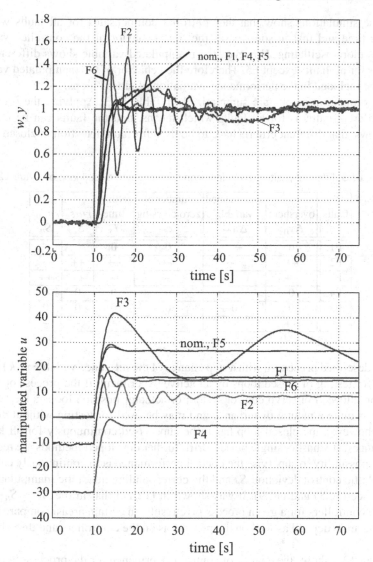

Fig. 12.2. Simulation of closed-loop behavior for a step change of the reference variable. The controlled process consists of a 2nd order actuator with small Coulomb friction, a 2nd order linear process and a 1st order sensor. The controller is an optimized PID-controller, [12.2]. The size of the fault is 10 % of the nominal value.

F1	offset sensor signal	F4	offset manipulated variable
F2	increased process gain	F5	increased sensor time constant
F3	increased actuator friction	F6	increased sensor gain

These simulations show that the controller compensates for all faults with increasing time and that in some cases the time behavior of the controlled variable becomes more oscillating. However, the manipulated variable shows different time behavior for all faults except F5. Therefore the behavior of the manipulated variable should be included in the performance evaluation of closed loops.

The simulation results are summarized in Table 12.3. Applying the 6 criteria, gives different patterns for all faults which means, that the faults can be isolated. This shows, that different criteria have to be used for fault detection in closed loops.

Table 12.3. Effect of positive 10 % faults on the simulated closed loop of Figure 12.2

faults	overshoot Δy_{0s}	manipulated variable Δu_∞	number of zero crossing κ_e	settling time T_s	rms_e S_e	rms_u S_u
F1	0	--	0	0	0	+
F2	++	--	++	++	++	--
F3	++	0	-	++	++	++
F4	0	--	0	0	0	0
F5	0	0	0	0	0	0
F6	++	-	+	+	-	-

The described procedure with a step change of the reference variable $w(k)$ can be applied for testing of closed loops by an active experiment, if the process operation condition allows. For servo control system or actuator position loops a step change of $w(k)$ is quite a natural disturbance and may be used for testing without difficulties. If, however, the closed loop has to be supervised continuously for all kind of disturbances, stochastic, single pulses, drift, further developed methods are required.

One possibility to supervise the control performance is to continuously calculate the quadratic control deviation S_e and the corresponding quadratic manipulation effort S_u, see (12.8) and (12.9). It was shown in [12.12] that in the $S_e - S_u$-plane different controllers for a given process give results in certain areas, compare Figure 12.3. The areas depend as well on the process as on the controllers and disturbances.

Figure 12.4 shows the change of control performance for the process

$$G_p(s) = \frac{y(s)}{u(s)} = \frac{K_p}{s^2 + 2D\,\omega_0 + \omega_0^2} \qquad (12.10)$$

$$K_p = 0.2 \quad \omega_0 = 5\frac{1}{s} \quad D = 0.9$$

by deviations of the process parameters from the nominal ones for which the controller was optimized, [12.2]. For larger process parameter changes the performance criteria leave an area which can be defined as a tolerance zone around the nominal point.

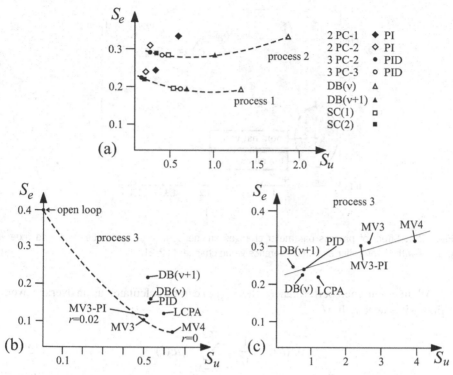

Fig. 12.3. Mean-squared control deviation S_e and mean-squared manipulation effort S_u for different controllers and processes: (a) and (c) step changes of the reference value w; (b) stochastic disturbances n at the process input, [12.11], [12.12]. Process 1: low pass process, $m = 3$, $d = 1$; process 2: non-minimum phase process, $m = 2$; process 3: low pass process, $m = 2$. DB: deadbeat controller; MV: minimum-variance controller; SC: state controller; LCPA: linear controller, pole assignment

Therefore a tolerance zone $\Delta S_e, \Delta S_u$ around the nominal performance criteria S_{en}, S_{un} can be defined. In addition, the control performance ratio

$$\eta_{eu} = S_e/S_u \qquad (12.11)$$

can be used. Hence, general applicable performance criteria for closed-loop supervision are

$$S_e(k) \geq S_{en} + \Delta S_e \ \text{ AND } \ S_u(k) \geq S_{un} + \Delta S_u \qquad (12.12)$$

Calculating the autocorrelation function (ACF) of the control deviation

$$R_{ee}(\tau) = \frac{1}{N} \sum_{k=1}^{N} e(k)\, e(k - \tau) \qquad (12.13)$$

indicates the kind of external or internal excitation of the closed loop. By this way more or less colored disturbances (noise) or periodic signals can be detected with less calculation effort.

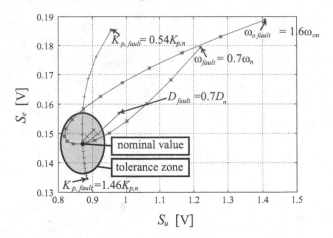

Fig. 12.4. Effect of process parameter changes on the control performance for a 2nd order process with P-controller and step reference value change, [12.2]

All the mentioned performance criteria (pc) can be calculated as an average over a time window length M

$$pc(k) = \frac{1}{M} \sum_{i=k-M}^{k} pc(i) \tag{12.14}$$

or by an average with exponential forgetting

$$pc(k) = \alpha \; pc(k-1) + (1-\alpha) \; pc(k) \tag{12.15}$$

12.3 Methods for the detection of oscillations in closed loops

As a partial solution to closed-loop performance supervision, the automatic detection of different kinds of oscillations is of importance. Oscillations are usually the result of too high controller actions or of nonlinearities like friction, backlash or saturation, e.g. in actuators or the process itself, or of signal quantization and external periodic disturbances.

A relatively simple method by calculating the integral of the absolute error (IAE) between successive zero crossings

$$\text{IAE} = \int_{t_{i-1}}^{t_i} |e(t)| \, dt \tag{12.16}$$

was proposed by [12.5]. If

$$\text{IAE} \geq 2 \, a/\omega_i \tag{12.17}$$

then a load-disturbance is likely to be present, where, e.g. $\omega_i = 2\pi/T_I$. T_I is the integral time of a PID-controller and a is an oscillation amplitude, e.g. 1 %. If the number of the load-detections become high, $n_i \geq n_{lim}$, an oscillation is likely to happen with, e.g. $n_{lim} = 10$.

Another possibility is to calculate the cross-correlation function (ccf) between the process input and output, [12.9], [12.8].

$$R_{uy}(\tau) = \frac{1}{N} \sum_{k=1}^{N} u(k)\, y(k - \tau) \qquad (12.18)$$

Oscillating external disturbances and an unstable loop lead to a phase shift of about π and therefore the ccf becomes an even function. Static friction, on the other side, results in a phase shift of about $\pi/2$, and therefore in an odd ccf.

To distinguish between sinusoidal signals and other periodic signals like rectangular, trapezoidal or triangle oscillations, a Fourier analysis can be performed. Except a peak at the basic frequency ω_1 further peaks at $3\omega_1, 5\omega_1, \ldots$ appear for the oscillations stemming from the nonlinearities like friction or backlash, see Figure 12.5, [12.2]. The calculation of the magnitude of the Fourier analysis may be limited to some distinct frequencies.

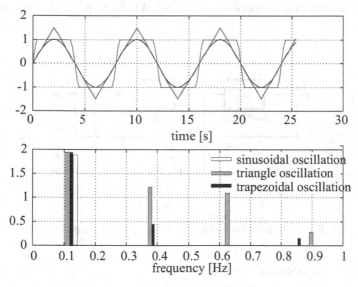

Fig. 12.5. Amplitude spectrum of different oscillations

12.4 Model-based methods for closed-loop supervision

As for a controller with fixed parameters, changes of the closed loop behavior result from process parameter changes and external disturbances. Another way to supervise control loops, is to observe changes of a *process model* and a *disturbance model.* The problem is then identical to that of information gaining for adaptive control. Especially the *parameter-estimation* method for model identification adaptive systems (MIAS), also called self-tuning controllers, can be applied here, e.g. [12.15] and [12.1]. However, then the identification conditions for closed-loop identification with discrete time models have to be satisfied if no external perturbations are used. These conditions are:

- model order and dead time exactly known;
- controller order larger or equal than the difference between model order and dead-time: $\nu \geq m - d$
- enough natural excitation of the process.

Based on the parameter estimates changes of the process are detected as a possible source for deterioration of closed-loop performance. Furtheron, the methods for the supervision of *adaptive control systems* can be applied, see [12.14], [12.15] and [12.6]. Also for adaptive controllers it is difficult to decide if output deviations come from disturbances or process parameter changes.

Now the application of *parity equations* in closed loop is considered. As shown in Figure 12.6 a residual r is generated by using a fixed process model and calculating a polynomial error or an output error, as described in Section 10.1, but now in closed loop.

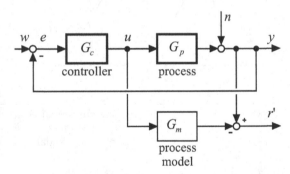

Fig. 12.6. Fault detection of closed loop with parity equations and output error r'

The *output error* then is

$$r'(s) = y_p(s) - y_m(s) = y_p(s) - G_m(s)\, u(s) \qquad (12.19)$$

By introducing (and omitting the Laplace variable s)

$$y_p = G_p\, u + n$$

$$u = \frac{G_c}{1 + G_c G_p}\, (w - n)$$

one obtains

$$r' = (G_p - G_m)\ u + n$$

$$= (G_p - G_m)\ \frac{G_c}{1 + G_c\ G_p}\ (w - n) + n$$

$$= \frac{G_c(G_p - G_m)}{1 + G_cG_p}\ w + \frac{1 + G_cG_m}{1 + G_cG_p}\ n \qquad (12.20)$$

Hence, if the model does not exactly agree with the real process, the residual depends on the inputs w and n. If, however, they agree, $G_p = G_m$, it holds

$$r'(s) = n(s) \qquad (12.21)$$

The residual then depends only on the external disturbance, or for additive faults, at the input and output, compare Figure 10.1 and (10.4):

$$r'(s) = G_m(s)\ f_u(s) + f_y(s) + n(s) \qquad (12.22)$$

Applying the *polynomial* or *equation error*, see (10.5),

$$r(s) = A_m(s)\ y_p(s) - B_m(s)\ u(s) \qquad (12.23)$$

leads with

$$G_p(s) = \frac{B_p(s)}{A_p(s)}\ \text{ and }\ G_c(s) = \frac{Q(s)}{P(s)}$$

$$y_p = \frac{B_p}{A_p}\ u + n$$

$$u = \frac{Q\ A_p}{P\ A_p + Q\ B_p}\ (w - n)$$

to

$$r = \frac{Q(A_m\ B_p - A_p\ B_m)}{P\ A_p + Q\ B_p}\ w + \frac{A_p(P\ A_m + Q\ B_m)}{P\ A_p + Q\ B_p}\ n \qquad (12.24)$$

If process and model agree, $A_p = A_m$, $B_p = B_m$ then

$$r(s) = A_m(s)\ n(s) \qquad (12.25)$$

Also here the residual depends only on external disturbances and for additive process input and output faults it follows

$$r(s) = B_m(s)\ f_u(s) + A_m(s)\ f_y(s) + A_m\ n(s) \qquad (12.26)$$

Hence, for exact agreement of process and model the output and the polynomial residual only depend on the disturbance and process faults in the same way as for the process in open loop, as comparison with (10.4) and (10.6) shows.

This means that the same procedure for fault detection with parity equations based on transfer functions can be applied for linear closed loops as for open loops. Therefore, for small disturbances n especially additive faults in the actuator, process and sensor can be detected, for example, sensor offsets, increased Coulomb friction or backlash in actuators. (The last two lead to direction-dependent residuals, $f_u = $ sign $\dot{u}$ or $f_u = $ sign u, see [12.13]). However, if disturbances n are becoming large, the threshold for the residuals have to be widened and then only large process faults can be detected. However, the residuals do not indicate deviations if the process oscillates because of wrong controller tuning or unstable closed-loop behavior if the model describes the process accurately.

This allows to use the parity equations also to *reconstruct the disturbance n* under the assumption that no faults and changes on the process side show up and there is a good agreement between process model and process. The disturbance signal is then

$$n(s) = r'(s) \quad \text{or} \quad n(s) = \frac{1}{A_m(s)}\, r(s) \tag{12.27}$$

In this way, the kind of disturbance can be observed, e.g. stochastic, deterministic, drift or periodic.

[12.2] applied on-line parameter estimation for the process model and residual generation by parity equations to the temperature control of a steam-heated heat-exchanger.

In summary, parity equations are suitable to be used in closed loop to detect faults in the process or extraordinary disturbances which may be the cause for changed control performance. Combination with methods for oscillation detection, Section 12.3, result in a good overall coverage of faults of closed loops.

Table 12.4. Fault-detection methods for closed loop in normal operation condition (no test signals) for some faults ($\checkmark$ means applicable)

Faults of closed loop	Signal-based methods							Process model-based methods	
	$\|e\|$	$u = u_{lim}$	IAE e	acf e	ccf u, y	Fourier anal.	S_e, S_u	param. estim.	parity equat.
sluggish behavior	$\checkmark$	$\checkmark$	$\checkmark$				$\checkmark$		
oscillations instability	$\checkmark$	$\checkmark$	$\checkmark$	$\checkmark$	$\checkmark$	$\checkmark$	$\checkmark$		
oscillations external disturbance	$\checkmark$	$\checkmark$	$\checkmark$	$\checkmark$	$\checkmark$	$\checkmark$	$\checkmark$		$\checkmark$
friction	$\checkmark$	$\checkmark$	$\checkmark$	$\checkmark$	$\checkmark$	$\checkmark$	$\checkmark$	$\checkmark$	$\checkmark$
backlash	$\checkmark$	$\checkmark$	$\checkmark$	$\checkmark$	$\checkmark$	$\checkmark$	$\checkmark$	$\checkmark$	$\checkmark$
sensor offset		$\checkmark$					$\checkmark$		$\checkmark$
sensor variance	$\checkmark$		$\checkmark$	$\checkmark$			$\checkmark$		$\checkmark$
controller detuned	$\checkmark$	$\checkmark$	$\checkmark$	$\checkmark$			$\checkmark$	$\checkmark$	

The discussion of several methods for fault detection in closed loops has shown, that it is difficult to decide if large deviations in closed loops are caused by large disturbances (stochastic, periodic, non-periodic, drift) or by faults in the controller (parameters), actuator, process dynamics or sensors. None of the discussed methods is able to detect all these faults. But by combining several detection methods a large portion of closed-loop faults is detectable and isolable. Table 12.4 shows the applicability of the discussed fault-detection methods. Depending on the expected faults, suitable methods should be combined to enable a diagnosis of the faults. A special issue on control performance monitoring is [12.10].

12.5 Problems

1) What are typical changes in the performance of a closed loop caused by not appropriately implemented and tuned controllers and faults of components?
2) What kind of faults can be observed for a closed loop, by monitoring the controlled variable and manipulated variable? Which faults can be differentiated?
3) List all methods for the detection of oscillations in closed loops.
4) Design a model-based fault detection with parity equations for a 2nd order linear process and a proportional controller according to Figure 12.6.

13

Fault detection with Principal Component Analysis (PCA)[1]

For large-scale processes, such as chemical plants, the development of model-based fault-detection methods require a considerable and eventually a too high effort. Then data driven analysis methods offer an alternative way. Especially methods based on multivariate statistical analysis and here especially Principal Component Analysis (PCA) and Projection to Latent Structures (PLS) have received attention, see, e.g. [13.9], [13.10], [13.1], [13.3].

These methods are attractive where the available process measurements are highly correlated but only a small number of events (faults) produce unusual patterns, [13.11]. When the process data are highly correlated, the original process data can be projected onto a smaller number of principal components (or latent variables), thus reducing the dimension of the variables. The PCA models are usually basically linear and static and are developed from a process in normal operation. However, they can be expanded to other situations.

13.1 Principal components

The basic idea of principal component analysis is to reduce the dimensionality of a data set considering a large number of interrelated variables, while retaining as much as possible of the variation present in the data set. This is achieved by transforming the measured data to a new set of variables, the *principal components*, which are uncorrelated. These principal components are ordered so that the first few retain most of the variation present in all of the original variables, [13.5], [13.6].

It is assumed that $\mathbf{x}$ is a vector of a large number of m random variables (measurement signals) and that the variance of the random variables and the structure of the covariances or correlations between the m variables are of interest. $\mathbf{x}$ can be input and output variables of a process.

Now, a reduced set of a considerable smaller number $r < m$ of variables is searched which preserve most of the information given in these variances and co-

---

[1] according to a presentation by Falko Haus, [13.4]

variances. This is obtained by a set of orthogonal vectors in the directions where most of the data variation occurs. Then, a few principal components are sufficient to capture the data variance, see Figure 13.1.

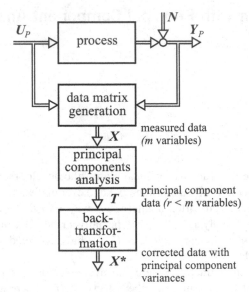

Fig. 13.1. Generation of a set of reduced number of variables **T** with only significant variables out of original $\mathbf{X}^T = [\mathbf{U}_P^T \ \mathbf{Y}_P^T]$ variables by principal component analysis. Back-transformation into original data coordinates $\mathbf{X}^*$ if required.

To illustrate the problem, $N = 8$ measurements of the two variables $x_1(k)$ and $x_2(k)$ are considered, see Figure 13.2. The measurements are represented in two vectors

$$\mathbf{x}_1^T = [x_1(1), x_1(2), \dots, x_1(8)]$$
$$\mathbf{x}_2^T = [x_2(1), x_2(2), \dots, x_2(8)] \tag{13.1}$$

A data matrix

$$\mathbf{X} = [\mathbf{x}_1 \ \mathbf{x}_2] \tag{13.2}$$

is formed, containing all measured data within the coordinate system $(x_2|x_1)$. As Figure 13.2 shows, the measured data fluctuate in both the directions of the coordinates x_1 and x_2.

Now, a transformation to a new coordinate system $(t_2|t_1)$ is searched, in which the variances of the data are maximal in the direction of t_1 and second maximal in the direction of t_2, such forming the first and second principal component. A further condition is that $(t_2|t_1)$ forms an orthogonal coordinate system.

Figure 13.3 shows the results. Hence, the data matrix **X** is transformed into a new data matrix

$$\mathbf{T} = [\mathbf{t}_1 \ \mathbf{t}_2] \tag{13.3}$$

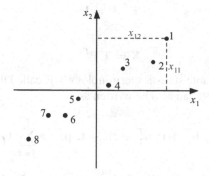

Fig. 13.2. Plot of 8 measurements of two variables

The variance of the data is now very large in the direction of t_1 and much smaller in the direction of t_2. Therefore one principal component t_1 approximates the variance of the data sufficiently and one variable t_2 can be neglected.

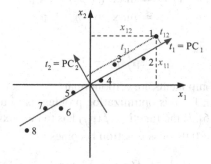

Fig. 13.3. Plot of 8 measurements of two variables with principal components

The general task consists in transforming a data matrix with m variables $x_i(k)$

$$\mathbf{X} = [\mathbf{x}_1, \mathbf{x}_2 \ldots \mathbf{x}_m] \tag{13.4}$$

with N measurements $k = 1, 2, \ldots, N$ into a new data matrix

$$\mathbf{T} = [\mathbf{t}_1, \mathbf{t}_2 \ldots \mathbf{t}_r] \tag{13.5}$$

with also N measurements, but smaller dimension $r < m$. This can be obtained through a transformation matrix $\mathbf{P}$

$$\mathbf{T}_{[N \times r]} = \mathbf{X}_{[N \times m]} \, \mathbf{P}_{[m \times r]} \tag{13.6}$$

$$\mathbf{P} = [\mathbf{p}_1, \mathbf{p}_2 \cdots \mathbf{p}_r] \tag{13.7}$$

As this transformation is a rotation matrix or orthonormal, it holds

$$\mathbf{P}^T \mathbf{P} = \mathbf{I} \tag{13.8}$$

Therefore also

$$\mathbf{X} = \mathbf{T} \mathbf{P}^T \tag{13.9}$$

is valid. In the multivariable statistics terminology $\mathbf{T}$ is called the *score matrix* and $\mathbf{P}$ the *loading matrix*. (13.9) can also be written as

$$\mathbf{X} = \mathbf{t}_1 \, \mathbf{p}_1^T + \mathbf{t}_2 \, \mathbf{p}_2^T + \ldots + \mathbf{t}_r \, \mathbf{p}_r^T = \sum_{j=1}^{r} \mathbf{t}_j \, \mathbf{p}_j^T \tag{13.10}$$

To find now the elements $\mathbf{p}_j$ of the transformation matrix $\mathbf{P}$ which leads to maximal variances a stepwise optimization has to be solved. For each step j with

$$\mathbf{t}_j = \mathbf{X} \, \mathbf{p}_j \tag{13.11}$$

a maximal variance of data $\mathbf{t}_j$ means

$$\begin{aligned} \max \mathbf{t}_j^T \mathbf{t}_j &= \max \left(\mathbf{X} \, \mathbf{p}_j \right)^T \left(\mathbf{X} \, \mathbf{p}_j \right) \\ &= \max \mathbf{p}_j^T \mathbf{X}^T \mathbf{X} \, \mathbf{p}_j \end{aligned} \tag{13.12}$$

under the constraint (13.7)

$$\mathbf{p}_j^T \, \mathbf{p}_j = 1 \tag{13.13}$$

which means that the components are orthonormal.

A standard approach for this optimization problem is to use the method of Lagrange multipliers, [13.6]. If the function $f(\mathbf{p}_j)$ has to be maximized under the condition $g = \mathbf{p}_j^T \, \mathbf{p}_j - 1 = 0$ the loss function becomes

$$V = f(\mathbf{p}_j) - \lambda_j \, g(\mathbf{p}_j) \tag{13.14}$$

where λ_j is the Lagrange multiplier. This leads to

$$V = \mathbf{p}_j^T \mathbf{X}^T \mathbf{X} \, \mathbf{p}_j - \lambda_j (\mathbf{p}_j^T \, \mathbf{p}_j - 1) \tag{13.15}$$

and

$$\frac{dV}{d\mathbf{p}_j} = 2 \mathbf{X}^T \mathbf{X} \, \mathbf{p}_j - 2\lambda_j \, \mathbf{p}_j = 0 \tag{13.16}$$

or

$$\left[\mathbf{X}^T \mathbf{X} - \lambda_j \, \mathbf{I} \right] \mathbf{p}_j = 0 \tag{13.17}$$

With

$$\mathbf{A} = \mathbf{X}^T \mathbf{X} \tag{13.18}$$

it holds

$$\left[\mathbf{A} - \lambda_j \, \mathbf{I} \right] \mathbf{p}_j = 0 \tag{13.19}$$

Hence, this is a classical eigenvalue problem. $\mathbf{A}$ is proportional to the correlation matrix or covariance matrix for zero mean variables of the measured data, λ_j is an eigenvalue and $\mathbf{p}_j$ an eigenvector of the matrix $\mathbf{A}$. From (13.19) it follows

$$\mathbf{p}_j^T \mathbf{A} \mathbf{p}_j = \mathbf{p}_j^T \lambda_j \mathbf{p}_j$$

and inserting in (13.12) yields for the maximal variance

$$\max \mathbf{t}_j^T \mathbf{t}_j = \max \mathbf{p}_j^T \lambda_j \mathbf{p}_j \qquad (13.20)$$

Therefore maximal eigenvalues λ_j give maximal variance for coordinates $\mathbf{t}_j$.

The procedure to determine the transformation matrix $\mathbf{P}$ and the new variable $\mathbf{T}$ is, compare Figure 13.1:

1) Calculation of the "correlation matrix"

$$\mathbf{A} = \mathbf{X}^T \mathbf{X}$$

with zero mean variables $E\{x_i(k)\} = 0$ and $E\{x_i^2(k)\} = 1$, [13.13].

2) Calculation of the eigenvalues λ_j of the matrix $\mathbf{A}$ and the eigenvectors $\mathbf{p}_j$ of

$$(\mathbf{A} - \lambda_j \mathbf{I}) \mathbf{p}_j = 0 \quad j = 1, \ldots, m$$

3) Selection of the largest (most significant) eigenvalues λ_j and corresponding eigenvectors $\mathbf{p}_j$, $j = 1, \ldots, r$, leading to the approximation

$$\mathbf{X}' = \mathbf{t}_1 \mathbf{p}_1^T + \mathbf{t}_2 \mathbf{p}_2^T + \ldots + \mathbf{t}_r \mathbf{p}_r^T$$

4) Determination of the transformation matrix $\mathbf{P}$

$$\mathbf{P} = [\mathbf{p}_1 \ \mathbf{p}_2 \cdots \mathbf{p}_r]$$

5) Calculation of the new data matrix

$$\mathbf{T} = \mathbf{X}' \mathbf{P} \ [\mathbf{t}_1 \ \mathbf{t}_2 \ldots \mathbf{t}_r]$$

with $\mathbf{t}_j = \mathbf{X}' \mathbf{p}_j$. The result is a new data matrix $\mathbf{T}$ with all original data but a reduced number $r < m$ of coordinates or variables, i.e. the principal components. The principal component data matrix $\mathbf{X}'$ carries approximately the same information on the variances of the variables as the original data matrix $\mathbf{X}$.

6) Back-transformation in the original data coordination system yields

$$\mathbf{X}_{[N \times m]}^* = \mathbf{T}_{[N \times r]} \mathbf{P}_{[r \times m]}^T = \mathbf{X}_{[N \times m]} \mathbf{P}_{[m \times r]} \mathbf{P}_{[r \times m]}^T \qquad (13.21)$$

By back-transformation in the original data coordination system one obtains the original variables with only significant variances, i.e. insignificant noise effects have been removed.

The principal component analysis was until now described for steady-state (static) behavior with fluctuating variables. By expanding the data by delayed samples for discrete-time models

$$\mathbf{X} = \left[\mathbf{x}_{1(k)}, \mathbf{x}_{1(k-1)} \cdots \mathbf{x}_{2(k)}, \mathbf{x}_{2(k-1)} \cdots\right] \tag{13.22}$$

or by derivatives for continuous-time modelling

$$\mathbf{X} = [\mathbf{x}_1, \dot{\mathbf{x}}_1 \ldots \mathbf{x}_2, \dot{\mathbf{x}}_2 \ldots] \tag{13.23}$$

PCA can be developed also for dynamic processes.

Example 13.1: Principal component analysis of a linear first order process

Figure 13.4 show a dynamic first order process with transfer function

$$G(s) = \frac{y(s)}{u(s)} = \frac{K}{1 + T s}$$

and normal distributed white output noise $n(t)$. The data matrix is

$$\mathbf{X} = [\mathbf{u} \, \dot{\mathbf{y}} \, \mathbf{y}] = \begin{bmatrix} u(0) & \dot{y}(0) & y(0) \\ u(1) & \dot{y}(0) & y(1) \\ \vdots & \vdots & \vdots \\ u(N) & \dot{y}(N) & y(N) \end{bmatrix}$$

For input excitation a sinusoidal function was chosen

$$u(t) = u_0 \, \sin \omega_1 \, t$$

The used parameters for simulations are

$$\begin{aligned} K &= 1 & T &= 5\,\text{s} \\ u_0 &= 1 & \omega_1 &= 1 \ 1/\text{s} \end{aligned}$$

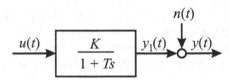

Fig. 13.4. Scheme of a linear process with output noise

Figure 13.5 depicts the (measured) simulated data and shows that the data fluctuate around a plane with fluctuations in all three directions of the coordinate system $[y, \dot{y}, u]$.

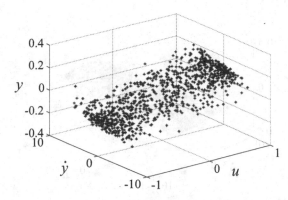

Fig. 13.5. Measured (simulated) data of a first order dynamic process with original data

Table 13.1. Eigenvalues and eigenvectors of data matrix $\mathbf{X}^T \mathbf{X}$

Eigenvalues	$\lambda_1 = 5847$	$\lambda_2 = 491$	$\lambda_3 = 6.37$
Eigenvectors	$\mathbf{p}_1 = \begin{bmatrix} 0.0531 \\ 0.9985 \\ 0.0105 \end{bmatrix}$	$\mathbf{p}_2 = \begin{bmatrix} 0.9940 \\ -0.0539 \\ -0.0956 \end{bmatrix}$	$\mathbf{p}_3 = \begin{bmatrix} -0.0960 \\ -0.0054 \\ 0.9954 \end{bmatrix}$

The calculated eigenvalues and eigenvectors determined with the correlation matrix $\mathbf{X}^T \mathbf{X}$ are given in Table 13.1.

As $\lambda_3 \ll \lambda_1$ and $\lambda_3 \ll \lambda_2$, the eigenvalue λ_3 can be neglected. The transformation matrix becomes

$$\mathbf{P} = [\mathbf{p}_1 \ \mathbf{p}_2] = \begin{bmatrix} 0.0531 & 0.9940 \\ 0.9985 & -0.0539 \\ 0.0105 & 0.0956 \end{bmatrix}$$

Back-transformation with (13.21) results in the data shown in Figure 13.6. Compared to Figure 13.5, the data are now concentrated in one plane indicating a reduction of noise effects.

□

13.2 Fault detection with PCA

For fault detection with the method of principal component analysis different possibilities can be used, Figure 13.7. Direct application of change detection methods on the transformed variable $\mathbf{t}_j$ lead to the mean, compare Chapter 7,

$$\mu_j(k) = E\left\{t_j(k)\right\} \tag{13.24}$$

and variances

$$\sigma_j^2(k) = E\left\{[t_j(k) - \mu_j(k)]^2\right\} \quad j = 1, 2, ..., r \tag{13.25}$$

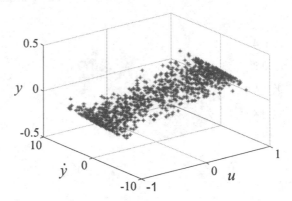

Fig. 13.6. Measured (simulated) data of a first order system with two principal components only

However, the interpretation of the results may be a problem and if new variances occur in variables which have been neglected by the PCA, they will not be detected, [13.7]. Therefore the change detection can also be applied to the back-transformed variables $\mathbf{X}^*$

$$\mu_i(k) = E\left\{x_i^*(k)\right\} \tag{13.26}$$

$$\sigma_i^2(k) = E\left\{[x_i^*(k) - \mu_i(k)]^2\right\} \quad i = 1, 2, ..., m \tag{13.27}$$

Here, observed deviations can be directly linked to the process variables $x_i(k)$.

A further way is to generate residuals between original and back-transformed variables.

$$r_i(k) = x_i(k) - x_i^*(k), \quad i = 1, 2, ..., m \tag{13.28}$$

and determine their mean and variances

$$\begin{aligned} \mu_{ri}(k) &= E\left\{r_i(k)\right\} \\ \sigma_{ri}^2(k) &= E\left\{[r_i(k) - \mu_{ri}(k)]^2\right\} \end{aligned} \tag{13.29}$$

see Figure 13.7c. These measures describe the differences between the present data $x_i(k)$ and their principal component analyzed values $x_i^*(k)$. Significant deflections of the variables $x_i(k)$ can be detected by exceeding a threshold with methods of change detection described in Chapter 7.

The discussed PCA procedure was described for one-block computation, i.e. the data have first to be stored and then to be transformed and residuals to be generated. This is directly applicable for time windowed data. The next example shows how PCA can be applied for fault detection in real-time.

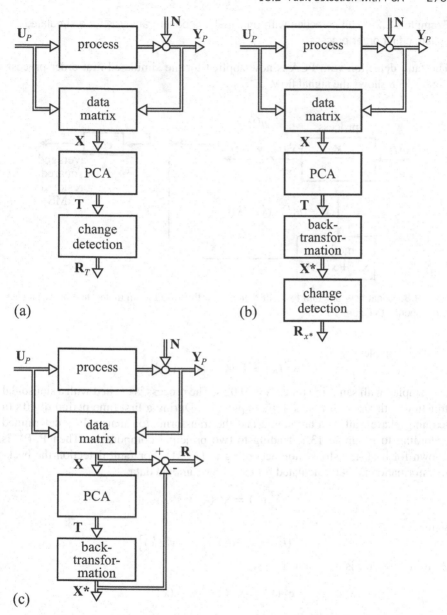

Fig. 13.7. Process fault detection with principal component analysis: (a) residuals by change detection of variables **T**; (b) residuals by change detection of variables **X***; (c) residuals by comparing variables **X** with **X***

Example 13.2: Fault detection with principal component analysis for a simulated dynamic first order process

The fault detection with PCA is now applied for the simulated first order process. Figure 13.8 shows the signal flow.

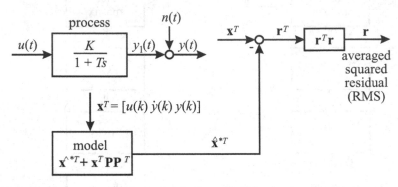

Fig. 13.8. Signal flow diagram for fault detection with PCA of a simulated first order process. (Corresponds to Figure 13.7c)

The variables
$$\mathbf{x}^T(k) = [u(k)\ \dot{y}(k)\ y(k)]$$
are sampled with sampling time $T_0 = 0.02$ s. The process is excited with a sinusoidal function with $\omega_1 = 1\ 1/s$ as in Example 13.1. During a first time period of 10 s (a learning phase), all data are stored and the transformation matrix $\mathbf{P}$ is determined according to Example 13.1, leading to two principal components. Then $\mathbf{P}\ \mathbf{P}^T$ is known for back-transformation according to (13.21). For fault detection the back-transformation is then calculated for each new sampled data

$$\mathbf{x}^{*T}(k) = \mathbf{x}^T(k)\ \mathbf{P}\ \mathbf{P}^T$$

with

$$\mathbf{x}^{*T}(k) = [u^*(k)\ \dot{y}^*(k)\ y^*(k)]$$

A residual vector is determined

$$\mathbf{r}^T(k) = \mathbf{x}^T(k) - \mathbf{x}^{*T}(k)$$

and a squared residual quantity

$$r'(k) = \mathbf{r}^T(k)\ \mathbf{r}(k) = \sum_{i=1}^{3} r_i^2(k)$$

then shows deviations caused by faults. Figure 13.9 shows simulations for changes (faults) of the time constant, the gain and output offset during the time period $t_1 = 10$ s to $t_2 = 20$ s.

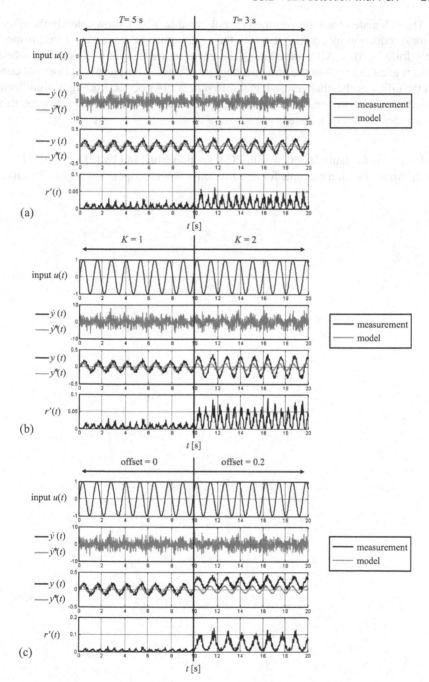

Fig. 13.9. Time history of signals without faults for $t < 10\,\mathrm{s}$ and with faults for $t > 10\,\mathrm{s}$: (a) change of time constant $T = 5\,\mathrm{s} \;\rightarrow\; 3\,\mathrm{s}$; (b) change of gain $K = 1 \;\rightarrow\; 2$; (c) output offset $y_{\mathrm{off}} = 0 \;\rightarrow\; 0.2$

The calculated back-transformed output variable $y^*(k)$ shows clearly the effect of noise reduction by comparison with the signal during the learning phase respectively for $t < 10$ s. After introducing the changes, the output $y(k)$ of the process deviates accordingly. But the principal component-based output $y^*(k)$ does not contain the effect of the changes such that differences $r(k)$ occur, leading to significant deviations of the squared residual quantity $r'(k)$. However, based on this quantity the changes cannot be isolated.

□

Examples for fault detection with PCA are published in [13.12], see also [13.8], [13.2]. An application example for an automotive wheel suspension is given in Chapter 22.

14

Comparison and combination of fault-detection methods

The comparison of the different methods for fault detection is not easily performed because the final practical results depend on many aspects, like kind of the process, kind of disturbances, open or closed loop, nonlinearities, experience of the designer, etc. For some classes of processes the structure and at least some parameters are relatively well known, like for electrical, mechanical or hydraulic processes. For other classes only rough models are available, as, e.g. for many process industries (cement, chemical, mineral, metal and biochemical processing). In addition, many processes change their behavior continuously because of different operating points, wear and ageing, such that all methods with constant models are problematic.

However, under specified conditions comparisons are possible, see, e.g. [14.3] and [14.1].

14.1 Assumptions of model-based fault detection

Model-based fault-detection methods use residuals which indicate changes between the process and the model. One general assumption is, that the residuals are changed significantly so that a detection is possible with regard to the mostly inherent stochastic character. This means that the offset of the residuals after the appearance of a fault is large enough and lasts long enough to be detectable. This may be called a "significant change". The various fault-detection methods are characterized by their underlying assumptions summarized in Table 14.1. All considered model-based methods require of course that the process can be described by a mathematical model. As there is almost never an exact agreement between the process and its model, the kind and size of model discrepancies are of primary interest. In the following an attempt is made to summarize some special features of the different methods:

a) Parity equations

- model structure and parameters must be known and must fit the process well;
- especially suitable for additive faults;

Table 14.1. Qualitative comparison of properties of different fault-detection methods for linear processes. SISO: single-input single-output. MIMO: multi-input multi-output

criteria	parity equations	state estimation		parameter estimation
		state observer	output observer	
assumptions				
model structure	exactly known	exactly known		known
model parameters	known, constant	known, constant		unknown, time-varying
disturbance models for unknown inputs	exactly known	exactly known		exactly known
noise	small	small		medium
stability of detection scheme	no problem	depends on design	no problem	no problem
excitation by the input	additive faults: no multiplicative faults: yes	additive faults: no multiplicative faults: yes		additive faults: no multiplicative faults: yes
detectable faults				
abrupt	yes	yes		yes
drift	yes	yes		yes
incipient	yes	yes		yes
single faults	yes	yes		yes
multiple faults	SISO: no MIMO: yes	SISO: no MIMO: yes		SISO: yes MIMO: yes
fault isolation	MIMO: yes	MIMO: yes		SISO: yes MIMO: yes
additive	yes	yes		yes
multiplicative	no	no		yes
general				
robustness parameter changes	problematic	problematic		unproblematic
nonlinear processes	many classes possible	limited		many classes possible
static processes	yes	no		straightforward
computational effort	small / medium	medium		medium / larger
closed loop	yes	yes		yes, external excitation

- in general multi-outputs required;
- very fast reaction after sudden faults;
- online real-time application possible for fast processes;
- computationally more intense than parity equations;
- no input signal changes required for additive faults (but then, some parameter changes are not detectable);
- some faults to be detected can be small, (e.g. additive faults and gain), some must be large, (e.g. time constants).

b) State observers, state estimation

- the model structure including parameters must be known rather accurately;
- especially suitable for additive faults;
- in general multi-outputs required;
- very fast reaction after sudden faults;
- on-line real-time application possible for fast processes, if not too many observers required;
- no input signal changes required for additive faults (but then some parameter changes, e.g. time constants, not detectable);
- some faults to be detected can be small, (e.g. additive faults and gain), some must be large;
- observers have very similar properties as parity equations.

c) Parameter estimation

- model structure to be known;
- especially suitable for multiplicative faults;
- also additive faults on the input and output signal can be detected;
- several parameter changes are uniquely detectable for one input and one output measurement;
- very small changes are detectable, which includes the detection of slowly developing as well as fast developing faults;
- on-line real-time application possible, even for fast processes;
- input excitation required for dynamic process parameters.

Parity equations and observer-based methods have partially almost identical properties, but parity equations are much simpler to design, to implement and to understand. They can also be easily expanded to nonlinear processes. Parity equations and observers are well suited for additive faults, but are not in general well suited for multiplicative faults. For multiplicative, i.e. parametric faults, parameter estimation is best suited. Also some additive faults can be modelled as unknown parameters. However, parameter estimation with dynamic models needs, in general, a dynamic input excitation. For static processes only measurements for different operation points are required.

A further essential difference is that parity equations and observer-based methods need more than one output measurement to detect and isolate several faults, but that for parameter estimation one input and one output are sufficient to detect and diagnose different faults.

In the case of abrupt faults state estimation and parity equations react faster than parameter estimation for the basic methods described above. This is due to the fact that parameter estimation is intended to estimate constant values and to remove the influence of disturbances with time. If, however, parameter estimation is designed for time varying parameters by a forgetting factor or by including a dynamic state model for the parameters (resulting in a Kalman-filter type estimator) it is able to follow rapidly abrupt parameter changes on cost of disturbance rejection. Also state estimation can be designed for better disturbance elimination on cost of rapid state changes following. Hence, the property to follow abrupt changes rapidly depends for both parameter and state estimation on the design.

Of course, a large influence on all methods have the *assumptions on fault modelling*. As considered in Chapter 5, the modelling of faults needs the understanding of the many kinds of real physical faults and their mapping to the used mathematical models and fault-detection methods, compare the remarks on fault modelling at the end of Section 5.2.3, where it was stated that many faults are of multiplicative nature and that additive faults are applicable to some sensor faults and actuator faults, see also the examples in Part V.

14.2 Suitability of model-based fault-detection methods

For *single-input single-output processes* (SISO) the results can be summarized as follows. As *parameter estimation* is especially suitable for multiplicative faults, this detection method can be primarily recommended for corresponding faults in the processes and faults which change the dynamics of actuators and sensors. But also additive faults at the input and output can be included in the parameter estimation, as for static actuator and sensor faults. *State estimation* and *parity equations* have their advantages for additive faults and are therefore feasible for corresponding faults in the sensors, actuators and in some cases for processes.

State observers for SISO processes can be applied if the faults map into the observed state variables, like for leak detection of pipelines. However, the applicability of *output observers* and *parity equations* is rather limited for SISO systems as only one residual can be generated which does not allow to isolate different faults.

For *single-input multi-output processes* (SIMO) and *multi-input multi-output processes* (MIMO), compare Figure 5.2, the analytical redundancy between the measured inputs and outputs increases. This is advantageous for parity equations and output observers, because it allows to generate different residuals which can be made independent on certain inputs and faults, thus enabling fault isolation. However, on the other side, it is more difficult to obtain precise MIMO process models with all cross-couplings.

14.3 Combination of different fault-detection methods

The preceding discussions show that parameter estimation on the one side and state estimation and parity equations on the other side show advantages and disadvantages with regard to the detection of the various types of faults. Therefore, if all faults should be detectable, different detection methods should be combined properly in order to mainly make use of their advantages. As in most cases the model parameters are unknown anyhow, it is quite natural to apply parameter estimation first. Then following combinations of model-based detection methods result, [14.2], [14.3].

1. Sequential parameter estimation and parity equations

- parameter estimation to obtain the model;
- parity equations for change detection with less computations;
- parameter estimation (on request) for deep fault diagnosis.

2. Sequential parameter and state estimation

- parameter estimation to obtain the model;
- state estimation for fast change detection;
- parameter estimation (on request) for deep fault diagnosis.

3. Parallel parameter and state estimation

- for multiplicative and additive faults;
- depending on input excitation.

The way of combination depends very much on the process, the faults to be detected and the allowable computational effort.

In some cases also the integration of process model-based and signal model-based detection methods give a good overall information.

4. Parameter estimation and vibration analysis

- parameter estimation for parameter mapping faults;
- vibration analysis for other type of faults like unbalance, knocking, chattering.
  (This is especially attractive for rotating machines).

By this way of combining suitable detection methods the most relevant analytical symptoms can be generated and used for integrated fault diagnosis.

Figure 14.1 shows a scheme for the combination 2. Examples are shown in Part V.

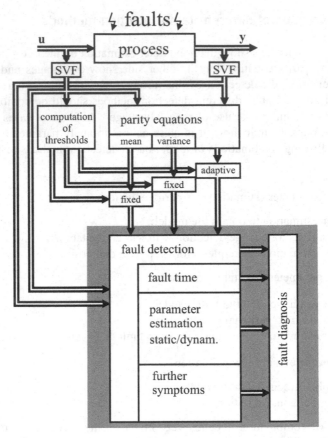

Fig. 14.1. Combination of parameter estimation and parity equations: parity equations are used to detect changes in the process on-line and real-time. After the residuals exceed thresholds, parameter estimation is applied to gain more information on the process and to allow a deeper fault diagnosis (online, close to real-time)(SVF: state variable filter, if required)

Part III

Fault-Diagnosis Methods

15

Diagnosis procedures and problems

15.1 Introduction to fault diagnosis

The fault diagnosis task consists of the determination of the fault type with as many details as possible such as the fault size, location and time of detection. The diagnostics procedure is based on the observed analytical and heuristic symptoms and the heuristic knowledge of the process, as shown in Figure 2.7.

Figure 15.1 summarizes the single steps as well for automatically measured variables as for human observation. In both cases a feature extraction and a detection of changes the normal or nominal situation takes place. Analytical and heuristic symptoms must then be brought in an unified symptom representation in order to perform the diagnosis.

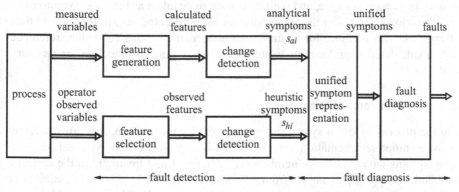

Fig. 15.1. General scheme for fault detection and fault diagnosis with analytical and heuristic knowledge

Note that *features* were described in Section 2.5 as extracted values from signal or process models, describing the status of the process, (e.g. parameters, state variables, parity equation errors or residuals) and that *symptoms* are unusual changes of the features from its normal or nominal values. In a fault-free case the symptoms are zero.

The inputs to the knowledge-based diagnosis procedure are all available symptoms as facts and the fault-relevant knowledge about the process. In more detail these are, compare Section 2.1:

a) Analytical symptoms

The analytical symptoms s_{ai} are the results of the limit checking of measurable signals, signal or process-model fault-detection methods and of change-detection methods, as described in Section 2.1 and Chapter 7.

b) Heuristic symptoms

Heuristic symptoms s_{hi} are the observations of the operating personnel in the form of acoustic noise, oscillations or optical impressions like colors or smoke, obtained by inspection. These empirical facts can usually only be represented in form of qualitative measures, e.g. as linguistic expressions like "little", "medium" or "much".

c) Process history and fault statistics

A third category of facts depends on the general status, based on the history (past life) of the process. This process history includes the past information of running time, load measures, last maintenance or repair. If fault statistics exist, (e.g. from "statistical process control") they describe the frequency of certain faults for the same or similar processes. Depending on the quality of these measures, they can be used as analytical or heuristic symptoms. However, the information on the process history in general is vague, and their facts have to be taken as heuristic symptoms.

The knowledge about the symptoms can be represented, *e.g*, in the form of data strings and can include, for example, number, name, numerical value, reference value, calculated confidence or membership value, time of detection, explanatory text, [15.1].

d) Unified symptom representation

For the processing of all symptoms in the inference mechanism, it is advantageous to use a unified representation. One possibility is to present the analytic and heuristic s_i symptoms with confidence numbers $0 \leq c(s_i) \leq 1$ and treatment in the sense of probabilistic approaches known from reliability theory, [15.1]. Another possibility is the representation as membership functions $0 \leq \mu(s_i) \leq 1$ of fuzzy sets, [15.4].

By these kinds of fuzzy sets and corresponding membership functions, all the analytic and heuristic symptoms can be represented in a unified way within the range $0 \leq \mu(s_i) \leq 1$. These integrated symptom representations are then the inputs for the diagnosis procedure, Figure 2.4 and Figure 15.1. Diagnosis knowledge-representation including a priori knowledge and various symptom representations is treated in Section 15.2.1.

e) Fault-symptom relationships

The propagation of faults to observable symptoms follows in general physical cause–effect relationships. Figure 15.2a shows that a fault in general influences events as intermediate steps, which then influence the measurable or observable symptoms, both by internal physical properties. The underlying physical laws, however, are mostly not known in analytical form, or too complicated for calculations. The fault diagnosis proceeds the reverse way. It has to conclude from the observed symptoms to the faults, Figure 15.2b. This implies the inversion of the causality. One cannot expect to reconstruct the fault-symptom chains solely from measured data, because the causality is not reversible or the reversibility is ambiguous, [15.2]. The intermediate events between faults and symptoms are not always visible from the symptoms behavior. Therefore, mostly structured knowledge has to be included, known from inspection of the process faulty behavior.

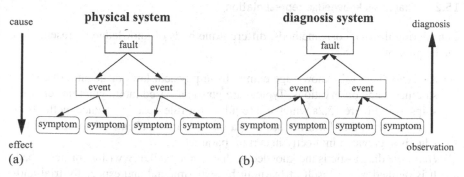

Fig. 15.2. Fault-symptom relationship: (a) physical system: from faults to symptoms; (b) diagnosis system: from symptoms to faults

If no information is available on the fault-symptom causalities, experimentally trained *classification methods* can be applied for fault diagnosis, see Chapter 16. This leads to an unstructured knowledge base. If the fault symptom causalities can be expressed in the form of if-then rules *reasoning* or *inference methods* are applicable. This is considered in Chapter 17. Figure 15.3 gives a survey on the diagnosis methods treated in the next two chapters.

15.2 Problems of fault diagnosis[1]

The main challenges of fault diagnosis are given by the knowledge representation, the introduction of prior knowledge, the typical symptom distributions and the data size and representation. These problems will be briefly introduced in the following.

---

[1] follows Dominik Füssel, [15.2]

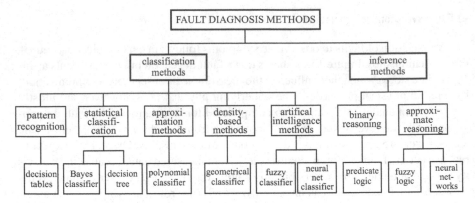

Fig. 15.3. Survey of fault-diagnosis methods

15.2.1 Diagnosis knowledge representation

Considering the diagnosis methods, different methods of knowledge representation can be present:

1) *Analytic* diagnostic knowledge comes from physical laws or quantitative measurements and observations. Typical are physical-based fault-symptom relationships as in Figure 15.2a which are required for rule-based inference methods or measurements that form the basis of a classification scheme where the knowledge is aggregated indirectly in certain parameters.

2) *Heuristic* diagnostic is the knowledge that is not explicitly written or described. It is defined as the result of learning by experimental and especially trial-and-error methods. It comes from the experience of operators and system engineers. Possible forms of knowledge representation are (following [15.1]):

   - Rules
   - Frames (object-oriented representations)
   - Predicate logics
   - Directed graphs (especially networks and tree structures)

   The heuristic knowledge is frequently expressed by inference mechanisms like the *forward* and *backward reasoning*.

15.2.2 Prior knowledge

In many applications, prior knowledge is present. It can come from experience or physical understanding of the processes. In some cases, it can even be quantitative knowledge of certain relationships. This can especially be advantageous within model-based fault detection and diagnosis. The reason can be found to be that physical parameters $\mathbf{p}_{phys}$ have a functional relationship to the model parameters, see Chapters 5.2, 9.4 and 23.4:

$$\theta_{model} = f(\mathbf{p}_{phys}) \tag{15.1}$$

The influence of faults to the physical parameters is usually known. With the knowledge of (15.1), one can utilize this information. Sometimes, only characteristics of the relationship (such as monotony) is known. An example for such a relation is the rotor resistance of a DC motor that depends on the physical variable "temperature". While the temperature is not a parameter of the motor model, one can nevertheless estimate its influence on the resistance of the motor wiring. Physical knowledge tells then that an overheating of the motor is indicated by a high resistance parameter.

Other prior knowledge might be more general such as information about similar or independent faults that can be used to structure the diagnosis system. In any case, all usable prior knowledge should be utilized.

15.2.3 Typical statistical symptom distributions

If a diagnosis system is to be built from experimental data, one has to consider the typical statistical data distributions that can occur. Such diagnosis systems rely on two types of symptoms:

1) Estimated model parameter changes
2) Deviations of model outputs and measured signals (residuals)

Faults that influence a physical parameter will be reflected by changes of the model parameters. Chapter 23.4 shows that the relationships (15.1) are frequently multi-linear functions of the kind

$$\theta_{model} = \sum_j c_j \prod_i p_{phys_{i,j}} \tag{15.2}$$

or can be approximated in certain regions as such. c_j denotes constant parameters and θ_{model}, p_{phys} the model and physical parameters respectively. Change of physical parameters lead to multiplicative deviations of the model output signals from the measured signals. The conclusion from this knowledge is the typical symptom distribution arising from multiplicative faults. The distributions can be favorable or unfavorable for some classification approaches, [15.2].

15.2.4 Data size

Another typical problem of the experimental fault diagnosis in form of classification methods is the size of available data sets. To have a statistically sufficient data base, one would have to evaluate a large number of data. That, however, is not only tedious to do: sometimes, faulty systems that could be used for measurements are simply not available or an artificial fault cannot be introduced for other reasons. A faulty system might be too dangerous to operate or too expensive to get. Numerical simulations will always only be a more or less adequate substitute because one can hardly simulate all effects of real faults. This problem is especially severe as the data is often highly-dimensional. This leads to extremely sparse data sets that are difficult to handle ("curse of dimensionality"). The diagnosis for a DC motor, for instance, is built from a 21-dimensional symptom space, see [15.2]. Algorithms that automatically build diagnosis systems must be able to handle that problem of sparse data.

15.2.5 Symptom representation

To derive a diagnosis system that is able to cope with different problems, one must reach a common symptom representation. Generally, two different sorts of information can be involved, as discussed in Section 15.1: *analytical information* and *operator-observed information*. This information can constitute as four different data types (following [15.5]):

1) Binary variables;
2) Multi valued variables;
3) Interval scaled variables;
4) Metric variables.

A *multi-valued variable* is given if its different states can be coded by characters. One has to distinguish if the characters constitute an order (like 1,2,3 or small, medium, large) or just different states must be named. The first kind can be translated into metric variables. The latter allows to replace multi valued variables by a set of binary variables. *Interval-scaled variables* contain a statement that refers to a certain interval (like 50%), but generally resemble *metric variables*. Since in practice metric variables can also not assume any value, there is practically no difference between these two variable types.

Binary variables, however, are different. They are often coded with a 0/1 scheme and then treated like metric variables. This is appropriate as long as not a too large number of binary variables hinders numerical methods. Some algorithms (such as clustering methods) can be influenced by such artificial variables and become unstable. Multi valued variables without ordering should always first be coded binary and then metrically. The difficulty of such conversions of data types is that the methods dealing with data are usually designed for one representation only and will deteriorate if such artificial numbers are involved.

The solution to this problem is the representation of all variables as *fuzzy sets* as mentioned in the last section. Figure 15.4 shows some examples for cases where the symptom s_i, either increases or decreases. Figure 15.4a, b has the advantage that only one membership function has to be processed, in contrast to Figure 15.4b where five membership functions for linguistically expressed changes have to be processed. This allows to process both, input from human operators as well as numerical data of different kinds, [15.3]. The disadvantage of the fuzzy representation is that the translation into fuzzy sets can yield an information loss through the projection into so-called fuzzy membership values. Care must therefore be taken to choose appropriate functions for the fuzzy sets. This issue will be addressed in Chapter 17.

15.3 Problems

1) Take a DC motor as an example and classify following symptoms into "analytical" or "heuristic": increase of armature resistance, vibrations, yellow or blue brush fire, smoke, reduced torque, reduced speed, high temperature.

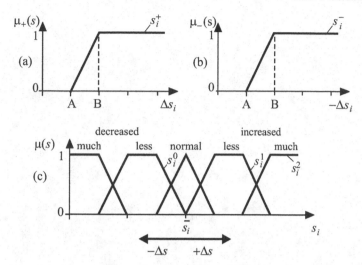

Fig. 15.4. Example for the unified symptom representation as membership functions in form of fuzzy sets

2) Use the symptoms of 1) and develop a fault-symptom tree for following faults: too large current, brushes without solid contact, plugged cooling channel, bearing fault.
3) Which analytical and heuristic diagnostic knowledge is available for an proportional acting electromagnetic solenoid actuator?
4) State observable symptoms for faults of a thermocouple as temperature sensor.

Fault diagnosis with classification methods[1]

The task of the diagnosis system is to separate n_f different faults

$$F_j , j \in \{1, \ldots, n_f\} \tag{16.1}$$

using n_s symptoms

$$s_i , i \in \{1, \ldots, n_s\} \tag{16.2}$$

The faults are combined into a fault vector

$$\mathbf{F} = [F_1 F_2 \ldots F_{nf}] \tag{16.3}$$

and the symptoms into a symptom vector

$$\mathbf{s} = [s_1 s_2 \ldots s_{ns}]^T \tag{16.4}$$

Nearly all methods compute a fault measure f_i for each fault class F_j. The decision about the most probable fault is then given by the fault with the maximum value f_j. In reality, however, not only the largest f_j is of relevance: unclear situations and measurement noise can create high values of multiple fault measures f_j. This can indicate uncertain decisions. Hence, all values of f_j are finally important for the diagnosis statement of the system.

In this and the next chapter two main classes of fault diagnosis are considered, the *classification methods* without structural knowledge and the *inference methods* with structural knowledge.

16.1 Simple pattern classification methods

This section will briefly review simple classification methods that are nevertheless common for fault diagnosis applications. The understanding of these methods lays ground for the more advanced methods of the following sections. If no structural

---

[1] compiled by Dominik Füssel, [16.7]

knowledge is available for the relation between the symptoms and the faults, classification or pattern recognition methods can be applied. Figure 16.1 shows that reference symptom vectors s_{ref} are determined for the faults F_j experimentally by learning or training. Comparison with the observed symptom s then determines the faults by *classification*.

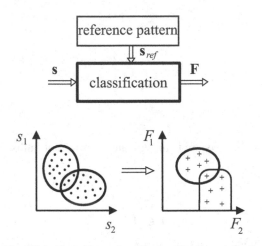

Fig. 16.1. Fault diagnosis with classification methods

Figure 16.2 summarizes the problem statement in more detail. The classification methods are used to represent the *diagnosis functional mapping* from the symptom space to the space s of *fault measures* f_j. The nomenclature t_j for the desired binary value of the f_j has been chosen following the equivalent term *target value* that comes from neural network learning terminology. There, the *desired output* of the network is the binary target value that the network needs to aim at producing. In Section 16.6 neural networks are discussed as diagnosis methods. In that context, the fault indicators become the target values of the network training.

The classifier needs to map the relationship between the computed symptoms (calculated from measurements) and the fault indicators f_j with binary desired values t_j. Since the classification methods usually compute a value between 0 and 1, an additional *maximum operation* is later needed to determine the fault diagnosis result.

The most commonly used classification methods are summarized in Figure 16.3. Historically, the statistical methods came first, followed by the density-based methods and general approximation approaches. The artificial intelligence methods were historically the latest to be developed for diagnosis problems.

16.2 Bayes Classification

The most well-known classification scheme is given by the so-called Bayes classification. The approach is based on reasonable assumptions about the statistical dis-

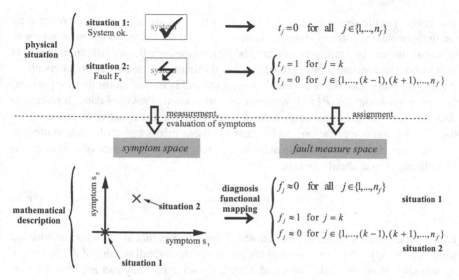

Fig. 16.2. General problem of fault diagnosis with pattern classification methods

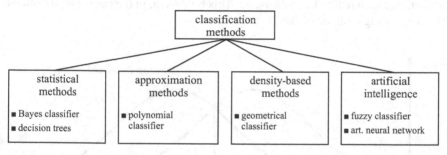

Fig. 16.3. Pattern classification methods

tribution of the symptoms. A common procedure is to assume Gaussian probability density functions, [16.8]:

$$p(\mathbf{s}) = \frac{1}{(2\pi)^{n_s/2}|\Sigma|^{1/2}} \exp\left(-\frac{1}{2}(\mathbf{s} - \mathbf{s}_0)^T \Sigma^{-1}(\mathbf{s} - \mathbf{s}_0)\right) \qquad (16.5)$$

This function is determined by its constants, the covariance matrix Σ and the centers $\mathbf{s}_0$. The common procedure is to determine maximum likelihood estimations for these parameters. The centers, for instance, are given by the mean values of the reference data. Other approaches use recursive parameter estimation methods or Bayes inference. Building a classification system from the probability density estimations requires the class specific densities. It can be shown that a minimum of wrong decisions is achieved if the maximum of the $p(F_j|\mathbf{s})$ is selected. This posterior probability can be calculated with the help of the Bayes-Law:

$$f_j(\mathbf{s}) = p(F_j|\mathbf{s}) = \frac{p(\mathbf{s}|F_j)P(F_j)}{p(\mathbf{s})}, \qquad (16.6)$$

The class specific densities $p(s|F_j)$ can be estimated from labelled reference data using those data points belonging to fault F_j, Figure 16.4. One of the methods mentioned above can be employed to gain $\hat{p}(s|F_j)$. Since only the maximum of the f_j is of interest, the denominator in (16.6) is not significant, because it does not depend on F_j. It only plays the role of a normalizing factor and is irrelevant for the comparison. The prior probabilities $P(F_j)$, however, are important. Provided enough reference data is available, they can be estimated from their frequency of occurrence in the data set. In many applications, unfortunately, these priors cannot be determined. If the reference data is created from experiments where the occurrence of the faults can be influenced, one should assume

$$P(F_j) = \frac{1}{n_f} \qquad (16.7)$$

unless experience suggests a better choice. This assumes that all faults occur with the same probability. The importance of the priors for the overall quality of the diagnosis system should not be underestimated. Carefully selected, they can improve the performance of a diagnosis system substantially. The common assumption of Gaussian distribution can in reality be problematic. This holds, e.g. in the case of distributions with overlapping fault areas, [16.7].

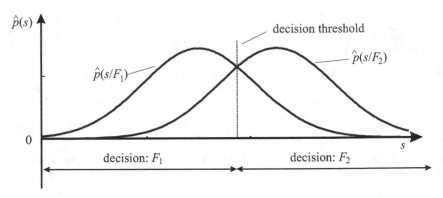

Fig. 16.4. Example for two class specific densities $p(s|F_j)$ (the statistical distribution of one symptom s belongs to two different faults F_1 and F_2)

A more realizable case is given if a histogram is used to generate a nonparametric estimation for $p(s|F_j)$. A necessary condition is then again the availability of sufficient amounts of data. For situations with a lack of enough data points, one can try using histogram methods with variable-size grids. This, again, is very similar to geometric classifiers that are subject of the following section.

16.3 Geometric classifiers

Geometric classifiers determine the class membership of a data point from its distance to reference data points. These reference data points are characterized by their symptom values $s_{ref,i}$ and the known class assignment (class F_1 or class F_2, etc.).

The simplest and most famous approach is the *nearest neighbor classification* that evaluates the Euclidean distance. If one wants to determine the class of the data point s_j, *for instance*, one has to compare the distances from this data point to all reference data points and determine the minimum:

$$\min_i (d_i) = \min_i \left(\sqrt{\|s_j - s_{ref,i}\|^2} \right) \quad i \in \{1, \ldots, n_{ref}\} \tag{16.8}$$

where n_{ref} denotes the number of reference data points. The class of the one $s_{ref,min}$ being closest to s_j is then taken as the class of s_j. Figure 16.5 pictures the concept.

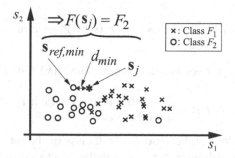

Fig. 16.5. Example of nearest-neighbor classification. The minimum distance to the point s_j is found to be to a reference data point belonging to F_2. Therefore, the result of the classification of s_j is F_2

The drawbacks of this method are obvious if for instance in the case of overlapping class regions only a few reference points exist. The resulting decision boundary is not smooth, hence, probably sub-optimal. A regularization can then be achieved if not the distance to just one reference point is evaluated, but rather k reference points s_{ref} are used. The k nearest neighbor approach utilizes a voting of the k closest data points. It is apparent that this method comes close to a local parameter-free probability density estimation. The class that most s_{ref} belong to is identical to the one which has the highest relative frequency of occurrence in the considered hypersphere centered around s. The size of the hypersphere is governed by the reference data density at s.

Commonly cited as a disadvantage of nearest neighbor classification is the need to store all reference data points. This problem is relieved in the view of modern computing and storage capabilities. Furthermore, there are techniques to reduce the number of necessary points to be stored by an intelligent selection strategy. One example is the so-called condensed nearest neighbor approach, [16.9].

16.4 Polynomial classification

The polynomial classification uses a special functional approximation for the posterior probabilities of the classes instead of the Gaussian functions assumed for the Bayes classification scheme. Employed are polynomials

$$\hat{p}(s|F_j) = f_j = a_{j,0} + a_{j,1}s_1 + a_{j,2}s_2 + \ldots + a_{j,n+1}s_1s_2 + \ldots \qquad (16.9)$$

defined by their parameters $\mathbf{a}_j = [a_{j,0} \, a_{j,1} \, \ldots \, a_{j,n_p}]^T$ with the maximal number of polynomial terms given by

$$n_{p,max} = \binom{n_s + o}{o} \qquad (16.10)$$

with o being the highest polynomial order. The coefficients are determined using a least squares approach with a loss function V_j:

$$V_j = \sum_{k=1}^{N}(t_j(\mathbf{s}(k)) - f_j(\mathbf{s}(k)))^2 \rightarrow \min. \qquad (16.11)$$

where k runs over all N reference data points. The $t_j(k)$ follow the usual 0/1 notation, i.e.

$$t_j(\mathbf{s}(k)) = \begin{cases} 1 \text{ if } \mathbf{s}(k) \text{ belongs to fault } F_j \\ 0 \text{ otherwise} \end{cases} \qquad (16.12)$$

This optimization is logical, because the fault class of the reference data points is known. Hence, the probabilities, here equal to the target values, can only be 0 or 1.

The polynomial classifier is used in the following way:

- evaluate all polynomials (one per fault class);
- find the maximum out of the computed f_j: the data point belongs to the corresponding polynomial.

Figure 16.6 shows a typical decision boundary for a two-class problem. The decision boundary is given be the line of equal polynomial values $\hat{p}(s|F_1) = f_1(s)$ and $\hat{p}(s|F_2) = f_2(\hat{s})$. The decision boundary line itself is in general composed of polynomials of the same order like the f_j.

The parameters $\mathbf{a}_j$ of the polynomials minimizing (16.11) can directly be solved for because the error measure is linear w.r.t. the parameters:

$$\mathbf{a}_j = \left(\mathbf{\Psi}^T\mathbf{\Psi}\right)^{-1}\mathbf{\Psi}^T\mathbf{t}_j \qquad (16.13)$$

with

$$\mathbf{t}_j = [t_j(\mathbf{s}(1)) \, t_j(\mathbf{s}(2)) \, \ldots \, t_j(\mathbf{s}(N))]^T \qquad (16.14)$$

and

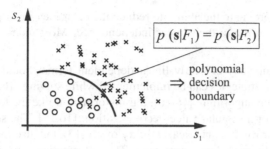

Fig. 16.6. Example of the decision boundary of a polynomial classifier

$$\Psi = \begin{pmatrix} 1 & s_1(1) & s_2(1) & \dots & s_1(1)s_2(1) & \dots \\ 1 & s_1(2) & s_2(2) & \dots & & \\ \vdots & \vdots & \vdots & \vdots & & \vdots \\ 1 & s_1(N) & s_2(N) & \dots & & \end{pmatrix} \qquad (16.15)$$

The simplest case is given with $o = 1$. The resulting classification system is popular and known under different names. One is the *linear discriminant technique* frequently used in statistics, especially in medical diagnosis applications. Such a system can solve linearly separable problems. These are all those problems for which the classes can be separated with a line (in 2-D), or a plane or hyper-plane for multidimensional problems respectively. This plane is retrieved from the linear polynomial classifier for two classes by setting (16.9) equal to 0.5. For the case of Gaussian distributions with diagonal covariance matrices, different approaches will lead to the identical result, regardless whether they are based on pseudo-inverse solutions like (16.13), nonlinear optimization techniques like the gradient descent method or perceptron learning. The achieved solution coincides with the optimal decision boundary given by the Bayes-law.

For the general polynomial classifier, the biggest problem is to choose the appropriate order o and select the correct polynomial terms. The polynomial should usually not be complete. Especially with $o > 1$ and larger n_s, a complete polynomial would in most applications be unnecessarily large. The following has to be considered in choosing the polynomial:

- a large number of terms can easily create overfitting with bad generalization behavior;
- a small number of terms might lead to systems not flexible enough for the distinction of the classes;
- the wrong selection of the a_j can lead to numerical problems. Linearly dependent column vectors of Ψ make a solution with (16.13) infeasible.

Typical polynomial orders are 1 (which equals the linear discriminant techniques) and $o = 2$ or $o = 3$. Higher orders are not necessary in most applications. The selection of only a subset of all possible polynomial terms can be based on two principles:

1) Selection according to the ability to reduce the error measure V_j from (16.11);
2) Selection according to their linear independence. Most independent terms are selected first.

A removal of completely or nearly linear dependent terms should be done in any case. The selection strategies can be implemented while solving (16.11) for instance with a Gauss-Jordan algorithm, [16.16], a transformation like the Orthogonal Least Squares, [16.14], or a singular value decomposition, [16.15]. The selection process does not only improve the numerical stability of (16.13), but also indicates inappropriately chosen or meaningless symptoms **s**. This information can be of great help in designing the symptoms or reducing the complexity of the system.

16.5 Decision trees

Originating in the social sciences are different types of decision trees that are used to classify data. Similar approaches are also common in the classification of botanical species. The system basically relies on a series of questions that have to be answered and depending on the answer the next question narrows the species more and more until the exact plant is determined. Typical for biological problems are binary features that can be answered without ambiguity. The collection of all questions forms the complete decision tree. One can picture a whole set S_0 of data tuples being subdivided into two sets S_{11} and S_{12} by a decision $\mathcal{D}_1$. The two sets are then again broken down into more sets forming a tree. Ideally, the splitting is finished if the sets contain solely a single class of data. Then, a further division is not necessary. The class information of the remaining set is assigned to a leaf of the tree. The use of the tree is straightforward: From the top a new data point is confronted with the decisions until it reaches a leaf and is classified according to the leaf's class membership.

Figure 16.7a shows such a tree for the distinction of the two faults F_1 and F_2 using two continuously distributed symptoms s_1 and s_2. The decisions are binary but based on a continuous variable. The example also shows that the tree is not complete, i.e. it does not have an identical size in all branches. The resulting symptom space segmentation can be seen in Figure 16.7b.

How is such a tree structure built if not from prior knowledge of an expert? For the case of fault diagnosis, this involves two questions: Which symptom is to be chosen and, following the example, which value is an appropriate threshold to yield a sensible division into subsets? The procedure is based on a single-step optimal strategy: One tries to implement the decision that results in subsets with maximum "purity", meaning that the sets should contain data of most similar type, ideally of only one single class membership. Instead of the purity one calculates a measure of impurity, the entropy of the set based on the statistical definition of an entropy index:

$$i_{entropy}(\mathbf{P}) = -\sum_j P_j \log P_j \qquad (16.16)$$

The occurrence probability P_j of the fault F_j in the resulting set is usually replaced by its relative frequency of occurrence. This is a valid procedure for larger data sets.

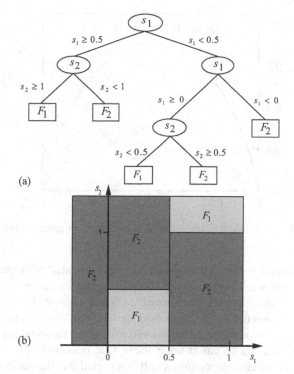

Fig. 16.7. Decision tree for the distinction of two classes F_1 and F_2: (a) decision tree; (b) resulting partitioning in a $s_1 - s_2-$ plane. The symptoms s_1 and s_2 are continuous variables

One can describe the tree growing as making the decisions that extract most information from the data sample. A finally pure leaf of the tree contains no information.

Figure 16.8 shows the behavior of the entropy index function. Displayed is the case of two classes. The abscissa is the probability of the first of the classes, P_1. P_2 is then directly given by $1 - P_1$. Figure 16.8 also displays a second function, the Gini-index, that is sometimes used instead of (16.16), being defined as:

$$i_{gini}(\mathbf{P}) = 1 - \sum_j P_j^2 \qquad (16.17)$$

This impurity measure is calculated faster because is does not rely on the evaluation of a logarithm. Its behavior is very similar to (16.16) as can be seen in Figure 16.8. Both functions are symmetric with regard to $P_1 = 0.5$, which is clear since the behavior must not depend on the naming of the data, and both functions vanish for pure sets (i.e. $P_{1,2} = 0$ or $P_{1,2} = 1$). The maximum impurity is reached for $P_1 = P_2 = 0.5$.

The decision tree is constructed by choosing the decisions that minimize the entropy of the sets on the next tree level. The algorithms are optimal in a single step, but do not necessarily lead to an overall optimal tree, that is, one that is of minimum size. A search for the overall optimal decision tree is not feasible with

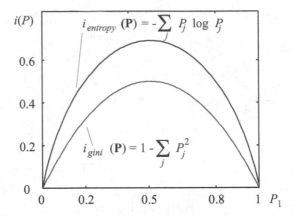

Fig. 16.8. Entropy measure function and Gini-index for a simple two-class problem.

normal computational power. The algorithms are augmented with various growing and pruning approaches. They weigh the complexity of the resulting structure against its classification performance to yield sensible compromises without overfitting.

Originally, decision trees were developed for decisions based on non-continuously valued data. The problem with continuous variables is the theoretically infinite number of decisions that can be based upon them. One possibility is to discretize the variables by means of interval techniques. It is reported that the necessary computational time is not a principle problem.

The more serious problem of the approach is the fact that it is only optimal in one step. In certain data configurations it can happen that a reduction of the index $i(\mathbf{P})$ through a subdivision is not possible which normally terminates the algorithm. It can, however, be beneficial to carry through the split and further subdivide the resulting sets. The average entropy of the resulting sets in the second level can be much lower than original set although the intermediate split does not decrease the entropy measure.

The decisions of standard trees are uni-variate resulting in axes-orthogonal splits in the input space. While it is possible to approximate any decision boundary with axes-orthogonal segments, this is not always desirable because it requires a large tree structure with many decisions. This explains why the methods work best with relatively weakly correlated inputs. There are methods utilizing multi-variate splits, but they are increasingly difficult to handle since the number of possibilities increases drastically.

Decision trees are common in medical diagnosis applications. It seems that the intuitive, simple scheme that can be trusted and understood is important for this area of application.

16.6 Neural Networks for fault diagnosis

In the following, the two most important neural network topologies will briefly be reviewed. These are multi-layer perceptron networks (MLP) and networks composed of radial basis functions (RBF). These networks differ from the aforementioned diagnosis approaches as they represent simple static function approximations, not restricted to any special kind of algebraic function.

While the Bayes Classifier assumes a special function (Gaussian) to be parameterized, the neural networks are designed to match an arbitrary function by reducing an appropriate error measure V, usually defined as a sum-of-squares of the errors similar to (16.11). The mapping function of the network depends on a number of weights $\mathbf{w}$ that contain the information of the network. Network training is done by adapting the weights to minimize V, see also Section 9.3.3. Neural networks of the aforementioned types have been shown to be universal approximators, meaning they can fit any function to an arbitrary accuracy, provided their structure is sufficiently large. As diagnosis tools they are trained with exactly the same target values as for instance the polynomial classifier which means that they also approximate the class-conditional posterior probability.

An interesting result of the sum-of-squares error function has been pointed out in [16.3]: The trained network mapping is given by the conditional average of the target data:

$$f_j(\mathbf{s}; \mathbf{w}^*) = E\{t_j/s\} = \int t_j\, p(t_j|\mathbf{s})dt_j \qquad (16.18)$$

with $\mathbf{w}^*$ representing the weights at the minimum of the error function. $p(t_j|\mathbf{s})$ denotes the probability of occurrence of the target value t_j given $\mathbf{s}$ as the symptom values. $E\{\ldots\}$ denotes the expectation value.

If the network is trained with as many outputs as there are faults to be distinguished and the target values are again chosen as binary values according to (16.12), this expression transforms to:

$$f_j(\mathbf{s}; \mathbf{w}^*) = \int t_j \left(\sum_{l=1}^{c} \delta(t_j - \delta_{jl}) P(F_j|\mathbf{s}) \right) dt_j \qquad (16.19)$$

with δ_{jl} being the Kronecker delta. $P(F_j|\mathbf{s})$ is the probability that $\mathbf{s}$ belongs to fault F_j. Since the delta function is zero everywhere else apart from 0, the integral can easily be computed to yield:

$$f_j(\mathbf{s}; \mathbf{w}^*) = P(F_j|\mathbf{s}) \qquad (16.20)$$

meaning that the outputs of the network correspond to the Bayesian posterior probabilities of the fault classes given the symptom pattern $\mathbf{s}$. This suggests that the network structures trained with a labelled set of reference data and sum-of-squares error functions have the potential to learn the statistically optimal decisions given by the intersections of the posterior probabilities while not being constrained to an assumption about the probability density functions.

It has to be stressed, however, that this result presupposes that the optimal network weights $\mathbf{w}^*$ at the minimum of $V(\mathbf{w})$ have been found. This requires not only an appropriately sized network to be able to fit the probability function, but also the solution of problems like overfitting and decent convergence of the optimization algorithm made use of.

In the field of fault diagnosis, neural networks are frequently employed: In a survey about half of the publications utilizing a classification procedure for fault diagnosis, relied on neural networks, [16.10]. Since efficient tools for network training and implementation have become easily available, it is likely that neural networks are used in more than half of the applications today. They provide a means to achieve decent classification results with relatively moderate design effort.

16.6.1 Multi-layer perceptron networks

The structure of a multi-layer perceptron network is shown in Figure 16.9. It uses each of the symptoms as one input and calculates one output per fault. Since the output represents a probability, its domain should be the interval $\{0; 1\}$. That can easily be ensured by utilizing a sigmoidal activation function in the output layer of the network. That is unusual for most other applications of MLP networks that commonly require an unlimited continuous output and hence employ a linear function in the last network layer as treated in Section 9.3.3 for process identification.

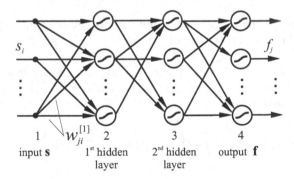

Fig. 16.9. Multi-layer perceptron network for fault diagnosis. The typical configuration has symptoms s_i as inputs and one output f_j per fault class and sigmoidal activation functions in the output layer.

The network layer first performs a projection of the data by a weighted sum of the input values and a bias. If one denotes the output of the first neuron in the first hidden layer with z_{11} it is calculated as:

$$z_{11} = \gamma(a_{11}) \tag{16.21}$$

with $\gamma(\cdot)$ being the activation function of the network. Its input a_{11} is given by:

$$x_{11} = \sum_{i=1}^{n_s} = w_{ij}^{[1]} s_i + w_{0j}^{[1]} \tag{16.22}$$

This projection is similarly performed in each of the layers of the network. It makes the MLP a good choice for high-dimensional problems since a small number of hidden neurons effectively removes irrelevant information from the inputs. On the other hand, a configuration with a high input dimensionality leads to large numbers of free parameters which is problematic if the reference data base is not sufficiently large.

The MLP should be used in cases where the shapes of the class boundary are complex and a large number of data points is available. Since it must be trained with a nonlinear optimization technique, its training time can become relatively long. With the availability of modern and highly efficient training algorithms, however, this is not a serious problem any more. Even networks with multiple hidden layers can be trained relatively fast. Examples of MLP networks for the discrimination of different fault situations can be found [16.1], [16.11], [16.19], [16.13], [16.12], [16.23], [16.24] and [16.26]. In [16.25], a set of neural networks is used with each network trained to distinguish only one of all possible faults from the nominal situation. The advantages of this approach are smaller network sizes with simpler functions to be learned. The overall number of parameters, however, is most likely larger than in a single network configuration.

In some applications, [16.2], the MLP network is not only utilized for the diagnosis but at the same time for generating symptoms from the measurement signals directly. In this case, the typical fault detection/fault diagnosis nomenclature is not applicable. The task of the neural network is then a direct mapping from measurements to faults. This concept might be appealing at first sight due to its simplicity; it is however feasible only for simple static systems. Furthermore, the often highly specific domain knowledge of the system designer is completely lost.

A large problem of the MLP networks is the difficult extrapolation behavior, as the data sets in diagnostic applications are not always complete and also relatively sparse compared to other classification areas like image processing. This creates a need for a diagnosis system working outside the trained symptom domain. This is especially problematic for MLP networks. The neurons use *global* activation functions that contribute significant shares of the network output over the whole input domain. If the network is trained with many degrees of freedom to fit a complex decision boundary, these neurons will create completely unpredictable outputs in the extrapolation region, even close to the training area. This problem was first formulated in [16.11] and it was suggested to use local approximating networks like the radial-basis function networks instead. The problem will occur in particular if a high-dimensional input space is sparsely populated, leaving vast areas without training data.

16.6.2 Radial-basis function networks

The second most widely used networks for diagnosis are radial-basis function networks. Most frequently, non-normalized basis functions with a singleton in the output layer are used, compare Figure 16.10. The advantage of the configuration is the simple optimization of the output layer weight that are linear in the output error, hence a deterministic single-step optimization can be used as described in Section 9.3.3.

Consequently, the research has concentrated on the more difficult problem of placing the basis functions. The standard procedure is to cluster the input data and locate basis functions at the cluster centers. Typical approaches are based on k-means clustering or Kohonen feature maps to group the input data.

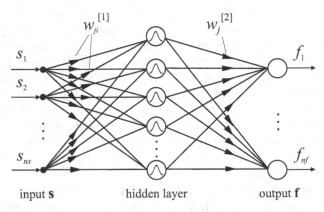

input **s**   hidden layer   output **f**

Fig. 16.10. Radial-basis function network for fault diagnosis.

Both, clustering as well as Kohonen feature maps are based on the minimization of a distance from the data points to reference vectors. An iterative procedure creates the reference vectors that represent a set of data points. The approach does usually not use information about the class membership of the data points and is thus called *unsupervised*. Hence it can happen that overlapping data from different classes are assigned to a single basis function. The network will then hardly be capable of differentiating the classes by adapting only the output layer weights.

This problem has lead to the development of various constructive algorithms that successively create new basis functions. They are typically driven by the performance of the network and refine the output by adding new locations for basis functions. Examples are the approach developed by [16.6], or the LOLIMOT-algorithm developed by [16.14]. An interesting approach is connected to the decision trees discussed in Section 16.5. By first learning a division of the input domain via a uni-variate decision tree algorithm, one derives a useful segmentation of the input space. Placing the basis functions one per segment is reported to yield a network with a superior classification performance compared to the decision tree as classifier alone.

The alternative to constructive algorithms is a procedure that first creates too many basis functions and then uses a sensible way of removing unnecessary ones. One of those approaches is known as *global ridge regression* and is realized with an additional regularization term in the least-squares error function (16.11):

$$V_j^* = \sum_{k=1}^{N} (t_j(\mathbf{s}(k)) - f_j(\mathbf{s}(k)))^2 + \alpha \sum_{i=1}^{m} \left(w_i^{[2]} \right)^2 \qquad (16.23)$$

The $w_i^{[2]}$ are the m output layer weights of the radial basis function network, α a constant controlling the strength of the regularization. The f_j are again the network outputs that can be between zero and one. The desired network outputs are the $t_j(\mathbf{s}(k))$. They are known fault indicators having either a value of 0 or 1. In the context of neural networks, the term *target value* is common for this desired network output.

Since the optimization tries to minimize V_j^*, weights that are not relevant for the performance of the network will be driven to zero. In a second step, the corresponding neurons can be pruned from the network yielding a compact, yet well performing network. The procedure should prune only one neuron at a time and repeat the optimization.

One major difference between MLP and RBF networks can be seen in the extrapolation behavior of RBF networks. Since each basis function influences the output of the data only in a limited region of the input space, RBF networks are said to be *locally approximating*. For diagnosis applications, that can be favorable: Observed data points outside the coverage of the training data do not significantly activate any basis function. Consequently, the outputs assume a value close to zero. This can be interpreted as a *rejection*, i.e. the outputs signal that a new situation has occurred and no definite conclusion can be drawn. This does not give a completely satisfactory answer but avoids at least a wrong diagnosis.

It can also serve for an on-line training of the network. In [16.4] it is suggested to augment the network with basis functions if a data point occurs that does not belong closely enough to any of the priori known fault classes, i.e. basis functions. This property is calculated by an appropriate measure to evaluate the distance of the point from known input areas. Similar concepts are common for constructive algorithms to locate the basis functions of radial basis function networks.

The local approximation of the network furthermore allows an interpretation of the system to some degree: The second layer weights $w_j^{[2]}$ belong to a well-defined area of the input domain. Taking result (16.19) into account explains the weights as a somewhat smoothed estimate of the corresponding posterior probabilities $p(F_j|\mathbf{s})$. Examples of radial-basis function networks for diagnosis are the fault diagnosis of industrial plants, [16.17], chemical reactors, [16.20], [16.22] and a two-link manipulator, [16.21].

Radial-basis function networks with an intelligent basis function placement strategy are suitable for problems with many reference data points because the optimization of the weights in the last layer can be performed very efficiently. They are generally less critical due to the local optimization behavior, but require more storage for the network parameters than a comparably performing MLP network.

16.6.3 Clustering and self-organizing networks

Occasionally, clustering approaches or the closely related Self-Organizing-Maps are used for the diagnostic classifier design, [16.5]. These methods are un-supervised and rely on the assumption that training data points that are near to each other in the symptom space will also belong to the same fault class. They find clusters in the data by evaluating neighborhood measures and employing competitive strategies. The next step is the determination of the majority of the class membership of the data points for each cluster. This gives labels for the clusters. When a data point is to be classified, one determines which cluster it belongs to and assigns the corresponding label.

This "clustering-labelling" approach, however, is not as powerful as the supervised neural network methods described above. Especially difficult decision boundaries with overlapping classes are hard to distinguish since the method does not use the class information to build the network or adapt any weights. The class information is only employed during the labelling. As initialization procedures for radial basis function networks or to determine initial membership functions in fuzzy logic based systems these clustering methods are nevertheless very useful, see Section 17.2.

An overview of other neural network techniques for fault diagnosis and some practical considerations for their use can be found in [16.18] and [16.19].

16.7 Problems

1) Compare the a priori assumptions of different classification methods, like Bayes, geometric and polynomial classifiers. Use as example Figure 16.5.
2) What are the differences between geometric classifiers and classifiers based on decision trees?
3) Specify the input and output variables for the classification with neural networks.
4) Which classification methods should be used if only few symptoms data (5) are available for each fault?

17

Fault diagnosis with inference methods

For some technical processes, the basic relationships between faults and symptoms are at least partially known in form of causal relations: fault → events → symptoms.

Figure 17.1a shows a corresponding causal network, with the nodes as states and edges as relations. The establishment of these causalities follows the fault-tree analysis (FTA), proceeding from faults through intermediate events to symptoms (the physical causalities) or the event-tree analysis (ETA), proceeding from the symptoms to the faults (the diagnostic forward-chaining causalities), see, e.g. [17.31] and Section 4.2. To perform a diagnosis, this qualitative knowledge can now be expressed in form of rules

$$\text{IF} \ \langle \text{ condition} \rangle \ \text{THEN} \ \langle \text{ conclusion} \rangle$$

The condition part (premise) contains facts in the form of symptoms s_i as inputs, and the conclusion part includes events e_k and faults f_j as a logical cause of the facts. If several symptoms indicate an event or fault, the facts are associated by AND and OR connectives, leading to rules in the form

$$\text{IF} \ \langle \ s_1 \ \text{AND} \ s_2 \ \rangle \ \text{THEN} \ \langle \ e_1 \ \rangle$$
$$\text{IF} \ \langle \ e_1 \ \text{OR} \ e_2 \ \rangle \ \text{THEN} \ \langle \ f_1 \ \rangle$$

compare Figure 17.1b.

For the establishment of this heuristic knowledge several approaches exist, see [17.18], [17.58].

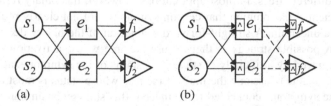

Fig. 17.1. Fault diagnosis using inference methods: (a) causal network; (b) fault symptom tree

17.1 Fault trees

Fault trees are established as a graphical tool for the visualization of the relationships in reliability and diagnosis, [17.50], representing *binary relationships*. They are directed graphs showing the fault situation at the top and symptoms and conditions below. Elements of the tree are logic connections, events and symptoms. An example is shown in Figure 17.2.

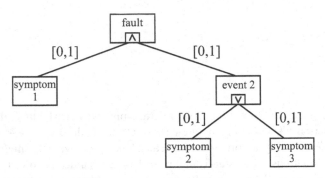

Fig. 17.2. Scheme of a fault tree for binary symptoms (states) [0, 1]

Fault trees are a good and intuitive graphical tool for displaying the binary relationships that will lead to failures. The hierarchical structure supports the human comprehension. They are common for analysis and diagnosis in safety-critical applications. Quantitative failure probabilities can also be derived from the fault trees. They require information about the failure probabilities of the individual elements. By combining these with the relationships from the fault trees it is possible to calculate the probability of a system failure. The fault tree is usually important during the early system design phase in identifying the critical subsystems that contribute most to the system failure probability. Examples of fault trees can for instance be found in [17.5] and [17.16]. The fault trees shown there for an industrial robot are not used for reliability analysis but rather for the diagnosis, i.e. to visualize the diagnostic reasoning from symptoms to faults. It is then necessary to design one fault tree for each of the n_f faults. The leaves of the tree are composed of the n_s available symptoms. The decision which fault has occurred is ideally a simple binary evaluation of the different trees. In most applications, however, this binary representation is not sufficient. It is common that certain symptoms are not clearly recognized or that they are uncertain. A straight-forward evaluation of the binary decision would not work. A possible strategy is then to use the tree whose activation could most easily be achieved by artificially changing the status of the symptoms at the leaves. It seems reasonable to assume the fault of the tree whose status is most active from checking the symptoms connected to it. Indeed, this strategy is similar to a pattern matching: The symptom pattern is assumed that is the closest to a known fault pattern. Viewing the problem from this perspective, however, suggests using a different procedure. Instead of trying to laboriously match tree structures, one should translate the problem to a functional one: The symptom occurrences are inputs and the activation level of the different faults are the outputs of the functional mapping that has to be determined. That strategy is followed by the other diagnosis methods presented in this chapter. It represents the standard procedure of any inference method.

As discussed the symptoms and events are considered as binary variables, and the condition part of the rules can be calculated by Boolean equations for parallel-serial-connection, see, e.g. [17.5], [17.16]. However, this procedure has not proved to be successful because of the *continuous nature of faults and symptoms*.

The fault detection of *discrete event systems* on the basis of discrete events only is treated, e.g. in [17.57]. The addition of the time of the events improves the situation, [17.34]. Model-based methods for fault-detection of timed discrete event systems are described in [17.52], [17.56] and [17.53]. An example for the application of the fault diagnosis with binary fault trees of a discrete-event system is shown in the following example.

Example 17.1: Discrete-event fault detection for a pump-valve-filter system

As an example for a *discrete-event system* a plant consisting of an electrical driven pump, electromotor actuated valve, a filter and a fluid storage is considered, see Figure 17.3. Measurements are the fluid pressure after the pump p_1, the end position of the valve s_1 and the fluid mass flow $\dot{m}_1$. A sequential controller switches on the pump ($U_1 = 1$) and after a few seconds opens the valve ($U_2 = 1$), independent on the pressure measurement p_1. It is assumed that the controller functions correctly. The corresponding signals are shown in Figure 17.4.

a) Fault detection with discrete-event amplitude

The measured outputs are just binary values 1 if in normal operation, obtained by limit switches, otherwise 0. Hence, in normal operation as well the inputs as the outputs indicate the discrete event 1.

The task now consists in detecting faults in the components pump, valve, filter and the applied sensors. Figure 17.5 shows a fault symptom tree for the case of three measurements. If the binary state of these measurements is 111 the plant is correctly operating. Based on physical-logic inspection of the binary states of the sensors, a fault-symptom table can be established, see Table 17.1. Out of six possible faults all faults can be detected, four of them can be isolated, but two are not isolable. The number of sensors is now reduced to two. Table 17.2 shows the results for different sensor combinations. Then five faults are possible. If the flow sensor $\dot{m}_1$ is included, all five faults are detectable, but only two are isolable and three cannot be isolated. If the flow sensor $\dot{m}_1$ (which measures the main output) is not used, but only the sensors p_1 and s_1, the filter fault F5 is not detectable, and four faults are detectable, but not isolable. Using only the flow sensor, Table 17.3, allows to detect four out of four possible faults, but they cannot be isolated. Table 17.4 summarizes the results for different measurements.

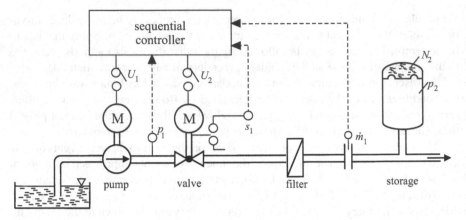

Fig. 17.3. Pump-valve-filter-storage plant

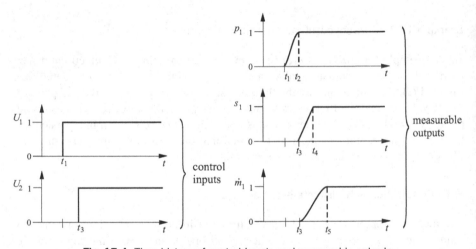

Fig. 17.4. Time history of control inputs and measurable outputs

Table 17.1. Fault symptom table for discrete events and three measurements, p_1, s_1 and $\dot{m}_1$

fault	symptoms			isolable	not isolable
	p_1	s_1	$\dot{m}_1$		
no fault	1	1	1	–	–
F1	0	1	0	x	
F2	0	1	1	x	
F3	1	0	0	x	
F4	1	0	1	x	
F5	1	1	0		x
F6	1	1	0		x

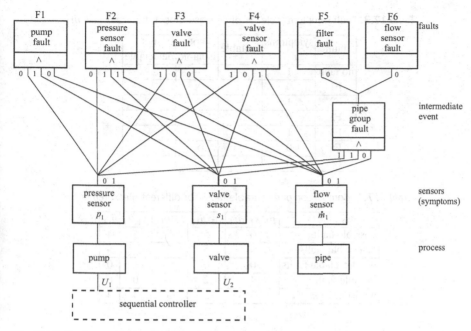

Fig. 17.5. Fault symptom tree with binary events for the discrete-event system Figure 17.3 and 17.4 and the case of three limit switch measurements for pressure, valve and flow

Table 17.2. Fault symptom table for two measurements, p_1 and $\dot{m}_1$, s_1 and $\dot{m}_1$, p_1 and s_1

fault	symptoms p_1	symptoms $\dot{m}_1$	isolable	not isolable	fault	symptoms s_1	symptoms $\dot{m}_1$	isolable	not isolable
no fault	1	1	–	–	no fault	1	1	–	–
F1	0	0	x		F1	1	0		x
F2	0	1	x		F2	–	–	–	–
F3	1	0		x	F3	0	0	x	
F4	–	–	–	–	F4	0	1	x	
F5	1	0		x	F5	1	0		x
F6	1	0		x	F6	1	0		x

fault	symptoms p_1	symptoms s_1	isolable	not isolable	not detectable
no fault	1	1			x
F1	0	1		x	
F2	0	1		x	
F3	1	0		x	
F4	1	0		x	
F5	1	1			x
F6	–	–	–	–	–

Table 17.3. Fault symptom table for one measurement, the mass flow $\dot{m}_1$

fault	symptoms $\dot{m}_1$	isolable	not isolable	not detectable
no fault	1	–	–	–
F1	0		x	
F2	1	–	–	–
F3	0		x	
F4	1	–	–	–
F5	0		x	
F6	0		x	

Table 17.4. Comparison of detectable faults for different measurements

sensors	$p_1\ s_1\ \dot{m}_1$	$p_1\ \dot{m}_1$	$s_1\ \dot{m}_1$	$p_1\ s_1$	$\dot{m}_1$
possible faults	6	5	5	5	4
detectable faults	6	5	5	4	4
not detectable faults	0	0	0	1	0
isolable faults	4	2	2	0	0
not isolable faults	2	3	3	4	4

Hence, it is important that the main output signal of the pump system, the mass flow $\dot{m}_1$ (or at least pressure p_2) is available as measurement. Then with all sensor configurations, either one, two or three sensors, all possible faults are detectable. However, with one sensor they cannot be isolated, i.e. cannot be diagnosed. But with increasing number of sensors more faults can be isolated, but not all of them.

b) Fault detection with discrete-event time intervals

A further information source for fault detection are the time intervals between reaching the discrete events. By measuring the time intervals between the time instants indicated in Figure 17.4 the fault symptom table shown in Table 17.5 results. If, for example, the pump does not deliver fluid or has too slow dynamic response after the switch command $U_1 = 1$ the measured time interval $t_2 - t_1$ is not within a pre-specified threshold and therefore generates a 0. The Table indicates that with four time intervals five faults can be isolated and two not, as in Table 17.1. This includes a CPU-program fault for not switching U_2 at the programmed time instant.

If the time intervals for nominal behavior is assumed to be very large, then the same information is gained as with the discrete events of the measured signals, but related to each other. Therefore the information which is additionally obtained with the time intervals depends on the time behavior or dynamics of the components which is not considered by observing the discrete event amplitudes only. Hence, faults which influence the dynamics of the components are additionally included by the time interval observations.

Combining the time interval fault detection of Table 17.5 with the discrete event amplitude fault detection by using three sensors as of Table 17.1 allows to isolate also too slow dynamics of four components. Combination with the case of two sensors or one sensor improves the isolability of faults, especially for measurement with p_1, s_1 and $\dot{m}_1$. Hence, this example shows that symptom trees with binary events are well suitable for discrete-event systems.

Table 17.5. Fault symptom table for measured time intervals between discrete events

fault		symptoms				isolable	not isolable
		$t_3 - t_1$	$t_2 - t_1$	$t_4 - t_3$	$t_5 - t_3$		
no fault		1	1	1	1		
F1	pump	1	0	1	0	x	
F2	pressure sensor	1	0	1	1	x	
F3	valve	1	1	0	0	x	
F4	valve sensor	1	1	0	1	x	
F5	filter	1	1	1	0		x
F6	flow sensor	1	1	1	0		x
F7	CPU-program for U_2	0	1	0	0	x	

□

17.2 Approximate reasoning

For the rule-based fault diagnosis of *continuous processes* with *continuous variable symptoms*, methods of approximate reasoning are more appropriate than binary decisions. Based on the available heuristic knowledge in the form of *heuristic process models* and *weighting of effects*, different diagnostic forward and backward *reasoning strategies* can be applied. Finally the diagnostic goal is achieved by a *fault decision* which specifies the type, size and location of the fault as well as its time of detection, [17.25], [17.26].

A review of developments in reasoning oriented approaches for diagnosis was given in [17.27]. As major areas of interest, medicine, [17.54], [17.49], and engineering, [17.14], can be observed. Engineering research especially with regard to reliability analysis of nuclear power stations, aero space systems and electrical equipment started much earlier and followed in many cases the concept of fault tree analysis, [17.12], [17.24], [17.11], [17.10], [17.5], [17.23], [17.31], [17.15], [17.9]. On the other side, artificial intelligence (AI) offered new methods for the treatment of cause-effect relations, [17.43], and for diagnostic problem solving, [17.36], [17.44], [17.45], [17.58]. The development in the area of artificial intelligence (AI) was oriented initially to medical diagnosis and then extended to technical processes. Therefore, also for technical diagnosis only heuristic symptoms were considered. Then more sophisticated diagnostic reasoning strategies were developed by increasing the level of abstraction see, e.g. [17.47] and [17.30] using the causalities as deep logic interdependencies. The fault-symptom trees known from engineering and the AI strategies for treating causalities can be brought together for fault diagnosis. Especially the analytical symptoms generated by the model-based fault detection then allow to perform a deep fault diagnosis by pinpointing on the possible physical fault origins (roots).

17.2.1 Forward chaining

By using the strategy of forward chaining a rule, the facts are matched with the premise and the conclusion is drawn based on the logical consequence (Modus ponens). Therefore with the symptoms s_i as inputs the possible faults F_j are determined using the heuristic causalities.

In general the symptoms have to be considered as uncertain facts. Therefore a representation of all observed symptoms as confidence functions $c(s_i)$ or membership functions $\mu(s_i)$ of fuzzy sets in the interval $[0,1]$ is feasible, especially in unified form as described in Section 15.1. For a short introduction into fuzzy logic see Chapter 23.3.

a) Approximate reasoning with fuzzy logic

With the structure of a fault-symptom tree, obtained by knowledge acquisition via the event-tree analysis (ETA) of the process, a fuzzy rule based system with multiple levels of rules can be established, as shown in Figure 17.6.

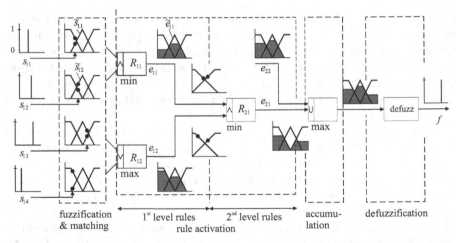

Fig. 17.6. Signal flow in a fuzzy-rule-based system for fault diagnosis with two levels of rules and max/min operations and singletons as inputs

The symptoms s_i are now represented by fuzzy sets $\tilde{s}_i^0, \tilde{s}_i^1, \tilde{s}_i^2, \ldots$ with linguistic meanings like "normal", "less increased", "much increased", etc. The general procedure within the fuzzy IF-THEN rule based system, [17.35], [17.64], [17.13], then follows:

- *inference*: The fuzzy IF-THEN rule expresses a fuzzy implication relation between the fuzzy sets of the premise and the fuzzy sets of the conclusion. If the rule is interpreted by the Mamdani-implication and the compositional rule of inference by [17.63] is applied, following steps can be distinguished:
  - matching of the facts with the rule premises (determination of the degrees of fulfillment of the facts to the rule premises → fit values);
  - if the rule contains logic connections between several premises by fuzzy AND or fuzzy OR the evaluation is performed by t-norms or t-conorms (The result gives then the degree of fulfillment of the facts to the complete premise);
  - the evaluation of the resulting conclusion follows the max-min composition, [17.63] (the degree of fulfillment of the premise restricts the fuzzy set of the conclusion)
- *accumulation* of all conclusion fuzzy sets if several rules contribute to the output: A union operation yields a global fuzzy set of the output;
- *defuzzification* to obtain crisp outputs: various defuzzification methods can be used, as, e.g. center of gravity, maximum-height, and mean of maximum, to obtain a crisp numerical output value. For a fault diagnosis, however, the defuzzification is often replaced by a simple maximum operation. For the conditional part of the IF-THEN rules relatively simple fuzzy-logic operations are obtained by max-min composition to obtain the most possible fault.

$$\text{Fuzzy - AND} : \mu(\eta) = \min[\mu(\xi_1), \ldots, \mu(\xi_v)] \tag{17.1}$$

$$\text{Fuzzy - OR} : \mu(\eta) = \max[\mu(\xi_1), \ldots, \mu(\xi_v)] \tag{17.2}$$

These operations agree with corresponding operations of Boolean logic. However, some information is not used as only the minimal or maximal value is taken. Another possibility is to use the prod-sum-operation

$$\text{Fuzzy - AND} : \mu(\eta) = \mu(\xi_1)\,\mu(\xi_2)\ldots\mu(\xi_v) \tag{17.3}$$

$$\text{Fuzzy - OR} : \mu(\eta) = 1 - \prod_{i=1}^{v}(1 - \mu(\xi_i)) \tag{17.4}$$

In this case all values are represented in the result. Note also the similarity to equations of probability theory.

The NOT-Operation is in both cases

$$\text{NOT} : \mu(\eta) = 1 - \mu(\xi) \tag{17.5}$$

The dimensions of the fuzzy rule based system is given by

- number of symptoms $i = 1, 2, \ldots, n_s$;
- number of levels $l = 1, 2, \ldots, P$;
- number of rules per level $v = 1, 2, \ldots, N$;

- number of faults $j = 1, 2, \ldots, n_f$.

The overall dimension may therefore blow up strongly even for small components or processes. Therefore the software implementation is important. Mainly two procedures to perform the reasoning are known:

- *Sequential rule activation* from level l to level $l + 1$ (horizontal procedure);
  - matching of symptoms $\tilde{s}'_{li}$ with the premises of rules of level $l = 1$ (including their logical operations) to determine the events $\tilde{e}'_{lv}$;
  - matching of events $\tilde{e}'_{lv}, l = 2, 3, \ldots, P$, with the premises of rules of level l to determine the events $\tilde{e}'_{(l+1),v}$
  - matching of events $\tilde{e}_{(l+1)v}$ with the fault's fuzzy subsets f'_j;
  - accumulation of fault fuzzy subsets to obtain the global possibility for each fault
  - fault decision by maximal possibility
  - defuzzification to obtain a crisp size of the most possible fault.
- *Multiple rule activation* from all symptoms to all faults (vertical procedure), by using the fuzzy relational equation

$$\mathbf{F} = \mathbf{S} \circ \mathbf{R} \qquad (17.6)$$

proposed by [17.61][17.60].
  - chaining $\tilde{s}_{1i} \rightarrow \tilde{e}_{1v} \rightarrow \ldots \tilde{e}_{lv} \rightarrow \ldots \tilde{f}_j$ by multiple composition to obtain overall relations $R_{s1i \times fj}$ in form of a matrix $\mathbf{R}$;
  - matching of $\mathbf{R}$ with all current symptoms $\mathbf{S}'$ by using the fuzzy relational equation $\mathbf{F} = \mathbf{S} \circ \mathbf{R}$;
  - accumulation to obtain the global possibility for each fault;
  - fault decision by maximal possibility;
  - defuzzification to obtain a crisp size of the most possible fault.

In both approaches a defuzzification can be performed after each level or not.

A similar approach using fuzzy relation equations was proposed earlier by [17.29] using derivations of directly measurable variables as symptoms and stating the fuzzy relation equation

$$\mathbf{F} = \mathbf{S} \circ \mathbf{R}^* \qquad (17.7)$$

The knowledge for the establishment of this equation stems from a fault-tree analysis (FTA). Then the inverse problem of (17.6) has to be solved, by calculating the possibilities of the faults $\mathbf{F}$, e.g. by using Sanchez's operator, [17.48]. [17.21] use the relational fuzzy equation (17.6), for a diagnosis after a failure has happened (post mortem diagnosis) and searching for more or less binary events (symptoms) as causes. Also in this case the inverse problem is solved.

Indeed, in many diagnosis application cases the knowledge acquisition first follows a fault-tree-analysis (FTA), by proceeding from assumed faults through events to symptoms by physical causality inspection according to Figure 15.2b. If the knowledge is complete, a graphical representation like Figure 17.6 can be obtained directly, from which (17.6) can be established. By this way the solution of the inverse problem can be circumvented. Then, also basic fuzzy-logic software tools can be used directly.

To reduce the *computational effort*, simplifying assumptions may help. Following simplifications are possible:

- singletons for the symptoms: matching is reduced to direct fuzzification of crisp values;
- singletons for the events and faults: the conclusions are singletons with a height of the matching degree between facts and rule premise;
- universal fuzzy sets for the events and faults: the conclusion is a universal set with height of the matching degree between facts and rule premise. But then the size of the event or fault cannot be determined.

In the cases of singletons or universal sets for events and faults the conclusions are identical with the membership degree of the facts with the logical operation in the premise. This means the whole procedure reduces to an aggregation of the facts with appropriate t-norms and t-conorms, and chaining of the results, [17.25]. Fuzzy fault diagnosis can be expanded to time dependent faults and symptoms, including incipient and intermittent faults and multiple faults, [17.60].

b) Simplified fuzzy logic reasoning for fault diagnosis

The basic fuzzy logic operations for given continuous variable symptoms are summarized in Chapter 23.4. For typical fault-diagnosis applications, the standard fuzzy system can be reduced. The main reason for that is the desired output of the diagnostic. Instead of an arbitrary fuzzy set of a continuous variable, the output is a fault measure representing a gradual measure for the possibility of the corresponding fault. If the observed symptoms are far apart from the linguistically defined pattern, this fault measure will be close to zero, whereas a perfect match will yield a fault measure of one. This means that the higher the fault measure becomes, the more likely the corresponding fault situation has occurred. The possibility of that event increases with the fault measure. The reduction of the rule consequences to a statement which fault has occurred can be represented by a singleton value which is scaled by the rule fulfillment. Therefore, no other output membership functions are necessary and also the defuzzification is not required. The resulting, simplified fuzzy logic system structure can be seen in Figure 17.7.

c) Probabilistic fault-symptom reasoning

Another possibility to cope with uncertain facts is to assign probabilities $P(\xi_i)$ to the symptoms (events) and $P(\eta_k)$ to the events (faults). Based on the causal tree a simplified Bayesian network can be assigned, including conditional probabilities $P(\xi_i, \eta_k)$, [17.43].

IF s_1 small AND s_2 medium THEN Fault f_1

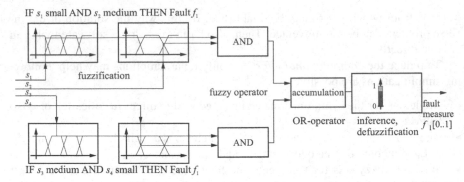

IF s_3 medium AND s_4 small THEN Fault f_1

Fig. 17.7. Simplified fuzzy logic system for fault diagnosis by using singletons as fault measures in the range $0 \ldots 1$

With the assumption that the symptoms ξ_i are statistically independent among each other and the symptoms ξ_i are statistically independent of the events η, it holds for an AND connection, [17.60]

$$P(\eta, \xi_1 \text{ AND } \xi_2) = P(\xi_1 \text{ AND } \xi_2 | \eta) \ P(\eta)$$
$$= P(\xi_1) \ P(\xi_2) \ P(\eta) \qquad (17.8)$$

Similarly one obtains for an OR-connection

$$P(\eta, \ \xi_1 \text{ OR } \xi_2) = \\ [P(\xi_1) + P(\xi_2) - P(\xi_1) \ P(\xi_2)] \ P(\eta) \qquad (17.9)$$

If in addition the events are assumed to have happened surely, $P(\eta) = 1$, one obtains for the AND-connection

$$P(\eta, \xi_1 \text{ AND } \xi_2 \text{ AND} \ldots \xi_v) = \\ P(\xi_1) \ P(\xi_2) \ldots P(\xi_v) \qquad (17.10)$$

and the OR-connection

$$P(\eta, \xi_1 \text{ OR } \xi_2 \text{ OR} \ldots \xi_v) = 1 - \prod_{i=1}^{v} (1 - P(\xi_i)) \qquad (17.11)$$

The similarity of the formulas to fuzzy-logic operations with the prod-sum operation, e.g. (17.3), (17.4), is obvious. However, the assumptions on statistical independence do not take into account the existing causalities of the fault-symptom trees.

17.2.2 Backward chaining

The strategy of backward chaining assumes the conclusion as known and searches for all relevant premises (modus tollens). This is especially of interest if the symptoms are not complete. Therefore the concluded events and faults are displayed to the operator after forward chaining with all known symptoms. A refinement of the diagnosis can then be achieved by selecting the most plausible events and faults as hypothesis and applying backward chaining by asking for missing symptoms. Then forward chaining is restarted. This procedure is implemented best within an interactive dialogue and repeated until terminated by the operator, [17.15], [17.16]. Other chaining strategies are Establish-Refine, Hypothesize and Test, Depth and Width searching, etc., see [17.22]. If all symptoms are already taken into account during forward reasoning, backward reasoning can be applied for the validation of diagnosed faults, [17.61]. The fault diagnosis considered until now assumed that all symptoms appear simultaneously and do not change with time. Further cases are therefore dynamically developing faults and symptoms, (e.g. either incipient or intermittent faults) and therefore dynamic fault trees, [17.41], [17.60] and also multiple faults.

17.2.3 Summary and comparison

The last two chapters have introduced some important approaches for fault diagnosis ranging from fault tree analysis to neural networks and approximate reasoning with fuzzy logic. Each method has its advantages and disadvantages. A summary of the characteristics of the methods is given in Table 17.6, [17.19].

Table 17.6. Comparison of the different fault diagnosis methods and their main characteristics. The performance evaluation is based on experience and assumes a typical fault-diagnosis problem with some explicit knowledge and a more or less complete training data set of medium complexity. Naturally, there are always examples for that one or the other method performs better or worse. Therefore, the rating here is preliminary. $++/--$: strongly pos./neg., $+/-$: pos./neg., o: average; expl.: explicit knowledge base, impl.: implicit knowledge base.

Method	Classification							Inference	
	Bayes	Nearest Neighb.	Polynominal	Decis. Tree	MLP-NN	RBF-NN	Cluster-ing	Fault Tree	Fuzzy Logic
Know-ledge Base	impl.	impl.	impl.	impl.	impl.	impl.	impl.	expl.	expl.
Design Effort	+	++	+	−	o	o	+	−−	−
Trans-parency	−	−−	−−	o	−−	−	−	++	++
Perfor-mance	−	+	+	o	+	+	−	−	+

It is apparent that two different sources of information can be used: Expert or domain (structured) knowledge on the one hand and measured data from fault experiments on the other. Each of the methods presented make use of one of the two.

This leads either to highly transparent systems that are tedious to design or too non-transparent data-based classifiers. It becomes clear that in reality a mixture of both sources of information resides. To assess them in parallel having benefit from reference data but also incorporate prior knowledge would be an improvement. The combinations of neural and fuzzy systems is such an approach and will be treated in the next section.

17.3 Hybrid neuro-fuzzy systems[1]

Combinations of neural network learning strategies with fuzzy logic elements have been discussed for about 15 years. The main objective of the approach is to overcome the *knowledge acquisition bottleneck* faced by humans while designing the knowledge base of a traditional fuzzy expert system. The neural network training techniques of neuro-fuzzy systems can handle the information retrieval from data using optimization techniques. The fuzzy system representation on the other hand provides the intuitive understanding of the resulting system and establishes the possibility of integrating expert knowledge. Since the two approaches have a different knowledge representation, their combination can be a persuasive way to fuse information from different sources, namely human experts and experimental data. Neuro-fuzzy systems can *generate new rules* from data or they can *refine existing rules* by adapting parameters within them.

The main application area of neuro-fuzzy systems is in control and decision support systems. In control, a strong focus of the neuro-fuzzy research has shifted towards the approximation of nonlinear functions, especially for modelling and system identification. In the context of fault diagnosis, however, the ability to build decision support systems is more important.

The following sections assume basic knowledge of fuzzy logic systems. If an introduction into the main terms is required, the reader is referred to the Chapter 23.3.

17.3.1 Structures

The combinations of neural networks and fuzzy logic are manifold. One can use neural networks as a part of a fuzzy inference system, for instance to map membership functions, see Figure 17.8. It is also possible to use a neural network to represent certain important parameters - such as parameters defining the fuzzy membership function - of a fuzzy inference system. On the other hand, a fuzzy rule base can be employed to specify parameters of a neural network that are difficult to determine, such as structural parameters like the number of hidden neurons. There, the fuzzy system automates the heuristics that would be utilized from a human expert experienced in designing neural network topologies.

---

[1] compiled by Dominik Füssel, [17.19]

Figure 17.9 visualizes some of the mentioned concepts. Furthermore, any combination of a serial or parallel set-up of a fuzzy system with a network is possible which are called general neuro-fuzzy systems, [17.3]. Here the term is used in a narrower context limited to those systems where the neural network and fuzzy inference engine are not completely separate.

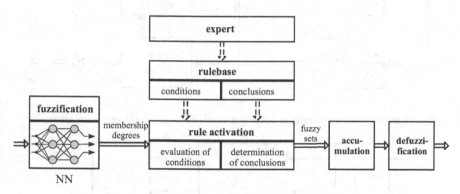

Fig. 17.8. Example of neuro-fuzzy system: A neural network (NN) is used for the determination of fuzzy membership functions in a fuzzy inference system.

The most interesting unions of fuzzy systems with neural networks are those with a *structural equivalence* of the two systems. These systems are generally referred to as *hybrid neuro-fuzzy systems*. They consist of a network structure with usually fixed number of layers where every layer fulfills a specific function that can be expressed as part of a linguistic representation. This means that the hybrid neuro-fuzzy system is both at the same time: a fuzzy inference system and a neural network. It has the ability to extract fuzzy rules from data by inspecting the result of a neural network-type training algorithm. Some examples of the neuro-fuzzy structure will be shown in the following sections.

A relatively complete overview of rule-learning methods with neuro-fuzzy structures can be found in [17.37], [17.8].

a) Generic hybrid neuro-fuzzy model

The different hybrid neuro-fuzzy systems that can be found in the literature vary in their structure and learning algorithms. In the following, a simple, generic hybrid neuro-fuzzy model will be outlined to show the main parts of such systems and to explain their function.

A hybrid neuro-fuzzy network is basically composed of the same elements as a standard fuzzy inference engine. This is depicted in Figure 17.10. The first task is a *fuzzification* that is performed in a fuzzification layer of the network. This is followed by a *neural processing unit* that consists of one or more layers of neurons that constitute the logics of the fuzzy rules. Common elements are for instance units computing a t-norm (logic "and") by a multiplication or minimum operation.

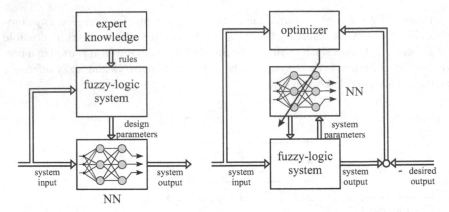

Fig. 17.9. Training configurations utilizing a combination of neural and fuzzy elements

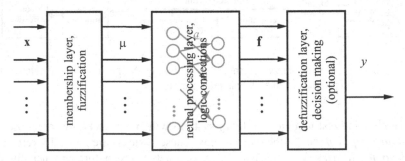

Fig. 17.10. Generic neuro-fuzzy structure for regression and classification (decision support).

The activations of the rules are then transmitted to the last element, the *defuzzification*. Here, one has to clearly distinguish two situations: Systems used for classification and those employed for regression. A regression problem is the approximation of a usually continuously valued one- or higher-dimensional function. For the one-dimensional case, this refers to a function $y = f(\mathbf{x})$, where y assumes any value in the continuous output domain. Regression neural network therefore usually possess a defuzzification with output membership functions or singletons and network elements that perform for instance a weighted sum, which in the case of singletons is equivalent to a center of gravity defuzzification.

Classification problems, on the other hand, are mappings from the continuous, discrete or mixed input domain to a discrete output domain. To distinguish them from the regression problems, in the following a different input/output notation is used: Considering the application of fault diagnosis, the inputs are again noted s_i (for symptoms) and the outputs f_j (fault measures). For the n_f class problem, a n_f-dimensional output domain is created. The indication of the classes is simply given by a 0/1 scheme. This means, that the desired output values assume only 0 or 1. If a training data point belongs to class 3, the values of all but dimension 3 are zero, whereas the third output equals one. During normal use, the classification system is expected to deliver an equivalent statement. Since the inputs to the network might still be continuous, this can only be realized by some hard threshold or decision function. It is common to use a maximum decision acting on the activations of the rules, i.e. the results of the neural processing units. This maximum filters out the strongest rule activation, that in turn becomes the classification statement of the system. Hence, the defuzzification of such hybrid neuro-fuzzy systems is only a maximum operation. It is common, however, to even omit that maximum operation or associate it with a later processing step. This is beneficial, because it is often useful to access the activations of the non-winning rules to make a statement about the reliability of the maximum decision.

Translated into fuzzy logic, rules of regression systems have a structure like:

$$R_l : \quad \text{IF} \dots \text{THEN } y \text{ is large} \tag{17.12}$$

whereas a classification rule looks like:

$$R_l : \quad \text{IF} \dots \text{THEN class } \# 1 \tag{17.13}$$

For classification rules output fuzzy attributes do not exist. The maximum decision will directly act on the activation of the rule premises. The conclusion of the rule with the strongest activation will be the classification statement.

It is important to note, however, that most concepts of neuro-fuzzy regression systems easily translate into classification problems. Therefore, the concepts being presented in this section apply for both types.

In the following, some example hybrid neuro-fuzzy approaches will be given to clarify the typical structure and elements of such systems. The selection of the examples is made to show some differing concepts and architectures present in neuro-fuzzy systems.

b) Mamdani and Singleton neuro-fuzzy systems

A typical example of hybrid neuro-fuzzy systems is the structure suggested by [17.32]. It is known as the *Fuzzy Adaptive Learning Control/Decision Network* (FALCON). The network structure maps Mamdani-type fuzzy rules like:

$$R_j : \quad \text{IF } s_1 \text{ is } A_1 \text{ AND } s_2 \text{ is } A_2 \dots \text{ AND } s_n \text{ is } A_n \text{ THEN } y_k \text{ is } B_k \tag{17.14}$$

A set of rules like (17.14) using identical conclusions can further be combined with a logical OR. Figure 17.11 shows the overall structure of the FALCON network.

This neuro-fuzzy system has typical properties: The tasks of the network layers can be understood linguistically. The network is not equipped with the weights that store the information in traditional connectionist models. Here, all links have weights with unit value. Only the links leading to the defuzzification layer can also be interpreted as having weights if the parameters of the membership functions are taken as such. The network is furthermore not fully interconnected as a normal MLP would be.

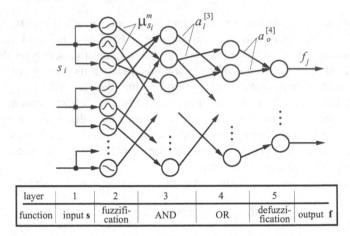

Fig. 17.11. Hybrid neuro-fuzzy system FALCON following [17.32]

The same neuro-fuzzy structure is implemented by [17.46], who apply it for a neuro-fuzzy diagnosis. The only structural difference to the method described above is the use of the maximum as OR-operator. For the diagnosis of incipient motor faults, the use of FALCON is also reported by [17.2].

A comparable structure was also proposed by [17.39]. Triangular membership functions are used in their *Neuro-Fuzzy Classification System* (NEFCLASS). The inference employs MinMax-operators. It is, however, alternatively suggested to compute a weighted sum as t-conorm in the last layer. The weights are fixed to one. This leads to a simple summation of the rule activations as an OR-operator.

A different network structure is proposed by [17.3]. Here, the distinction between the logic AND and OR is broken up. Instead, the *System for Adaptive Rule Aquisition with Hebbian Learning* (SARAH) uses a so-called ANDOR-neuron model that is able to represent both, a t-norm as well as a t-conorm, see Chapter 23.3. It is a parametric composite of the bounded difference operator (t-conorm) and the bounded sum operator (t-norm). Its feature is a continuous change between a logic AND and a logic OR by means of a parameter α. The operator is then approximated by a sigmoidal function. The resulting neuron resembles a standard sigmoidal perceptron.

The function of the individual neurons is determined during training. The input is fuzzified by Gaussian membership functions in the antecedent layer of the SARAH structure. Within the output layer, the rule activation scales the associated output singletons. This yields a crisp output value.

c) Takagi-Sugeno neuro-fuzzy systems

The second large class of neuro-fuzzy systems that has increased in importance over the recent years is given by the networks with a Takagi-Sugeno (TS) structure. The difference to the networks described above lies in the consequence of the fuzzy rules. Instead of a fuzzy attribute or singleton, the consequence is given as a linear or nonlinear function of the inputs. The correspondence to a rule like (17.14) is then:

$$R_l : \text{IF } s_1 \text{ is } A_1 \text{ AND } s_2 \text{ is } A_2 \; \ldots \; \text{AND } s_{n_s} \text{ is } A_{n_a} \text{ THEN } f_j = f_{l,j}(\mathbf{s})$$
$$(17.15)$$

The most frequent case is the linear function in the consequence:

$$R_l : \text{IF } s_1 \text{ is } A_1 \text{ AND } s_2 \text{ is } A_2 \; \ldots \; \text{AND } s_{n_s} \text{ is } A_{n_a}$$
$$\text{THEN } f_{l,j} = w_0 + w_1 s_1 + w_2 s_2 + \; \ldots \; + w_{n_s} s_{n_s} \qquad (17.16)$$

The overall output is then determined as the weighted sum over all consequences.

A frequently employed learning algorithms is the *Adaptive-Network-based Fuzzy Inference System* (ANFIS) proposed by [17.28]. The inputs are used twice: They are inputs of the fuzzification layer and at the same time inputs of the consequences. The membership functions are Gaussian exponential functions and the t-norm is a product operator. An important step is the normalization of the membership functions performed in the 4th layer.

A very similar structure is present in the *Local Linear Model Tree* (LOLIMOT) approach presented by [17.40]. The only difference to the ANFIS-structure is the use of multi-dimensional membership functions whereas ANFIS uses a grid-based structure of membership functions with explicit AND-operator. The two approaches further differ in their learning algorithms.

TS-fuzzy systems are usually employed for the approximation of nonlinear functions in general or, more specific for control, in regression problems as part of dynamic model identification. As tools for the dynamic modelling they have been used for fault detection with multiple models as well as for the generation of nonlinear parity equations. Design and applications of nonlinear parity equations from these models are described in [17.4], other examples of TS models in fault detection applications can be found in [17.33] and [17.51].

As function approximation schemes, TS systems are not typical classification tools. However, they can be used as local polynomial classification systems for fault diagnosis, [17.19].

17.3.2 Identification of membership functions

The first step in the training of most neuro-fuzzy systems is to create a set of membership functions that constitute the fuzzy attributes like "small" or "large". Many different approaches exist to create the fuzzy attributes. To characterize these, one has first to distinguish whether the membership functions are uni-variate or defined on the complete multi-dimensional input domain. In the first case, one will always end up with fuzzy systems based on some kind of grid considering the segmentation of the input space that is possible from the combination of the individual membership functions. Multi-dimensional functions offer a more general segmentation of the input domain - helpful for the classification performance, but impractical for a good understanding of the system.

The *design of the membership functions* can follow three basic principles:

- *Manual construction:* The first and simplest choice is to design the shape and number of the functions manually. This can be done purely following a grid-based approach or, usually preferable, from prior knowledge through visual inspection of the data or knowledge of the physical behavior of the input data. In the latter case, it is desirable to choose the parameters defining the functions so that the different classes can optimally be separated using the selected functions.
- *Concurrent rule and MSF construction:* Some of the algorithms design the rule base concurrently with the membership functions. This encloses in particular the various procedures relying on splits of the input domain. These methods will not be examined further in this section.
- *Batch data-based construction:* Mainly clustering approaches are used for data-based construction algorithms. Different competitive training algorithms are also applied.

Besides these general construction procedures, an error-based refinement of the functions is usually added during training of neuro-fuzzy systems.

The data-based construction of fuzzy attributes is most frequently implemented with a clustering approach. [17.3], [17.42] and [17.59] report the use of unsupervised input space clustering approaches. They base on the assumption that data points close to each other in the symptom space will also belong to the same class and hence can be described with a joint attribute or cluster. Once the clusters are found and every training data point is assigned to a cluster center, that prototype will form the base of the function with the spread of the points specifying the width.

Most frequently, the Fuzzy-C-Means (FCM), [17.6] algorithm, a fuzzified version of the hard c-means algorithm for clustering, is applied.

17.3.3 Identification of rules with predefined membership functions

The second step in the training of neuro-fuzzy systems is the identification of rules. The problem addressed with neuro-fuzzy systems is that conventional fuzzy models suffer from combinatorial rule explosion: In other words, the model complexity grows exponentially with the input dimension. Most problems though, inherently possess a much smaller complexity. Typically, a number of at most 50-100 fuzzy rules should suffice for many technical diagnosis applications. Besides the fact that the high complexity is usually not required, one should consider that the transparency that is the main reason for using neuro-fuzzy approaches, is lost if the system grows too complex. This means that highly complex problems are not appropriately tackled with neuro-fuzzy approaches and should consequently solved with other algorithms.

For the case that the membership functions are already determined, one observes two main general methodologies for finding the rules: *Sequential Backward Elimination (SBE)* and *Sequential Forward Selection (SFS)*.

a) Sequential backward elimination (SBE)

The approaches based on Sequential Backward Elimination start with all possible rule combinations that can be built with the membership functions defined. Then a selection process eliminates rules that do not apply for the particular problem. The approach is restricted to diagnosis problems with not too many symptoms. The reason is simple: Because of the combinatorial rule explosion, it is infeasible to work with higher dimensional problems. In general, assuming n_s symptoms (inputs), each with n_a fuzzy attributes being defined on the input domain, will lead to $n_{\text{premise}} = n_a^{n_s}$ possible premise terms. Denoting the number of faults (or output membership functions in the case of regression problems) with n_f, gives a total of $n_{\text{rules}} = n_f n_{\text{premise}}$ possible rules. Even if the rules are limited to n_p premise terms, one has to consider

$$n_{\text{premise}} = \binom{n_s}{n_p} \cdot n_a^{n_p} \qquad (17.17)$$

possible rule premises. For the relatively moderate case of 5 fuzzy attributes, 10 symptoms and 3 faults this yields about 30 million possible rules.

A typical example of the SBE approaches is realized in the learning of the SARAH network. [17.3] reports the following training scheme: A matrix $\mathbf{R}$ is created with n_{premise} rows and n_f columns. This means that all possible rules are represented by an entry in the matrix $\mathbf{R}$. All rules are possible. The matrix elements are initialized with zero. The training requires that input membership functions exist. The algorithm now iterates through the training data. For each data point, the membership values of all fuzzy attributes and from them the rule premise activations are computed. The strongest activation wins. In a similar way, the strongest output attribute is determined and the corresponding entry in $\mathbf{R}$ is then increased by a certain value. The procedure repeats until no data points are left. It is referred to as *Hebbian Learning*.

A nearly identical method was suggested by [17.32] for the rule structure determination of the neuro-fuzzy system FALCON from Figure 17.11. Instead of explicitly forming a matrix containing the association strengths, it is indicated to act directly on the connection weights of the neural structure.

The underlying idea of the mentioned and also many other learning approaches is to find *concurrent activations* in the membership domain occurring *in parallel* to the given output activation. The rules or weights reflecting that input/output behavior are then strengthened or created in the first place. Competing rules that would contradict can be reduced. A threshold of some kind finally decides which rules are kept and which discarded. Since the approach is based on the association strength of the activations of the input and output patterns, it will be referred to as *activated learning*.

b) Sequential forward selection (SFS)

In contrast to the methods utilizing SBE there are the algorithms that successively build up the rule base. The Sequential Forward Selection methods start with an empty rule base and iteratively augment new rules to model the training data. The main distinction between the different SFS algorithms are then the criteria to choose a rule and stop the rule growing process.

The SFS approaches have to deal with contradictory or redundant rules, just as the SBE algorithms. Here, the same checks and procedures that were mentioned in the previous section can be applied.

In general, the rule construction process of the SFS methods can be driven by the iteration through all data points or by the minimization of an appropriate overall error measure. The advantage of the first procedure in contrast to the SBE methods is that the complexity for activated learning scales only with the number of data points. This makes small but highly dimensional data sets feasible.

The NEFCLASS approach for instance processes a new data point only if there is no existing rule that describes its input/output relationship. This way, the data points contributing new information are found. An activated learning then determines whether the activation is strong enough to justify a new rule. [17.38] suggests different learning criteria: "Simple" rule learning utilizes the data with the best classification performance, while "best per class" proceeds with one best point per class.

c) Identification of rules concurrent with membership functions

Structurally different from the methods described above are the approaches that create the rule structure and membership functions at the same time. They are usually based on an overall error measure. It is normally applied with a least-squares objective function. Using all training data points, the overall objective is determined. As long as a further reduction of it is desired, more rules and membership functions are created to better fit the training data. This usually means some sort of segmentation of the input space or placing of locally valid basis functions which will continue until the necessary granularity of data description is reached.

This concept is for instance realized in the Neural Gas approach by [17.17]. This method determines the centers of radial basis function networks by placing new basis functions at locations where the error is the highest. A similar objective is followed be the LOLIMOT-learning that is based on a recursive splitting of the input domain, thereby concurrently creating the fuzzy input segmentation.

While the LOLIMOT approach and other RBF networks are difficult interpreted linguistically, other approaches use the error-based method more directly to create fuzzy systems: [17.1] presented a hyperbox approach for fuzzy classification. The approach is simple: The data of the two different classes is fitted each into a separate hyperbox. If two boxes overlap, the overlap is removed by creating smaller boxes within the overlapping region. This continues until any overlap is removed. A projection of the boxes creates the fuzzy attributes.

17.3.4 Optimization methods

Due to the equivalence of neuro-fuzzy systems with standard neural networks, they can be optimized with the identical methods that are employed for normal neural networks. Depending on the structure of the neuro-fuzzy systems, common linear and nonlinear optimization methods are used.

Since almost all authors report the use of least-squares objective functions, the optimization becomes especially simple for those parameters that influence the objective linearly. This usually applies for the neuro-fuzzy systems with a radial basis function equivalent structure such as the LOLIMOT network utilizing singleton or linear function output layers.

Most structures, however, change parameters of the first and second layers corresponding to hidden relevance weights or membership function parameters. The FALCON network for instance, establishes a backpropagation learning of the membership layer weights. A standard least-squares objective function is built from the network output vector $\mathbf{f}$ and the desired target vector $\mathbf{t}$ for each of the k data points.

[17.32] state that the parameter optimization is performed relatively easily compared to classical neural networks because the clustering approach for finding initial membership functions guarantees a good starting point for the parameters.

The NEFCLASS approach also tunes parameters of the membership functions only. A difference to the approach above is the selective optimization. Instead of a parallel optimization of all MSF parameters, only a few membership functions are adjusted in each optimization step: A single training pattern is presented to the network and its outputs computed. Each rule unit with an output error > 0 will then propagate its error back to the rule neuron. Since the conjunction operator in the rule unit is a minimum, one can easily identify the membership function activation that dominates the others. It will be the one with the smallest activation level. Only the parameters of that membership function will be adjusted to yield a higher or smaller activation for the present data sample depending on the sign of the error.

The SARAH network differs from the ones above as all parameters of it are subject to an optimization. [17.3] reports the use of an unconstrained backpropagation algorithm.

The ANFIS as well as the LOLIMOT structure are equipped with parameters that influence the output error linearly. The weights of the linear function in the consequence layer (17.16) can be optimized in a single-step pseudo-inverse optimization. The antecedent parameters of the ANFIS network are subsequently optimized with a nonlinear optimization (gradient descent). The LOLIMOT algorithm differs from that as the antecedent layer parameters determining the positions of the MSF are found with a simple, iterative splitting procedure.

More details on this survey are given in [17.19]. There also conditions for the transparency, i.e. useful rule examination and interpretation are stated, like selection of membership functions, local behavior, weight constraints and a small rule basis.

17.3.5 Self-learning classification tree (SELECT)

a) Select structure

The SELECT approach according to [17.20], [17.19] combines ideas from the field of decision trees with the adaptation properties of neural networks and the interpretation of fuzzy logic. Its basic element is a simple neuro-fuzzy structure that consists of a fuzzification layer similar to those discussed in the previous sections and a neural

AND-operator that is implemented in a straightforward manner as an approximated bounded-sum operator. From the discussion of the previous section, a learning procedure with a sequential forward selection (SFS) strategy seemed advisable for the application of fault diagnosis. Also, a constrained optimization is implemented to retain the desired transparency.

Figure 17.12 shows this basic element composed of the operator and the fuzzy attribute layer. The pictured operator is equivalent to the ANDOR-operator from Figure 23.4 in Chapter 23.3 with $\alpha = 1$. Due to the sigmoidal approximation of the threshold, this results in a soft version of the conjunction. The output can assume values close to one even if not all inputs are equal to one.

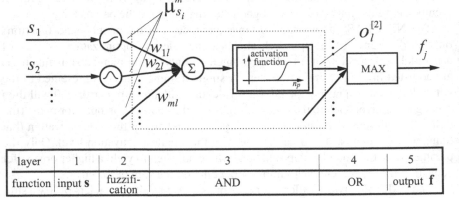

layer	1	2	3	4	5
function	input **s**	fuzzification	AND	OR	output **f**

Fig. 17.12. Basic processing element (rule R_l) and OR-operation for output class F_j of the SELECT-structure. From the left, the inputs s_i are fed into the fuzzification layer. This yields membership values $\mu_{s_i}^m$ that describe how strongly the symptom s_i belongs to the fuzzy attribute A_i^m. Multiple membership values are then fed into the AND-operator by computing a weighted sum. The activation function is then applied to this sum. Finally, the outputs of the AND-operators that describe a rule with the same rule conclusion, are combined in the OR-operator ("accumulation"). The output f_j of the SELECT-system is a measure for the corresponding fault class F_j.

The connections from the fuzzy attribute m to the operator neuron l are equipped with weights w_{ml} that in combination with the operator can be interpreted as relevance weights. This yields an output activation given by

$$o_l^{[2]} = \frac{1}{1 + e^{-\lambda\left(\sum_{m=1}^{n_p} w_{ml}\mu_{s_i}^m - 0.75 n_p\right)}} \tag{17.18}$$

Here, n_p is the number of inputs into the operator. λ controls the slope of the function, thereby determining the "degree of fuzziness" of the operator. A large λ brings the operator closer to the binary equivalent. The threshold of the sigmoidal function at $0.75 n_p$ was empirically chosen to yield a fuzzy version of the AND-operation. The symbol for the membership value is an abridged notation. In fact, the chosen fuzzy attribute depends not only on s_i and m but rather on the individual rule premise. Since the fuzzy attribute can be one out of the $n_a(s_i)$ attributes defined for the symptom s_i, a more precise notation would be:

$$\mu_{s_i}^m = \mu_{s_i}^{\kappa(m)} \tag{17.19}$$

where κ is from $[1; n_a(s_i)]$. For simplicity reasons, however, the abridged version is kept for the remainder of this chapter.

The specialty of the SELECT-approach is the tree structure composed of AND-operators. This creates an equivalent representation as a tree composed out of fuzzy rules. The nodes of the tree consist of AND-operators while the leaves contain a consequence of the rule, see Figure 17.13. The output of the operator determines the rule fulfillment $o_i^{[2]}$ of the consequence. Multiple leaves can contain an identical consequence. The degrees of fulfillment of the rules are then combined by an appropriate t-conorm (OR-operation) such as the maximum yielding the output f_j of the SELECT structure.

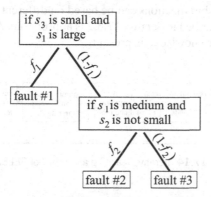

Fig. 17.13. Tree representation of a SELECT system. In this simple tree there are no OR-operators necessary, because there are no two rules with the same conclusion

The processing of the tree is performed in a sequential manner: After evaluating the top node, a first neuron output $o_1^{[2]}$ is determined. The next node is then multiplied with $(1 - o_1^{[2]})$, so that its activation is given by $o_2^{[2]} = (1 - o_1^{[2]})o_2'^{[2]}$, where $o_2'^{[2]}$ is the output from (17.18). This multiplication is continued so that the sum of all activations $o_i^{[2]}$ will not exceed 1.

For large trees that tend to create small outputs because the sigmoidal function yields an output $o_i'^{[2]} > 0$ even during small activations, the multiplication is omitted. Otherwise, the multiplication would lead to an ever decreasing output of the subsequent neurons.

The steps for the determination of a SELECT system from data and prior knowledge consist of the design of membership functions, followed by a structure learning based on prior knowledge and finally a neural network-type optimization. These steps will be explained below.

b) Learning algorithm

The learning approach is based on the observation that a number of faults must be distinguished by relatively many symptoms that span a sparsely populated symptom space due to few experimental data. This makes sequential backward elimination approaches unsuitable. Another observation is that not all symptoms are necessary to isolate every fault situation. Instead, some faults are easily identified with only one or two significant symptom reactions. This favors systems that utilize an adaptive complexity depending on the difficulty of the decision.

The following learning approach has been inspired by *decision tree learning algorithms*. As such, it is relatively unique for the area of fault diagnosis and neuro-fuzzy systems. The procedure is iterative and relies on the reduction of the training data set. Figure 17.14 shows a flowchart of it. The main steps membership function design, candidate rule selection, and rule evaluation are explained in the following.

Design of Membership Functions
The design of membership functions can be based on different principles. Generally, any of the methods described in Section 17.3.2 is appropriate. A manual construction is advantageous if prior knowledge is present.

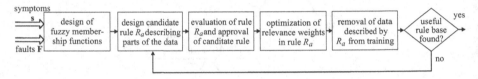

Fig. 17.14. Iterative learning algorithm of SELECT.

For the SELECT approach, a fuzzy clustering with the Fuzzy-C-Means (FCM) algorithm based on a one-dimensional projection of the training data proofed to work sufficiently well. Due to the usually nonuniform distribution of measured symptoms, however, a modified version, the *Degressive FCM* algorithm was developed. It is explained in [17.19]. Its main feature is to include a nonlinear transformation of the cluster membership to accommodate uneven data distributions. The result is a set of clusters that is usually finer around zero (symptom is not reacting) and broader towards the limits of the symptoms. After determining the cluster centers and assignments of the data points to them, adjacent and highly overlapping clusters were combined. A typical set of fuzzy attributes realized by the membership functions can be seen in Figure 17.15.

The functional form of the fuzzy attributes was chosen to be the product of two sigmoidals. This is advantageous to realize nonsymmetric functions and is simple to implement. The function is:

$$\mu_{s_i}^m(s_i) = \frac{1}{1 + e^{-a_{1,i,m}(s_i - b_{1,i,m})}} \frac{1}{1 + e^{-a_{2,i,m}(s_i - b_{2,i,m})}} \tag{17.20}$$

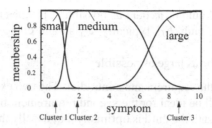

Fig. 17.15. Example of symptom membership functions from clustering. The three cluster centers are noted in the picture.

The four parameters $a_{1,i,m}, b_{1,i,m}, a_{2,i,m}, b_{2,i,m}$ determine the exact shape and location of the membership functions. The attributes at the limits of the data range were chosen to be single sigmoidals.

The approach to derive the parameters of (17.20) from the FCM clustering is as follows: After the cluster centers are found, the data points are uniquely assigned to the one cluster with the center v_m to that they have the highest fuzzy membership degree assigned to by the FCM method (usually the one that is the closest). From all N_m points that were assigned to one cluster v_m, the standard deviation of these points is estimated:

$$\sigma_m = \frac{1}{N_m - 1} \sum_{i=1}^{N_m} (s_i - v_m)^2 \tag{17.21}$$

Combined with the cluster centers, the parameters of the membership functions are chosen to fulfill two requirements:

- the membership must be at least $(1 - \varepsilon)$ at the cluster center, (e.g. $\varepsilon = 0.03$);
- the intersection of two membership functions is at the Bayes-limit, i.e. at the point that would be ideal to differentiate data from the two clusters if Gaussian data distributions are assumed. This point is defined by two neighbored Gaussian distribution probability densities having the same value 0.5.

This way, the attributes are already good separators for the data if the clusters cover points from different fault classes. An additional normalization of the functions is not necessary.

Selection of Candidate Rules

Candidate rules are rules that promise good help in distinguishing the data from different fault classes. They are found in a systematic way and evaluated for performance before they build the SELECT tree structure elements from Figure 17.12.

The basic problem of the selection of candidate rules is this: Given is a set S_t of training data points consisting of data from the faults $F_j, j = 1 \ldots n_f$. Which fuzzy rule $\mathcal{R}_a$ can be found that is activated by a subset S_a such that:

1) If $\mathcal{R}_a$ is evaluated with all training data S_t, the set S_a should yield a high rule fulfillment, while the rest, $\{S_t \setminus S_a\}$ should create an activation as small as possible.

2) The data from S_a should be as pure as possible. This means, all elements should belong to only one fault F_j. Typically, S_a should comprise all data from one particular fault.

3) The set S_a should be as large as possible.

It is clear that some of the requirements contradict. The two extremes are: S_a is only one data point. This will be ideal for the second requirement but bad for the third. If $S_a = S_t$ on the other hand, the third is optimally, but usually the second requirement not fulfilled.

A sensible way to arrive at a useful rule R_a is to successively tighten its condition, i.e. the rule premise of R_a, and check if the requirements 1-3 are fulfilled. This is done by taking more and more premise terms into account. This procedure makes R_a only as complex as necessary.

The selection of a candidate rule R_a is based on the training data set. With the membership functions, each data sample is assigned to the fuzzy attribute A_i^m that yields the highest membership function value $\mu_{s_i}^m$. That way, the training set is translated into a training set S_{ft} with fuzzy attributes instead of observed symptom quantities. The procedure continues as follows:

1) For each fault F_j, a candidate rule R_j is built. These rules start with a premise containing a single term like "IF s_3 large". These terms are selected as follows: For each symptom s_i, the most frequent entry in S_{ft} belonging to fault class F_j is found. The one symptom where the same term appears most often in S_{ft} is then chosen to build the starting rule for fault F_j.
   That way, the number of rules equals the number of faults F_j;

2) All rules R_j are evaluated with the method explained below. If a satisfactory rule R^* is found, it will be kept. If none of the rules performs good enough, their premise will be extended by an additional input dimension, i.e. by an additional term like: "AND s_5 small". Again, the performance is evaluated and, if necessary, the rule premises augmented up until a specified complexity $n_{p,max}$ is reached;

3) The chosen rule R^* forms a new neuron following Figure 17.12;

4) The correctly classified data set S_{cj} is removed from the training set. This means that the training data set decreases in size, which in turn makes the rule extraction for the subsequent neurons simpler;

5) The procedure then repeats with the smaller training set.

The learning is finished as soon as the training set is empty, no useful rule can be found, or a predefined complexity is reached. The learning will lead to a tree-like structure of rules that are evaluated sequentially.

Evaluation of Candidate Rules
For the evaluation of R_a with the conclusion F_j w.r.t. the requirements above, a threshold δ with $S_a = \{s \in S_t \mid o^{[2]}(s) > \delta\}$ is defined. This will lead to four subsets of the complete training data set S_t:

$$\begin{aligned}
S_{cj} &= \{\mathbf{s} \mid a(\mathbf{s}) > \delta, F(\mathbf{s}) = F_j\} \\
S_{mj} &= \{\mathbf{s} \mid a(\mathbf{s}) > \delta, F(\mathbf{s}) \neq F_j\} \\
S_{mo} &= \{\mathbf{s} \mid a(\mathbf{s}) < \delta, F(\mathbf{s}) = F_j\} \\
S_{co} &= \{\mathbf{s} \mid a(\mathbf{s}) < \delta, F(\mathbf{s}) \neq F_j\}
\end{aligned} \tag{17.22}$$

It becomes evident that S_{cj} denotes the data correctly classified with rule $\mathcal{R}_a$ as belonging to fault F_j and S_{co} the data correctly classified as not belonging to F_j. S_{mj} and S_{mo} are the sets of misclassified data. The definition of a classified data point is an output $o^{[2]}$ greater than δ. The *cardinalities*, i.e. number of the sets are denoted as $N_{cj}, N_{mj}, N_{co}, N_{mo}$. Additionally, $N_j = N_{cj} + N_{mj}$ and $N_o = N_{co} + N_{mo}$. The total number of training data point is given by $N = N_o + N_j$.

The classification performance of $\mathcal{R}_a$ can now be evaluated following the ideas of decision trees based on a measure of entropy. They can be used to evaluate the impurity of the data sets after applying the rule $\mathcal{R}_a$ compared to the impurity of the data set before applying the rule. Ideally, $\mathcal{R}_a$ should distinguish the data such that all data points from F_j are separated from the rest of the data. Ideally, that would yield pure data sets and a minimal entropy.

Instead of calculating the logarithms of the entropy, however, a simpler strategy was used: [17.55] suggests to work with the cardinalities $N_{cj}, N_{mj}, N_{co}, N_{mo}$. They state that a maximization of the following entropy

$$S_{Sun} = \frac{N_j}{N}\left(\left(\frac{N_{mj}}{N_j}\right)^2 + \left(\frac{N_{cj}}{N_j}\right)^2\right) + \frac{N_o}{N}\left(\left(\frac{N_{mo}}{N_o}\right)^2 + \left(\frac{N_{co}}{N_o}\right)^2\right) \tag{17.23}$$

is sufficient for all practical purposes. In the SELECT approach, however, (17.23) will not work, since it is symmetric, i.e. it will also be maximized if the rule $\mathcal{R}_a$ is fulfilled by all but data from F_j. Therefore, a modified measure was created:

$$S = \frac{N_j}{N}\left(-\left(\frac{N_{mj}}{N_j}\right)^2{}' + \left(\frac{N_{cj}}{N_j}\right)^2\right) + \frac{N_o}{N}\left(\left(\frac{N_{mo}}{N_o}\right)^2 - \left(\frac{N_{co}}{N_o}\right)^2\right) \tag{17.24}$$

It possesses the desired properties: It is maximal for a perfect separation and penalizes wrong rule fulfillment.

Besides S, also a simpler measure of the rule performance was heuristically found to be useful. It measures the difference $\Delta o_j^{[2]}$ of the mean rule fulfillment between data from class F_j and all other data:

$$\Delta o_j^{[2]} = \frac{\sum_{(s(k) \in S_a)} o_j^{[2]}(k)}{N_{cj}} - \frac{\sum_{(s(k) \notin S_a)} o_j^{[2]}(k)}{N_{co}} \tag{17.25}$$

where k denotes a running index over the training data set.

The following example will illustrate the rule selection and evaluation approach.

Learning example for SELECT

Figure 17.16 shows a simple two-dimensional problem with three fault classes F_j. For each dimension, two fuzzy attributes have been found. They are also shown in Figure 17.16. The three first *candidate rules* $\mathcal{R}_a$ are:

$$\mathcal{R}_1 : \quad \text{IF } s_2 \text{ small THEN fault } F_1$$
$$\mathcal{R}_2 : \quad \text{IF } s_1 \text{ large THEN fault } F_2 \qquad (17.26)$$
$$\mathcal{R}_3 : \quad \text{IF } s_1 \text{ small THEN fault } F_3$$

It should be emphasized that this set of rules does not suffice to solve the problem. It is only a selection of possibilities from which a first rule can be taken. The second rule $\mathcal{R}_2$ performs best and is kept as $\mathcal{R}^*$. The procedure removes all data belonging to class F_2 from the training because it is correctly described by $\mathcal{R}^*$. Again, new rules are created:

$$\mathcal{R}_1 : \quad \text{IF } s_2 \text{ small THEN fault } F_1$$
$$\mathcal{R}_3 : \quad \text{IF } s_2 \text{ large THEN fault } F_3 \qquad (17.27)$$

It should be noted that a rule for class 2 does not need to be designed any more. Considering the data, both rules perform equally well. The first is retained as $\mathcal{R}^*$, because there are more data samples from class F_1. Since the remaining data belongs only to F_3, there is no need to divide the data using an additional rule. The final SELECT rules are:

$$\text{IF } s_1 \text{ large THEN fault } F_2$$
$$\text{ELSE IF } s_2 \text{ small THEN fault } F_1 \qquad (17.28)$$
$$\text{ELSE fault } F_3$$

It should be noted that the use of the keyword for the alternative, "ELSE", shows up in the rules.

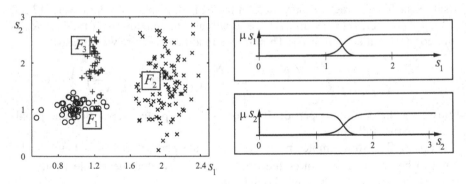

Fig. 17.16. Simple three-class problem. The membership functions from the clustering are shown on the right.

The example shows a characteristic of SELECT structures: The most informative premises are selected first. Furthermore, the system tends to create a smaller complexity than a standard parallel rule base would. The reason is that the sequential structure essentially "hides" complexity in the "ELSE" statement.

For comparison, the complete parallel rule set for the problem would be:

$$
\begin{aligned}
&\mathcal{R}_1^p : \quad \text{IF} \quad (s_1 \text{ small}) \quad \text{and} \quad (s_2 \text{ small}) \quad \text{THEN fault} \quad F_1 \\
&\mathcal{R}_2^p : \quad \text{IF} \quad (s_1 \text{ small}) \quad \text{and} \quad (s_2 \text{ large}) \quad \text{THEN fault} \quad F_3 \\
&\mathcal{R}_3^p : \quad \text{IF} \quad (s_1 \text{ large}) \quad \text{and} \quad (s_2 \text{ small}) \quad \text{THEN fault} \quad F_2 \\
&\mathcal{R}_4^p : \quad \text{IF} \quad (s_1 \text{ large}) \quad \text{and} \quad (s_2 \text{ large}) \quad \text{THEN fault} \quad F_2
\end{aligned}
\tag{17.29}
$$

The last step of the SELECT learning approach is an optimization of the weights w_{ij}. This is accompanied by a pruning process and will be shown in the next section.

c) Optimization of relevance weights

The rule extraction phase is followed by an optimization of the free parameters to increase the performance of the system. The optimization allows a fine-tuning to better adapt to the data distribution. This is advantageous because the rules are up to now based on the fuzzy membership functions from the one-dimensional clustering. By a fine-tuning of the relevance weights, one can influence the region that is selected with a rule to a certain extend. This optimization will be performed on the weights of multiple inputs simultaneously. In contrast to the one-dimensional clustering, it therefore can benefit from information concering the correlation of the symptoms.

The procedure follows the neural network training approaches. The objective is to minimize a least-squares function

$$
E = \sum_{k=1}^{N} (\mathbf{f}(k) - \mathbf{t}(k))^2 \longrightarrow \min.
\tag{17.30}
$$

where N is the number of training samples and $\mathbf{t}(k)$ denotes the target data. $\mathbf{f}$ are the fault measures. A 0/1 scheme is used for $\mathbf{t}$. Subject to the optimization are the relevance weights w_{ml}. The optimization is performed for each element individually: Instead of a parallel optimization of all parameters, only the weights that are connected to one AND-neuron are optimized. The data base used for this optimization is not always the complete training set. It rather follows the learning approach: In the same way as during learning, the training set is reduced from one rule to the next. Data samples that are correctly identified will not be included.

In addition to this data reduction, the data samples that lead to a strong rule fulfillment must be identified. This is done by evaluating the output $o_l^{[2]}$ of the AND-operator if the data belonging to its conclusion is presented. This subset of the training data and the data belonging to other classes are included into the optimization. The optimization drives the output

$$
o_l^{[2]}(s_i(k)) = \frac{1}{1 + e^{-\lambda \left(\sum_{i=1}^{n_s} w_{ml} \mu_{s_i}^m - (0.75 n_p) \right)}}
\tag{17.31}
$$

of the rules with n_p inputs towards 0 or 1.

In order to ensure the interpretation of the relevance weights, the following constraints were introduced:

$$0 < w_{ml} < 1 \qquad \sum_{m=1}^{n_p} w_{ml} = n_p \qquad (17.32)$$

For the optimization, the standard procedure in neural networks is to employ a back-propagation algorithm. The partial derivative of $o_l^{[2]}(s_i(k))$ can favorably be calculated with the known derivative of the sigmoidal function

$$f(x) = \text{sig}(x) = \frac{1}{1 + e^{-x}} \qquad (17.33)$$

The further procedure applying the softmax function, [17.7], is described in [17.19].

An important property of the optimization algorithm is the aforementioned sequential optimization with decreasing training set size. Instead of one massively parallel optimization as common for neural networks, here a series of smaller optimizations is performed. This is of great computational advantage. A typical optimization might include 10 individual optimization runs with 4 parameters instead of a parallel optimization with 40 free parameters. If the SELECT system consists of n_r operators with each having n_p inputs, it will create a computational need of the order $n_r \cdot \mathcal{O}(n_p^2)$ instead of $\mathcal{O}((n_r n_p)^2)$. This lets a typical optimization of the structure be performed in seconds to minutes on a standard Personal Computer and also reduces the problems associated with a high-dimensional but sparsely populated parameter space (known as *curse of dimensionality*).

Even more efficient and faster for large data sets than the nonlinear optimization with the softmax-transformation is an optimization based on a *linearization of the sigmoidal function* of the AND-neuron. The idea is to work with an error description that is linear with regard to the relevance weights. This enables the use of fast optimization algorithms. The idea of the approach is visualized in Figure 17.17. It shows the output $o_l^{[2]}$ as a function of the weighted sum of the inputs, here abbreviated with ρ. By using this approximation of the operator function, the problem is posed with an objective function that the weights w_{ml} enter linearly. The constraint that the weights add up to unity can also be included. The optimum is then found by a one-step procedure. The derivation of the method can be found in [17.19]. An optimization with the linearization showed results comparable to the nonlinear optimization.

The described SELECT method offers some possibilities to incorporate *prior knowledge*:

1) Manual placement of membership functions.
2) Manual augmentation of the rule structure with rules at the top. The weights of these manually added rules can be improved by the optimization.
3) Structural information about similar fault situations can help building a more hierarchical structure that is easier to interpret physically.

The manual introduction of fuzzy rules is especially useful if the fault–symptom relationships of some fault situations are known or can be derived from physical understanding of the process under consideration. This also enables to manually place

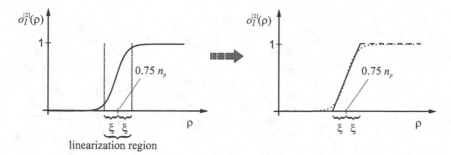

Fig. 17.17. Linearization of the ANDOR-operator. The region of the sigmoidal function with the transition from close to 0 to the region close to 1 is replaced by a straight line. This corresponds to the original AND-operator, see Chapter 23.3. The dashed lines indicate the region where the sigmoidal function is replaced by a straight line.

fuzzy rules in the tree, see [17.19]. Applications of this SELECT self-learning diagnosis methods are shown in Sections 20.1.5 and in [17.19] and [17.62]. The required computation time is relatively small, i.e. in the range of seconds to minutes.

17.4 Problems

1) Establish a fault tree of an illumination system, consisting of a power supply, cable, plug, fuse, switch and two lamps in parallel connection. The faults are defects of the components. Symptoms are 1, 2 or no lamp burning. Design a fault-diagnosis systems.
2) State the differences between fuzzy-logic and probabilistic reasoning.
3) In which cases is forward chaining as well as backward chaining feasible for fault diagnosis?
4) Design a fuzzy logic diagnosis system for a dust cleaner, consisting of a power switch, a universal motor with fan, a dust container, a vacuum pressure sensor, a flexible tube and a nozzle. Observable outputs are the noise, pressure, operation time since replacing the dust container, and suction power.
5) A DC motor driven drill machine shows irregular speed. Establish a fault-symptom tree and a fuzzy-logic diagnosis system if the observations are speed for idling and for load, brush fire and motion of the cable.

Part IV

Fault-Tolerant Systems

18
Fault-tolerant design

The improvement of reliability can be increased by two different approaches, *perfectness or tolerance*, [18.4]. Perfectness refers to the idea of avoiding faults and failures by means of an improved mechanical or electrical design. This includes the continued technical advancement of all components that increase the service life. During operation the intactness of the component must be maintained by regular maintenance and replacement of wearing parts. Methods that facilitate fault detection at an early stage allow for replacing the regular maintenance schedule with a maintenance-on-demand scheme, as discussed in Chapter 2 and 3.

Tolerance describes the notion of trying to contain the consequences of faults and failures thus that the components remain functional. This can be reached by the principle of *fault-tolerance*, see also Section 2.4. Herewith, faults are compensated in such a way that they do not lead to system failures. The most obvious way to reach this goal is *redundancy* in components, units or subsystems. However, the overall systems then become more complex and costly. In the following, various types of fault-tolerant methods are reviewed briefly, see [18.3].

Fault-tolerance methods generally use *redundancy*. This means that in addition to the considered module, one or more modules are connected, usually in parallel. These redundant modules are either *identical* or *diverse*. Such redundant schemes can be designed for hardware, software, information processing, and mechanical and electrical components like sensors, actuators, microcomputers, buses, power supplies, etc.

18.1 Basic redundant structures

There exist mainly two basic approaches for fault-tolerance, static redundancy and dynamic redundancy. The corresponding configurations are first considered for *electronic hardware* and then for other components. Figure 18.1a shows a scheme for *static redundancy*. It uses three or more parallel modules that have the same input signal and are all active. Their outputs are connected to a voter, who compares these signals and decides by majority which signal value is the correct one. If a triple modular-redundant system is applied, and the fault in one of the modules generates a wrong output, this faulty module is masked (i.e. not taken into account) by the two-out-of-three voting. Hence, a single faulty module is tolerated without any effort for specific fault detection, n redundant modules can tolerate $(n-1)/2$ faults (n odd).

To improve the fault tolerance also the voter can be made redundant, [18.8]. Disadvantages of static redundancy are high costs, more power consumption and weight. Furtheron, it cannot tolerate common-mode faults, which appear in all modules because of common fault sources.

Dynamic redundancy needs less modules at the cost of more information processing. A minimal configuration consists of two modules, Figure 18.1b and c. One module is usually in operation and, if it fails, the standby or back-up unit takes over. This requires fault detection to observe if the operation modules become faulty. Simple fault-detection methods only use the output signal for, e.g. consistency checking (range of the signal), comparison with redundant modules or use of information redundancy in computers like parity checking or watchdog timers. After fault detection, it is the task of the reconfiguration to switch to the standby module and to remove the faulty one.

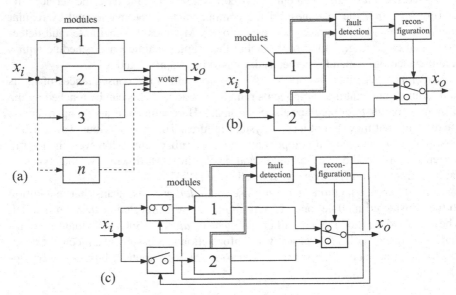

Fig. 18.1. Fault-tolerant schemes for electronic hardware: (a) static redundancy: multiple-redundant modules with majority voting and fault masking, m out of n systems (all modules are active);(b) dynamic redundancy: standby module that is continuously active, "hot standby"; (c) dynamic redundancy: standby module that is inactive, "cold standby"

In the arrangement of Figure 18.1b, the standby module is continuously operating, called "*hot standby*". Then, the transfer time is small at the cost of operational aging (wear-out) of the standby module.

Dynamic redundancy, where the standby system is out of function and does not wear, is shown in Figure 18.1c, called *"cold standby"*. This arrangement needs two more switches at the input and more transfer time due to a start-up procedure. For both schemes, the performance of the fault detection is essential.

Dynamic redundancy can be extended to two and more standby modules, thus tolerating two and more faults. Combinations of static and dynamic redundancy lead to *hybrid redundant schemes* to avoid the disadvantages of both ones on cost of higher complexity, [18.8].

Similar redundant schemes as for electronic hardware exist for *software fault-tolerance*, i.e. tolerance against mistakes in coding or errors of calculations, compare Table 18.1. The simplest form of static redundancy is repeated running ($n \geq 3$) of the same software and majority voting for the result. However, this only helps for some transient faults. As software faults in general are systematic and not random, a duplication of the same software does not help. Therefore, the redundancy must include diversity of software, like other programming teams, other languages, or other compilers. With $n \geq 3$ diverse programs, a multiple-redundant system can be established followed by majority voting as in Figure 18.1a. However, if only one processor is used, calculation time is increased and using n processors may be too costly.

Dynamic redundancy by using standby software with diverse programs can be realized by using recovering blocks. This means that in addition to the main software module, other diverse software modules exist, [18.8], [18.5].

For digital computers (microcomputers) with only a requirement for fail-safe behavior, a duplex configuration like Figure 18.2 can be applied. The output signals of two synchronized processors are compared in two comparators (software) which act on two switches of one of the outputs. In case of disagreement of the output signals one of the switches disconnects the output from the following components. This scheme covers both, hardware and software faults but is not fault tolerant. It is useful if the miss of the output brings the system in safe-state. (This fail-safe system is, e.g. used for ABS braking systems).

Fault-tolerance can also be designed for purely *mechanical and electrical systems*. Static redundancy is very often used in all kinds of homogeneous and inhomogeneous materials (e.g. metals and fibers) and in special mechanical constructions like lattice-structures, spoke-wheels, dual tires or in electrical components with multiple wiring, multiple coil windings, multiple brushes for DC motors and multiple contacts for potentiometers. This quite natural built-in fault-tolerance is generally characterized by a parallel configuration like in Figure 18.3a. However, the inputs and outputs are not signals but, e.g. forces, electrical currents or energy flows, and a voter does not exist. All elements operate in parallel and if one element fails (e.g. by breakage) the others take over a higher force or current, following the physical laws of compatibility or continuity. Hence, this is a kind of "stressful degradation".

Examples are two gears, two belts, two chains, two valves, two hydraulic cylinders, two power supplies or two electrical motors with each half load in normal operation. Further examples are the tandem-piston-system for hydraulic brakes or double magnet ignition for aircraft engines. Fault tolerance by redundant kinematics was

Table 18.1. Fault-tolerance methods for different systems

Systems	Fault-Tolerance Methods							
	Multiple redundant modules (static)				Multiple standby modules (dynamic)			
	Type	Modules	Selection	Fault masking	Type	Modules	Fault detection	Reconfiguration
Electronic Hardware	Static redundancy n–out-of-m	Identical $n \geq 3$	Majority logic voting	Yes	Dynamic redundancy: hot standby, cold standby	Identical $n \geq 2$	Simple output checks	Switch to standby module: output, input and output
Software	Multiple version programming	Repeated running	Majority logic voting	Yes	Recovering blocks (like cold standby)	Diverse program modules	Acceptance tests with outputs	Switch to standby program and back to recovery point
		Diverse programs $n \geq 3$	Majority logic voting	Yes				
Mechanical systems	Static redundancy	Identical $n \geq 2$	Physical compatibility	No	Dynamic redundancy: cold standby	Identical or diverse $n \geq 2$	Simple output checks	Switch to standby unit
Electrical systems	Static redundancy	Identical $n \geq 2$	Physical compatibility	No	Dynamic redundancy: cold standby	Identical or diverse $n \geq 2$	Simple output checks	Switch to standby unit
Mechatronic systems	–	–	–	–	Dynamic redundancy: cold standby	Identical or diverse $n \geq 2$	Sophisticated: inputs & outputs	Switch to standby unit

proposed by [18.9]. Mechanical and electrical systems with dynamic redundancy as depicted in Figure 18.1b, c can also be built. Hot standby, where the standby unit continuously operates in idle running is usually not used because of unnecessary wear or ageing and power consumption. It makes only sense if the interruption during the transfer has to be avoided.

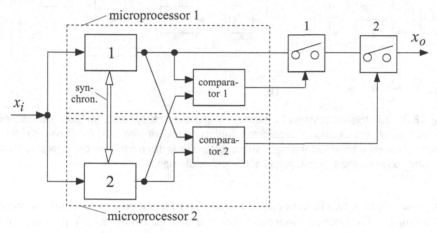

Fig. 18.2. Duplex microcomputer system with dynamic redundancy for fail-safe behavior

Therefore, mostly cold standby is meaningful, Figure 18.3b. Fault detection of the operating unit may be based on measured outputs like position, speed, force, torque, pressure, flow, voltage or current. However, then only large failures like complete break down can be detected. Examples are standby feedwater pumps for steam boilers or backup power generators. To improve fault detection also the input signals and other intermediate signals should be available. As this is rarely the case for pure mechanical components like linkages, gears, or drive chains or pure electrical components like amplifiers, cables, transformators dynamic redundancy can mainly be applied for electro-mechanical systems like speed or position controlled electromotor units, electro-hydraulic systems or electro-mechanical actuators.

Fault tolerance with dynamic redundancy and cold standby is especially attractive for *mechatronic systems* where more measured signals and embedded computers are already available and therefore fault detection can be improved considerably by applying process model-based approaches. Table 18.1 summarizes the appropriate fault-tolerance methods for different systems.

18.2 Degradation steps

Mainly because of costs, space and weight, a suitable compromise between the degree of fault tolerance and the number of redundant components has to be found. In contrast to fly-by-wire systems, for industrial, mobile and mechatronic systems,

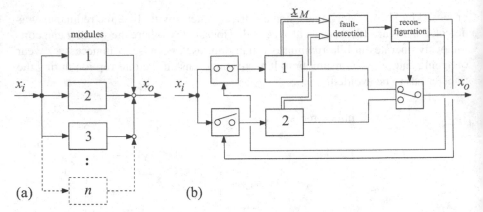

Fig. 18.3. Fault-tolerant schemes for mechanical and electrical systems: (a) Static redundancy for mechanical and electrical components: multiple redundant elements; (b) Dynamic redundancy for electro-mechanical and mechatronic systems: standby module which is inactive, "cold standby". $\mathbf{x}_M$: measured input, output and intermediate signals

only one single or two failures can be tolerated for hazardous cases, mainly because a safe state can be reached easier and faster. This means that not all components need very stringent fault-tolerance requirements. The following steps of degradation are distinguished:

- fail-operational (FO): one failure is tolerated, i.e. the component stays operational after one failure. This is required if no safe state exists immediately after the component fails;
- fail-safe (FS): after one (or several) failure(s), the component directly possesses a safe state (passive fail-safe, without external power) or is brought to a safe state by a special action (active fail-safe, with external power);
- fail-silent (FSIL): after one (or several) failure(s), the component is quiet externally, i.e. stays passive by switching off and therefore does not influence other components in a wrong way.

For vehicles, it is proposed to subdivide FO into "long time" and "short time". Considering these degradation steps for various components, one has to check first if a safe state exists. For automobiles, (usually) a safe state is stand still (or low speed) at a nonhazardous place. For components of automobiles, a fail-safe status is (usually) a mechanical back-up (i.e. a mechanical or hydraulic linkage) for direct manipulation by the driver. Passive fail-safe is then reached, e.g. after failure of electronics if the vehicle comes to a stop independently of the electronics, e.g. by a closing spring in the throttle or by actions of the driver via mechanical backup. However, if no mechanical back-up exists after failure of electronics, only an action by other electronics (switch to a still operating module) can bring the vehicle (in motion) to a safe state, i.e. to reach a stop through active fail-safe. This requires the availability of electric power.

Generally, a *graceful degradation* is envisaged, where less critical functions are dropped to maintain the more critical functions available, using priorities, [18.2]. Table 18.2 shows degradation steps to fail-operational for different redundant structures of electronic hardware. As the fail-safe status depends on the controlled system and the kind of components, it is not considered here.

Table 18.2. Fail behavior of electronic hardware for different redundant structures. FO: fail-operational; F: fail; FS: fail-safe not considered

Structure	Number of elements	Static redundancy		Dynamic redundancy		
		Tolerated faults	Fail behavior	Tolerated failures	Fail behavior	Discrepency detection
Duplex	2	0	F	0	F	2 comparators
				1	FO-F	fault-detection
Triplex	3	1	FO-F	2	FO-FO-F	fault detection
Quadru-plex	4	1	FO-F	3	FO-FO-FO-F	fault detection
duo-duplex	4	1	FO-F	–	–	–

For flight-control computers, usually a triplex structure with dynamic redundancy (hot standby) is used, which leads to FO-FO-FS, such that two failures are tolerated and a third one allows the pilot to operate manually, [18.7], [18.1], [18.6]. If the fault tolerance has to cover only one fault to stay fail-operational (FO-F), a triplex system with static redundancy or a duplex system with dynamic redundancy is appropriate. If fail-safe can be reached after one failure (FS), a duplex system with two comparators is sufficient. However, if one fault has to be tolerated to continue fail-operational and after a next fault it is possible to switch to a fail-safe (FO-FS), either a triplex system with static redundancy or a duo-duplex system, see Figure 18.4, may be used. The duo-duplex system has the advantages of simpler failure detection and modularity.

Figure 18.5 shows the improvement of the reliability for some of the discussed fault-tolerant structures with dynamic redundancy. For example, if a single module with failure rate $\lambda = 2 \cdot 10^{-4} h^{-1}$ (e.g. microcomputer) is used with MTTF $= 5 \cdot 10^3 h$ the triplex or duo-duplex systems improve the failure rate to $\lambda_{tot} \approx 2 \cdot 10^{-7} h^{-1}$ or MTTF$_{tot} = 5 \cdot 10^6 h$. Hence, the failure rate is improved by a factor 1000.

18.3 Problems

1) How many modules are required for fail-safe behavior with static and with dynamic redundancy?

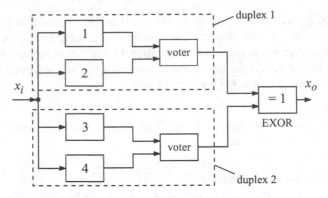

Fig. 18.4. Duo-duplex system with static redundancy

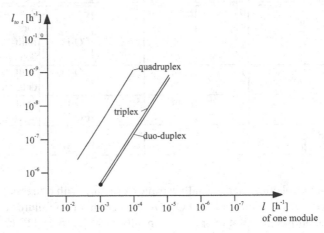

Fig. 18.5. Improvement of the reliability for some fault-tolerant structures, [18.7]. λ: failure rate $[h^{-1}]$; MTTF = $\frac{1}{\lambda}$ mean time to failure [h]; Example: one component: $\lambda = 2 \cdot 10^{-4}[h^{-1}] \rightarrow$ MTTF = $5 \cdot 10^3$ h; three components as triplex or duo-duplex: $\lambda = 2 \cdot 10^{-7}[h^{-1}] \rightarrow$ MTTF = $5 \cdot 10^6$ h

2) Design fault-tolerant schemes for FO-FO-F behavior with static and dynamic redundancy.
3) How is the reliability improved by a 2-out-of-3 static redundant system if the modules and the voter have a failure rate of $\lambda = 10^{-4}[h^{-1}]$? Calculate also the MTTF.
4) How is the reliability improved by a hot-standby duplex system, if the modules and the switch have a failure rate of $\lambda = 10^{-4}[h^{-1}]$ and the fault-detection system $\lambda = 10^{-3}[h^{-1}]$? Calculate also the MTTF.
5) Same problem as in 4) but cold-standby with a failure rate of the inactive module as $\lambda = 10^{-5}[h^{-1}]$.

19

Fault-tolerant components and control

High-integrity systems require a comprehensive overall fault-tolerance by fault-tolerant components and corresponding control. This means the design of fault-tolerant sensors, actuators, process parts, computers, communication (bus systems), and control algorithms. Examples of components with multiple redundancy are known for aircraft, space and nuclear power systems. However, lower cost components with built-in fault tolerance for general applications still have to be developed. In the following, some examples are given for sensors and actuators.

19.1 Fault-tolerant sensors

A fault-tolerant sensor configuration should be at least fail-operational (FO) for one sensor fault. This can be obtained by applying hardware redundancy with the same type of sensors or by analytical redundancy with different sensors and process models.

19.1.1 Hardware sensor redundancy

Sensor systems with static redundancy are realized, for example, with a triplex system and a voter, Figure 19.1a. A configuration with dynamic redundancy needs at least two sensors and fault detection for each sensor, 19.1b. Usually, only hot standby is feasible. Another less powerful possibility is plausibility checks for two sensors, also by using signal models (e.g. variance) to select the more plausible one, Figure 19.1c.

The fault detection can be performed by *self-tests*, e.g. by applying a known measurement value to the sensor. Another way uses *self-validating sensors*, [19.10], [19.6], where the sensor, transducer and a microprocessor form an integrated, decentralized unit with self-diagnostic capability. The self-diagnosis takes place within the sensor or transducer and uses several internal measurements, see also [19.19]. The output consists of the sensor's best estimate of the measurement and a validity status, like good, suspect, impaired, bad and critical.

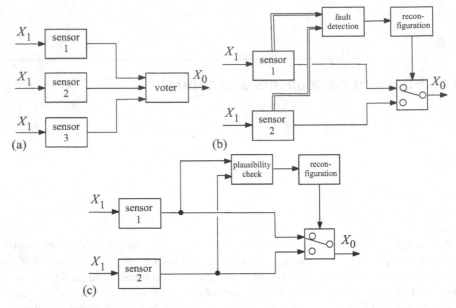

Fig. 19.1. Fault-tolerant sensors with hardware redundancy: (a) triplex system with static redundancy and hot standby; (b) duplex system with dynamic redundancy, hot standby; (c) duplex system with dynamic redundancy, hot standby and plausibility checks

19.1.2 Analytical sensor redundancy

As a simple example, a process with one input and one main output y_1 and an auxiliary output y_2 is considered, see Figure 19.2a. Assuming the process input signal u is not available but two output signals y_1 and y_2, which both depend on u, one of the signals, e.g. $\hat{y}_1$ can be reconstructed and used as a redundant signal if process models G_{M1} and G_{M2} are known and considerable disturbances do not appear (ideal cases).

For a process with only one output sensor y_1 and one input sensor u, the output $\hat{y}_1$ can be reconstructed if the process model G_{M1} is known, Figure 19.2b. In both cases, the relationship between the signals of the process are used and expressed in the form of analytical models.

To obtain one usable fault-tolerant measurement value y_{1FT}, at least three different values for y, e.g. the measured one and two reconstructed ones, must be available. This can be obtained by combining the schemes of Figure 19.2a and b as shown in Figure 19.3a. A sensor fault y_1 is then detected and masked by a majority voter and either $\hat{y}_1$ or $\hat{y}_{1u}$ is used as a replacement depending on a further decision. (Also, single sensor faults in y_2 or u are tolerated with this scheme.)

One example for this combined analytical redundancy is the yaw rate sensor for the ESP (electronic stability program) of vehicles, where additionally the steering wheel angle as input can be used to reconstruct the yaw rate through a vehicle model as in Figure 19.2b, and the lateral acceleration and the wheel speed difference of the right and left wheel (no slip) are used to reconstruct the yaw rate according to Figure 19.3a.

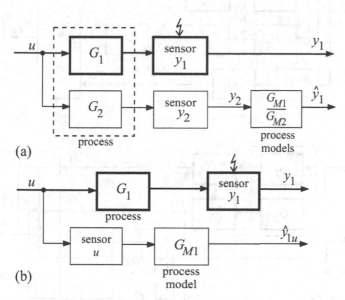

(a)

(b)

Fig. 19.2. Sensor fault-tolerance for one output signal y_1 (main sensor) through analytical redundancy by process models (basic schemes): (a) two measured outputs, no measured input; (b) one measured input and one measured output. G_i : $G_i(s)$ transfer functions.

A more general sensor fault-tolerant system can be designed if two output sensors and one input sensor yield measurements of the same quality. Then, by a scheme as shown in Figure 19.3, three residuals can be generated and by a decision logic, fault-tolerant outputs can be obtained in the case of single faults of any of the three sensors. The residuals are generated based on parity equations. In this case, state observers can also be used for residual generation, compare, e.g. the dedicated observers by [19.5]. (Note that all schemes assume ideal cases. For the realizibility, constraints and additional filters have to be considered.)

If possible, a faulty sensor should be fail-silent, i.e. should be switched off. However, this needs additional switches that lower the reliability. For both hardware and analytical sensor redundancy without fault detection for individual sensors, at least three measurements must be available to make one sensor fail-operational. However, if the sensor (system) has in-built fault detection (integrated self-test or self-validating), two measurements are enough and a scheme like Figure 19.1b can be applied. (This means that by methods of fault detection, one element can be saved).

Examples of fault-tolerant sensor systems are described in [19.11].

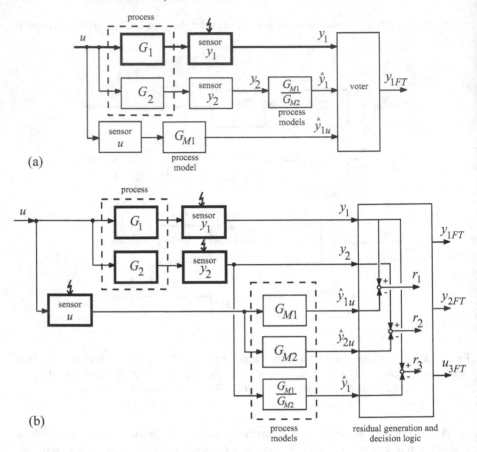

Fig. 19.3. Fault-tolerant sensors with combined analytical redundancy for two measured outputs and one measured input through (analytical) process models: (a) y_1 is main measurement, y_2, u are auxiliary measurements (combination of Figure 19.2a and b; (b) y_1, y_2 and u are measurements of same quality (parity equation approach)

19.2 Fault-tolerant actuators

Actuators generally consist of different parts: input transformer, actuation converter, actuation transformer and actuation element (e.g. a set of DC amplifier, DC motor, gear and valve, as shown in Figure 19.4a. The actuation converter converts one form of energy (e.g. electrical or pneumatic) into another form (e.g. mechanical or hydraulic). Available measurements are frequently the input signal U_i, the manipulated variable U_0 and an intermediate signal U_3.

Fault-tolerant actuators can be designed by using multiple complete actuators in parallel, either with static redundancy or dynamic redundancy with cold or hot standby (Figure 18.1). One example of static redundancy are hydraulic actuators for fly-by-wire aircraft where at least two independent actuators operate with two independent hydraulic energy circuits.

Another possibility is to limit the redundancy to parts of the actuator that have the lowest reliability. Figure 19.4b shows a scheme where the actuation converter (motor) is split into separate parallel parts. Examples with static redundancy are two servo-valves for hydraulic actuators, [19.22] or three windings of an electrical motor (including power electronics), [19.17], see also [19.11]. Within electromotor-driven throttles for SI engines, only the slider is doubled to make the potentiometer position sensor static-redundant, see Section 20.2.

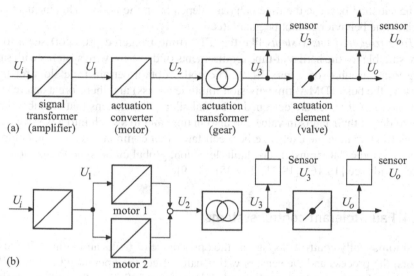

Fig. 19.4. Fault-tolerant actuator: (a) common actuator; (b) actuator with duplex drive

One example for dynamic redundancy with cold standby is the cabin pressure flap actuator in aircraft, where two independent DC motors exist and act on one planetary gear, [19.21], see [19.11].

As cost and weight generally are higher than for sensors, actuators with fail-operational duplex configuration are to be preferred. Then, either static-redundant structures, where both parts operate continuously, Figure 18.1a, or dynamic redundant structures with hot standby, Figure 18.1b, or cold standby, Figure 18.1c, can be chosen. For dynamic redundancy fault-detection methods of the actuator parts are required, [19.14]. One goal should always be that the faulty part of the actuator fails silent, i.e. has no influence on the redundant parts.

19.3 Fault-tolerant communication

As shown in Chapter 19.5 fault-tolerant management systems require also a fault-tolerant communication system between several electronic control units, sensors and actuators (nodes). This can be realized by a multiple bus system which has to cover hard real-time requirements. At least a dual bus system with two independent buses and independent power supplies must be realized. Both buses are then connected to all nodes, where several of them are also at least dual. Hence, a multiple access distributed real-time system results.

The CAN-Bus (Controller-Area Network) was developed in 1983 for automobiles as a serial bus system with high reliability of data transfer and high flexibility and extendibility. It is an *event-triggered* and therefore *asynchronous bus* with highest priority access, indicated by the nodes identifier. (CSMA/CA: carrier sense multiple access collision avoidance protocol). Usually only soft real-time requirements can be satisfied because the time behavior depends on the nodes. This means that a precise time behavior cannot be guaranteed.

Time-triggered bus systems like the TTP (time triggered protocol) seem to be more suitable for the hard real-time requirements of drive-by-wire systems with sampling times around some ms, [19.9]. The nodes obtain certain time slots for their access to the bus (TDMA: time division multiple access) and therefore a deterministic behavior. All nodes are designed to be fail-silent. This means that all subsystems have to detect their faults in value and also in time and to switch into a passive state. Means to guarantee the exchange between fail-silent components are, e.g. composability, periodic data transfer, fast fault detection, global clock synchronization. For more details see [19.16], [19.31], [19.25], [19.9].

19.4 Fault-tolerant control systems

For automatically controlled systems, the appearance of faults and failures in the actuators, the process and the sensors will usually affect the operating behavior. With feedforward control, generally all small or large faults influence the output variables and therefore more or less the operation. However, if the system operates with feedback control, *small additive* or *multiplicative faults* in the actuator or process are in general covered by the controller, because of the usual robustness properties. The controller can even be made very robust with regard to some known smaller changes or faults in the actuator or the process. But this means a trade off between good control performance and robustness against faults. This robustness property is a *passive controller fault-tolerance* because no active measures are undertaken. However, additive and gain sensor faults will immediately lead to deviations from the reference values. This holds also for feedforward control. For *large changes* in actuators, process and sensors, which exceed the robustness properties, the dynamic control behavior becomes either too sluggish or too less damped or even unstable. Then an *active fault-tolerant* control system is required to save the operation, Figure 19.5.

Active fault-tolerant control systems consist of fault-detection methods, a decision method and a reconfiguration mechanism with the goal that the operating behavior is hold in an acceptable way, compare the supervision loop, Figure 2.4. Depending on the fault, the fault-tolerant system may change to

- controller structure and parameters;
- used actuators;
- used sensors

taking into account a degradation of the normal performance. Of major importance is that severe faults are detected fast and reliable and the reconfiguration is settled in very short time. This imposes high requirements on the real-time capabilities.

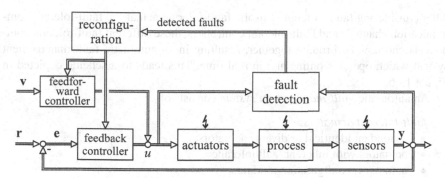

Fig. 19.5. Fault-tolerant control system: faults in actuators, process and sensors are detected and lead to a reconfiguration of the feedback and feedforward controller

Early publications on fault-tolerant control appeared especially for aircraft, e.g. [19.26], [19.27], [19.18], [19.3], [19.4], [19.29] or space-craft, [19.7], [19.2]. A more recent summary of reconfigurable flight control is given in [19.1] and [19.8], see also [19.28]. An application to fault-tolerant lateral control shows [19.30].

A survey on fault-tolerant control in general is given by [19.23]. Four areas are considered like fault detection, robust control, reconfigurable control and supervision, which manages the fault decision and selects the control configuration. Passive fault-tolerant control with fixed, robust controllers and active fault-tolerant control with variable structure, based on fault-accumulation, is distinguished. However, because of the many possibilities for the design and the individual practical requirements, the author states that only some structures and approaches could be described.

A further literature review on reconfigurable fault-tolerant control is [19.32]. They classify the literature according to control algorithms (e.g. pre-computed or online redesign) and application areas, like safety-critical, life-critical, mission-critical and cost-critical.

The stability analysis includes three stages, the fault-free period, the transient-period during reconfiguration and the steady-state after reconfiguration. Of further importance are the constraints in inputs and states, the uncertainties of FDD, the delay of reconfiguration, closed loop identification and real-time issues. The authors conclude that most publications focuss mainly on algorithmic design, neglecting the overall architecture and technical platform. Hence, this area of fault-tolerant control is in early state of development and needs more systematic research and practical realizations.

The following chapter describes the tasks of an automatic fault-management system, taking into account all components of control systems. The considered fault-tolerant control systems discussed in this chapter can be considered as a subset of a general fault-management system.

19.5 Automatic fault-management system

After considering fault -tolerant sensors, fault-tolerant actuators, fault-tolerant communication channels and fault-tolerant controllers, these different fault-tolerant components can now be brought together, resulting in an automatic fault-management system which operates online and in real-time. This leads to a scheme depicted in Figure 19.6.

An automatic *fault-management system* consists of:

1) *fault-tolerant actuators*:
   - redundant identical or diverse actuators;
   - actuators with inherent fault tolerance;
   - actuator reconfiguration module.
2) *fault-tolerant sensors*:
   - redundant identical or diverse sensors;
   - sensors with inherent fault tolerance;
   - virtual sensors based on analytical redundancy;
   - sensor reconfiguration module.
3) *active fault-tolerant controllers*:
   - redundant identical or diverse controller hardware;
   - redundant diverse controller software;
   - different controller structure and parameters
     - pre-designed for a priori-known faults;
     - redesigned or adaptive after fault detection
4) *fault-detection module*:
   - normal closed-loop operating signals are used to detect and isolate faults in the components (parity equations, observers, parameter estimation);
   - test signals are introduced either periodically or on request to improve fault detection and, if required, fault diagnosis;
   - indication of the degree of impairment and degree of safety criticality.
5) *fault-management module*:
   - decisions based on fault detection with an indication of the degree of impairment and effect on safety of the components;
   - reconfiguration strategies with
     - hard or soft reconfiguration;
     - change of operating conditions (setpoints, process performance);
     - closed loop or open loop (feedforward) operation.

The scheme in Figure 19.6 is an example with two manipulated and two controlled variables. If the normal *actuator 1 fails*, e.g. be getting stuck, another actuator 2 replaces its functions. This can be a second actuator of the same type or another actuator with a similar manipulation effect on the control variable (The passenger cabin pressure outflow valve is an example of a duplex actuator, see [19.20], [19.11]. Certain additional control surfaces for aircraft are designed as other, redundant actuators, e.g. ailerons or rudders.

In the case that the fault-detection system detects that *sensor 1* has a large fault or even fails totally a second sensor of same type is switched or some analytical redundancy with other sensors is used to generate a virtual sensor output 2, as described in Section 19.1. (Examples are the electrical throttle with a double potentiometer, see Chapter 20, or the model-based calculation of the yaw rate of automobiles from the lateral acceleration and wheel speed sensors, [19.12] or a horizontal and vertical gyro for the bank angle of aircraft, [19.23].

Depending on the reconfigured actuators or sensors, the controller structure and/or controller parameters of the *fault-tolerant controller* have also to be reconfigured.

The structure of Figure 19.6 also holds for *faults in the process* itself. If then the other actuator 2 or the other sensor 2 can be used to maintain the operation, the reconfiguration just selects the actuator-sensor configuration, adjusts the controller 2 and the reference variable w_2 accordingly. An example is a fluid 1 / fluid 2 heat exchanger: The outlet temperature of fluid 1 can be manipulated by changing the flow of fluid 2 instead of the temperature of fluid 2, if the plant allows.

The *controller structure and parameters* have to be adapted to the new process behavior. If the transfer behavior of the reconfigured actuator-process-sensor-system is known in advance, preprogrammed controllers have just to be switched. If it is not known, selftuning or adaptive control algorithms could be used. However, this adaptation must be supervised and properly excited with perturbation signals, see [19.13], what can be a problem, if a very fast recovery is required.

The task of the *fault-detection module* is to detect faults in all components, like actuators, sensors, controllers and the process as early as possible. A diagnostic capability is not necessarily required, because it is mostly enough information for the reconfiguration to know if the actuator or sensor has failed, independent on the causes.

It should be mentioned, that in the case of *closed-loop control* fault detection must be made under closed-loop conditions with all the problems discussed in Chapter 12. Because of the danger of a reconfigured replacement controller not functioning as expected with a replacement sensor, it is sometimes better not to reconfigure an alternative closed-loop control, but to apply a *feedforward control* without an output sensor substitute. This may result in a loss of control performance, but instability is avoided. This is, for example, used in engine control. In the case the oxygen-sensor (λ-sensor) fails, a stoichiometric air/fuel ratio is maintained, based on air-flow measurement and the setpoint of injected fuel mass.

Faults in the *controller hardware or software* can be detected as described in Chapter 12. Then new controllers as described above are applied or feedforward control is used.

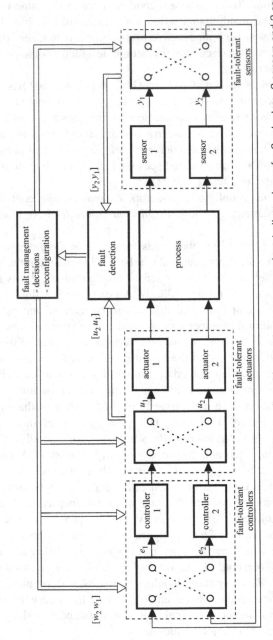

Fig. 19.6. Automatic fault-management system with reconfigurable actuators, sensors and controllers, shown for 2 actuators, 2 sensors and 2 controllers

The discussion on automatic fault management shows that there are many different possibilities. Therefore it is difficult to treat applicable methods generally and it is recommended to consider concrete cases.

An experimental investigation of fault detection in closed loop and reconfiguration to a redundant sensor is shown for the electrical throttle valve actuator in [19.24] and [19.15].

19.6 Problems

1) How can a fault-tolerant temperature measurement system be built with two sensors (thermocouple and resistance thermometer) and three sensors of the same type?
2) What are the differences of the degradation steps: fail-safe, fail-operational, fail-silent? Take an electrical driven elevator as example and explain the three steps for the electrical drive system, the position sensors and the control unit.
3) Design a fault-tolerant management system for the careful closing of elevator doors.
4) Describe the possibilities for the construction of fault-tolerant drives with two electrical motors.
5) Which kind of fault-tolerance possess hydraulic brakes for passenger cars? What are the degradation steps? How many faults can be tolerated? What kind of faults may happen? Which information is received by the driver after a fault?

Part V

Application Examples

20

Fault detection and diagnosis of DC motor drives

The theoretically developed methods for fault detection and diagnosis require an experimental testing with different kinds of technical processes. Most of the described methods assume ideal situations, as, for example, linear behavior or specific structures of nonlinear processes, precise measurements, small disturbances, stationary stochastic disturbances, constant parameters or open loop operation and modelling of faults. However, in practice frequently some of the simplifying assumptions are violated. Therefore it is of interest how robust the treated methods are with regard to these violations. As already discussed in Chapter 13, the suitability of the different methods depends on the behavior of the processes and real faults. Therefore some applications for two *DC motors, a circulation pump* and an *automotive wheel suspension* are shown in the following chapters, highlighting the advantages and disadvantages of the applied detection methods. Many more case studies and applications on, e.g. *electrical, pneumatic* and *hydraulic actuators, AC motors, pumps, machine tools, robots, heat exchangers, pipelines, combustion engines* and *passenger cars* will be treated in another book, [20.5].

20.1 DC motor

20.1.1 DC motor test bench

A permanently excited DC motor with a rated power of $P = 550$ W at rated speed $n = 2500$ rpm is considered, [20.3]. This DC motor has a two pair brush communication, two pole pairs, an analog tachometer for speed measurement and operates against a hysteresis brake as load, see Figure 20.1. The measured signals are the armature voltage U_A, the armature current I_A and the speed ω. A servo amplifier with pulsewidth-modulated armature voltage as output and speed and armature current as feedback allows a cascaded speed control system. The three measured signals first pass analog anti-aliasing filters and are processed by a digital signal processor (TXP 32 CP, 32 bit fpt, 50 MHz) and an Intel Pentium Host PC. Also the hysteresis brake is

(a)

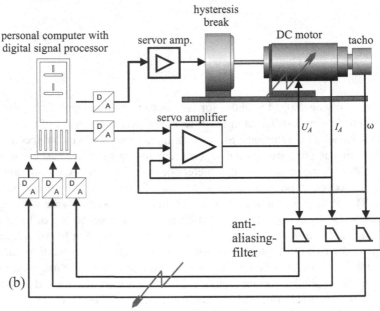

(b)

Fig. 20.1. DC motor test bench with hysteresis brake: a. test bench; b. scheme of equipment

controlled by a pulsewidth servo amplifier. Usually such DC motors can be described by linear dynamic models.

However, experiments have shown that this model with constant parameters does not match the process in the whole operational range. Therefore, two nonlinearities are included so that the model fits the process better. The resulting first-order differential equations are:

$$L_A \, \dot{I}_A(t) = -R_A \, I_A(t) - \Psi \, \omega(t) - K_B |\omega(t)| \, I_A(t) + U_A^*(t) \tag{20.1}$$

$$J \, \dot{\omega} = \Psi \, I_A(t) - M_{F1} \, \omega(t) - M_{F0} \, \text{sign} \, (\omega(t)) - M_L(t) \tag{20.2}$$

Figure 20.2 depicts the resulting signal flow diagram. The term $K_B |\omega(t)| I_A(t)$ compensates for the voltage drop at the brushes in combination with a pulsewidth-modulated power supply. The friction is included by a viscous and a dry friction term $M_{F1}\omega$ and $M_{F0}\text{sign}(\omega)$. The parameters are identified by least-squares estimation in the continuous-time domain. Table 20.1 give the nominal values. Most of them

$(R_A, \Psi, K_B, M_{F1}, M_{F0})$ influence the process gain, and the other two (L_A, J) the time constants. The data is measured with a sampling frequency of 5 kHz, and state variable filtered by a fourth-order lowpass-filter with Butterworth characteristic and a cut-off frequency of 250 Hz.

Table 20.1. Data of the DC motor

armature resistance	$R_A = 1.52\ \Omega$
armature inductance	$L_A = 6.82 \cdot 10^{-3}\ \Omega\ s$
magnetic flux	$\Psi = 0.33\ V\ s$
voltage drop factor	$K_B = 2.21 \cdot 10^{-3}\ V\ s/A$
inertia constant	$J = 1.92 \cdot 10^{-3}\ kg\ m^2$
viscous friction	$M_{F1} = 0.36 \cdot 10^{-3}\ N\ m\ s$
dry friction	$M_{F0} = 0.11\ N\ m$

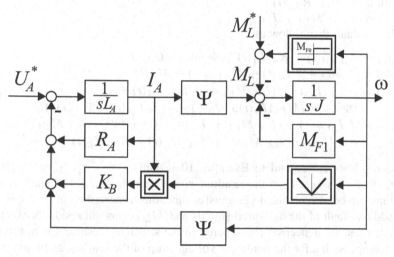

Fig. 20.2. Signal flow diagram of the considered DC motor

20.1.2 Parity equations

For the detection and isolation of sensor (output) and actuator (input) faults a set of structured parity equations with state-space models according to Section 10.2.1 is applied.

As the differential equations (20.1) and (20.2) are nonlinear, the design procedure for a linear parity space cannot be applied directly. But defining $U_A^* - K_A|\omega(t)|I_A$ as voltage input U_A and M_{F0} sign (ω) as load input M_L leads to a linear description. The linear state-space representation then becomes

$$\dot{\mathbf{x}} = \begin{bmatrix} \dot{I_A} \\ \dot{\omega} \end{bmatrix} = \begin{bmatrix} -\frac{R_A}{L_A} & -\frac{\Psi}{L_A} \\ \frac{\Psi}{J} & -\frac{M_F}{J} \end{bmatrix} \begin{bmatrix} I_A \\ \omega \end{bmatrix} + \begin{bmatrix} \frac{1}{L_A} & 0 \\ 0 & -\frac{1}{J} \end{bmatrix} \begin{bmatrix} U_A \\ M_L \end{bmatrix}$$

$$\mathbf{y} = \begin{bmatrix} I_A \\ \omega \end{bmatrix} = \begin{bmatrix} 1 & 0 \\ 0 & 1 \end{bmatrix} \mathbf{x}$$

(20.3)

A corresponding signal flow diagram is depicted in Figure 10.8.

An observability test reveals both outputs (I_A and ω) can also observe each other. This is a precondition for a parity space of full order (here: 2). Then, $\mathbf{W}$, (10.52), is chosen such that a set of structured residuals is obtained, where residual $r_1(t)$ is independent of $M_L(t)$, $r_2(t)$ of $U_A(t)$, $r_3(t)$ of $\omega(t)$ and $r_4(t)$ of $I_A(t)$, see also [20.3], [20.7], [20.1]

$$\mathbf{W} = \begin{bmatrix} R_A & \Psi & L_A & 0 & 0 & 0 \\ -\Psi & M_{F1} & 0 & J & 0 \\ \alpha & 0 & \beta & 0 & J\,L_A & 0 \\ 0 & \alpha & 0 & \beta & o & J\,L_A \end{bmatrix}$$

(20.4)

with $\alpha = \Psi^2 + R_A\,M_{F1}$
$\quad\quad \beta = L_A\,M_{F1} + J\,R_A$
The residuals then follow as:

$$r_1(t) = L_A\,\dot{I_A}(t) + R_A\,I_A(t) + \Psi\,\omega(t) - U_A(t)$$
$$r_2(t) = J\,\dot{\omega}(t) - \Psi\,I_A(t) + M_{F1}\,\omega(t) + M_L(t)$$
$$r_3(t) = J\,L_A\,\ddot{I_A}(t) + (L_A\,M_{F1} + J\,R_A)\,\dot{I_A}(t) +$$
$$\quad\quad (\Psi^2 + R_A M_{F1})I_A(t)J\,\dot{U_A}(t) - M_{F1}\,U_A(t) - \Psi\,M_L(t)$$
$$r_4(t) = J\,L_A\,\ddot{\omega}(t) + (L_A\,M_{F1} + J\,R_A)\,\dot{\omega}(t) + (\Psi^2 + R_A\,M_{F1})\,\dot{\omega}(t)$$
$$\quad\quad -\Psi\,U_A(t) + L_A\,\dot{M_L}(t) + R_A\,M_L(t) + R_A\,M_L(t)$$

(20.5)

(These residuals correspond to Example 10.3, with $r_1 = -r_4'; r_2 = -r_1'; r_3 = -r_3'; r_4 = -r_2'$, where r_1' is the residual r_1 in Example 10.3.) The same residual equations can be also obtained via transfer functions as described in Example 10.3. If an additive fault of the measured signals and M_L occurs, all residuals except the decoupled one are deflected. The scheme of the structured residuals is not touched by the compensation for the nonlinear voltage drop of the brushes, as its magnitude is small enough. Two parameters R_A and M_{F1}, however, depend on the present motor temperature. The behavior of R_A and its effect on residual r_1 is depicted in Figure 20.3. Therefore, the use of adaptive parity equations improves the residual performance, see Section 10.5.

The residuals are now examined with regard to their sensitivity to additive and parametric faults. As r_1 and r_2 comprise all parameters and all signals, it is sufficient to consider only these two, although r_3 or r_4 can also be taken. From (20.5) and (10.82) it yields

$$r_1(t) = -\Delta\,L_A\,\dot{I_A}(t) - \Delta\,R_A\,I_A(t) + \Delta\,\Psi\,\omega(t)$$
$$\quad\quad + L_A\,\Delta\,\dot{I_A}(t) + R_A\,\Delta\,I_A(t) + \Psi\,\Delta\,\omega(t) - \Delta\,U_A(t)$$
$$r_2(t) = -\Delta\,J\,\dot{\omega}(t) + \Delta\,\Psi\,I_A(t) - \Delta\,M_{F1}\,\omega(t)$$
$$\quad\quad + J\,\Delta\,\dot{\omega}(t) - \Psi\,\Delta\,I_A(t) + M_{F1}\,\Delta\,\omega(t) - \Delta\,M_L(t)$$

(20.6)

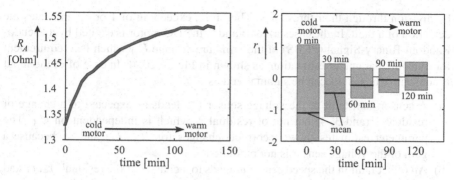

Fig. 20.3. Influence of the motor temperature on resistance R_A and residual r_1

In the presence of residual noise, e.g. of r_1 with a magnitude of about 1 [V] and an armature current of 3 [A], a resistance change must be at least 0.3 [Ω] in order to deflect the residual significantly. Therefore, the two linear parameters R_A and M_{F1} are selected to be tracked according to the single parameter estimation, see Section 10.5. λ is chosen to be 0.99.

20.1.3 Parameter estimation

The parameter estimation is based on the two differential equations (20.1), (20.2) in the simplified form of

$$\dot{I}_A(t) = -\hat{\theta}_1\, I_A(t) - \hat{\theta}_2\, \omega(t) - \hat{\theta}_3\, U_A(t) \qquad (20.7)$$
$$\dot{\omega}(t) = -\hat{\theta}_4\, I_A(t) - \hat{\theta}_5\, \omega(t) - \hat{\theta}_6\, M_L(t) \qquad (20.8)$$

with the process coefficients

$$R_A = \frac{\hat{\theta}_1}{\hat{\theta}_3};\ L_A = \frac{1}{\hat{\theta}_3};\ \Psi = \frac{\hat{\theta}_2}{\hat{\theta}_3}\ \text{and}\ \Psi = \frac{\hat{\theta}_4}{\hat{\theta}_6};\ J = \frac{1}{\hat{\theta}_6};\ M_{F1} = \frac{\hat{\theta}_5}{\hat{\theta}_6} \qquad (20.9)$$

Applying the recursive parameter estimation method DSFI (Discrete Square Root Filtering in Information Form), [20.6], with forgetting factor $\lambda = 0.99$ yields the parameters $\hat{\theta}_i$. Then all process coefficients can be calculated with (20.9). Experimental results with idle running ($M_L = 0$) resulted in standard deviations of the process coefficients in the range of $2\% < \sigma_\theta < 6.5\,\%$, [20.3].

20.1.4 Experimental results for fault detection

Based on many test runs, five different faults are now selected to show the detection of additive and multiplicative faults with parity equations and recursive parameter estimation, [20.2]. The time histories depict the arising faults at $t = 0.5$ s. The faults are step changes and were artificially produced. Figure 20.4 shows the parameter estimates and the residuals of parity equations. The residuals are normalized by division through their thresholds. Therefore, exceeding of 1 or -1 indicates the detection of a fault. In the cases a) to d) and f) the DC motor is excited by a Pseudo-Random-Binary-Signal (PRBS) of the armature voltage U_A which is a requirement for dynamic parameter estimation, as shown in Figure 20.4f. In case of e) the input is constant. The results can be summarized as

a) A sensor-gain fault of the voltage sensor U_A leads as expected to a change of residual 1 (and 3, 4) but not of residual 2, which is independent on U_A. The parameter estimates show (incorrect) changes for R_A, L_A and Ψ, because a gain of the voltage sensor is not modelled.

b) An offset fault in the speed sensor ω leads to a change of the residuals r_4, r_1 and r_2, but r_3 remains uneffected, because it is independent on ω. The parameter estimate of Ψ shows an (incorrect) change.

c) A multiplicative change of the armature resistance R_A yields a corresponding change of the parameter estimate $\hat{R}_A$. However, the residuals increase their variance drastically and exceed (incorrectly) their thresholds.

d) A change of the ratio of inertia is correctly given by the parameter estimate $\hat{J}$. But all residuals, except r_1, exceed their thresholds by increasing their variance.

e) The same fault in R_A as in c) is introduced, but the input U_A is kept constant. The parameter estimate $\hat{R}_A$ does not converge to a constant value and the parity estimation r_1 and r_4 change their mean, however, with large variance.

f) A brush fault leads to an increase of R_A and L_A but not of Ψ. The residuals show an increase of the variance.

Table 20.2 summarizes the effects of some investigated faults on the parameter estimates and parity residuals.

These investigation have shown:

1) *Additive faults* like the offsets of sensors are well detected by the parity equations. They react fast and do not need an input excitation for a part of the faults. However, they have a relatively large variance, especially if the model parameters do not fit well to the process;

2) *Multiplicative faults* are well detected by parameter estimation, also for small faults. Because of the inherent regression method the reactions are slower but smoothed. But they require an input excitation for dynamic process models.

Therefore, it is recommended to combine both methods, as shown in Section 14.3, Figure 14.1. The parity equations are used to detect changes somewhere in the process and if the fault detection result is unclear a parameter estimation is started, eventually by a dynamic test signal for some seconds. If the motor operates dynamically anyhow (as for servo systems and actuators) then the parameter estimation can be applied continuously, but with a supervision scheme, see [20.6].

[20.3] has shown that a considerable improvement can be obtained by continuously estimating the armature resistance with a single parameter estimation using parity equations in order to reach the temperature dependent resistance parameters, [20.4]. Furtheron, adaptive thresholds are recommended, to compensate for model uncertainties, see Section 7.5.

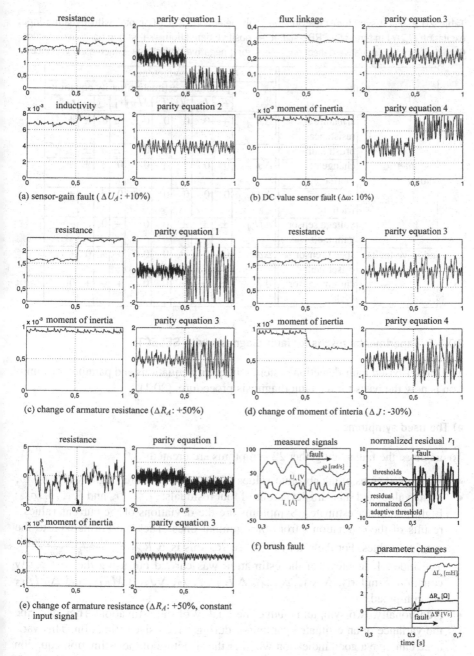

Fig. 20.4. Time histories of signals, residuals of parity equations and parameter estimation at fault occurrence

Table 20.2. Fault-symptom table for the fault detection of a DC motor with dynamic input excitation $U_A(t)$ in form of a PRBS-signal. $+$ positive deflection; $++$ strong positive deflection; 0 no deflection; $-$ negative deflection; $--$ strong negative deflection; $\pm$ increased variance

faults			symptoms								
			parameter estimation					parity equations			
			R_A	L_A	Ψ	J	M_{F1}	r_1	r_2	r_3	r_4
parametric faults	armature resistance	ΔR_A	$++$	0	0	0	0	$\pm$	0	$\pm$	$\pm$
	brush fault		$++$	$+$	$+$	0	0	$\pm$	0	$\pm$	$\pm$
	change of inertia	ΔJ	0	0	0	$++$	0	0	$\pm$	$\pm$	$\pm$
	change of friction	ΔM_{F1}	0	0	0	0	$++$	0	$\pm$	$\pm$	$\pm$
additive faults	voltage sensor gain fault	ΔU_A	$\pm$	$\pm$	$\pm$	0	0	$-$	0	$-$	$-$
	speed sensor offset fault	$\Delta\omega$	0	0	$-$	0	0	$+$	$+$	0	$+$
	current sensor fault	ΔI	$\pm$	$\pm$	$\pm$	0	0	$+$	$-$	$+$	0

20.1.5 Experimental results for fault diagnosis with SELECT

The model-based fault-detection system with parity equations and parameter estimation is now the basis for the fault diagnosis procedure, [20.1].

a) The used symptoms

To diagnose the faults, altogether 22 symptoms are created:

- Windowed sums of the absolute values of the three measured signals U_A^*, I_A, ω;
- Mean values and standard deviations of four residuals: $\bar{r}_1, \ldots, \bar{r}_4$ and $\bar{\sigma}_{r1}, \ldots, \bar{\sigma}_{r4}$;
- Eight parameter estimates. Symptoms are the deviations of the current values - results of the estimation - from the nominal ones. They are normalized to the nominal values. For the rotor resistance R_A this is $\Delta R_{A1} = \frac{(R_{A,nom.} - R_{A,est.})}{R_{A,nom.}}$. The index 1 denotes that the estimation was carried out using the first parity equation. Similarly, $\Delta R_{A4}, \Delta L_{A1}, \Delta L_{A4}, \Delta J_2, \Delta J_3, \Delta M_{F12}$, and ΔM_{F13} are computed;
- Additionally, two symptoms judge the quality of the estimation. They describe the variance of an estimated parameter during a recursive estimation. This variance can give a good indication whether the structure of the estimation equation is valid. A structural change of the system will result in a bad estimation result where the recursively estimated parameters fluctuate significantly. Two parameter estimations were chosen: Ψ and M_{F1}. Their estimation variances are denoted by $\sigma_{est.,\Psi}$ and $\sigma_{est.,MF1}$.

The symptoms serve to differentiate between 14 fault situations that can artificially be introduced on the test rig.

The DC motor diagnosis was performed by learning a SELECT tree from experimentally gained fault data. For the fault cases, typically 10-50 test cycle measurements for a parameter estimation were performed. The residuals were computed from the test runs. That way, each test run results in one data point in the symptom space. The membership functions were created with the degressive fuzzy-c-means method. To utilize a maximum of transparency and create a highly interpretable system, prior knowledge was used to structure the diagnosis system.

b) Incorporation of structural knowledge

In most applications, a certain amount of knowledge about the symptom behavior is present. Even if exact values for thresholds etc. are not known, there usually is some insight into the process like physical understanding of similar faults or similar effects of faults on certain symptoms. For the DC motor, this could be as simple as to use the windowed sums of the signals in order to to detect a broken sensor cable. This information is quite obvious, but its benefits are sometimes neglected, if a diagnosis system is designed with the aim to be solely learned from measured data. Hence, the task could be simpler if the designer used this information from the beginning.

Furthermore, the selection of the symptoms for the diagnosis becomes a matter of *robustness*. Some symptoms are affected by faults for which they are not an appropriate indicator. In an experimental environment, it is virtually impossible to gather enough measurements to adequately reflect every influence. Especially changes in the environmental conditions and long time changes due to wear are hardly captured in a limited time frame. This leads to diagnosis systems that work well under the experimental conditions but fail otherwise. The diagnosis of a fault should therefore be based on the *appropriate subset* of all available symptoms. Only the relevant ones should be selected.

Often, different faults can be categorized into larger groups if their effects on the process are similar. It is then advantageous to find a classification system for the larger groups first and later separate within them. This leads to the concept of a *hierarchical* diagnosis system.

Overall, it is proposed to use prior knowledge to *structure* the diagnosis system. The designer builds groups of faults and identifies the corresponding relevant symptoms to first differentiate between and later within them. The exact decisions can be found automatically if enough measured data is available.

If the set of all different fault situations F_i is denoted by

$$\mathcal{F} = \{F_1, F_2, \ldots F_r\} \tag{20.10}$$

and the available symptoms given by

$$\mathcal{S} = \{s_1, s_2, \ldots s_t\} \tag{20.11}$$

one can form meta-classes $\mathcal{C}_i$, $i = 1 \ldots m$ with

$$\mathcal{F} = \mathcal{C}_1 \cup \mathcal{C}_2 \cup \ldots \cup \mathcal{C}_m \tag{20.12}$$

In the DC motor diagnosis, for instance, such a meta-class is given by all faults on the mechanics of the motor. Such a hierarchy based on meta-classes requires at least $q = m + r$ decisions d_j, $j = 1 \ldots q$ assumed that no $\mathcal{C}_i$ is a single-element set. Each d_j is based on a subset $\mathcal{S}_{dj} \in \mathcal{S}$. The SELECT approach will then produce a system with p parameters where p is given by

$$p = \sum_{j=1}^{q} \mathrm{card} \left(\mathcal{S}_{dj} \right) \tag{20.13}$$

which is typically much less than a parallel network structure would result in. The usually larger number of parameters in parallel network configurations can lead to slower convergence and ill-conditioned optimization problems.

In addition to the structural knowledge, one can incorporate more detailed knowledge into the individual rules if desired.

c) Results

A total of 14 different fault situations are applied on the DC motor test bench:

- Change of rotor inductance or resistance F_{RA}, F_{LA};
- Broken rotor wiring (F_W);
- Failure of one the four brushes (F_B);
- Increased friction in the bearings (F_F);
- Offset on voltage, current or speed sensor signal ($F_{O,UA}$, $F_{O,IA}$, $F_{O,\omega}$);
- Gain change of voltage, current or speed sensor signal ($F_{G,UA}$, $F_{G,IA}$, $F_{G,\omega}$);
- Complete voltage, current or speed sensor failure (F_{UA}, F_{IA}, F_ω).

Repeated experiments with different faults were performed using a test cycle. The symptoms described in a) were computed for each of the experiments. Overall, the training set for the approach consisted of data from 140 experiments.

Figure 20.5 shows the resulting structure for the DC motor diagnosis. Details have been omitted to visualize the concept only. Each shaded block comprises a meta-class $\mathcal{C}_i$ of faults. Every branching of the tree is connected to a decision d_j learned with the SELECT approach, i.e. it contains a fuzzy rule. In each meta-class, a classification tree decides which individual fault has occurred based on a subset $\mathcal{S}_i$ of the symptoms.

The hierarchical decision tree proved to be highly suitable for the diagnosis. It achieved 98 % classification rate in a cross-validation scheme.

The groups of faults have been selected following basic understanding of the DC motor supervision concept. Firstly, the three total sensor breakdowns are different from other faults due to their strong effects on all symptoms. They form the first meta-class $\mathcal{C}_i$ and can be easily differentiate by the three windowed sums of the signals. These three symptoms accordingly form the set $\mathcal{S}_i$.

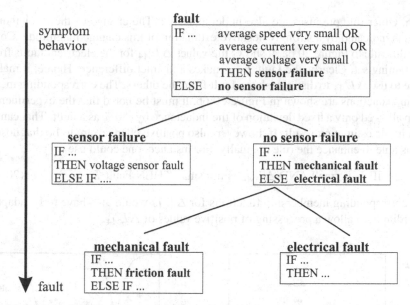

symptom
behavior

fault
IF ...   average speed very small OR
          average current very small OR
          average voltage very small
          THEN sensor failure
ELSE   **no sensor failure**

sensor failure
IF ...
THEN voltage sensor fault
ELSE IF ....

no sensor failure
IF ...
THEN **mechanical fault**
ELSE **electrical fault**

mechanical fault
IF ...
THEN **friction fault**
ELSE IF ...

electrical fault
IF ...
THEN ...

fault

Fig. 20.5. Hierarchical fault diagnosis system. Each block comprises a fuzzy classification tree

Since the motor can be understood as a combination of an electrical and a mechanical component, faults on these two parts were again treated separately, creating two more meta-classes, C_2 and C_3. Accordingly, the appropriate subsets of symptoms S_2 and S_3 for the diagnosis were selected. Basically, S_2 and S_3 consist of the residuals and parameter deviations connected to the corresponding meta-class. The diagnosis of electrical faults, for instance, is not based on parameter estimates of the mechanical parameters. Although some electrical faults may have an influence on the estimates of the mechanical parameters, this influence should not be used as the estimates are misleading and not reliable. Hence, S_2 does not contain ΔJ_2, ΔJ_3, ΔM_{F12} or ΔM_{F13}.

To give an example of the SELECT approach, the rules for the distinction of the electrical faults are given below:

IF $\bar{r}_1$ is small AND ΔL_{A4} is strongly negative THEN Fault F_{LA}
ELSE IF $\bar{r}_1$ is small AND $\bar{\sigma}_{r4}$ is medium THEN Fault F_{RA}
ELSE IF $\bar{r}_1$ is small AND $\bar{\sigma}_{r4}$ is large THEN Fault F_B
ELSE IF $\bar{r}_2$ is not small THEN Fault $F_{0,IA}$ (20.14)
ELSE IF $\bar{r}_1$ is small THEN Fault $F_{G,IA}$
ELSE IF $\bar{r}_1$ is large AND $\sigma_{est.,\psi}$ is not small THEN Fault $F_{0,UA}$
ELSE Fault $F_{G,UA}$

The relevance indices of the rule premises are not listed here. They also play a role for the exact decision boundaries.

Nevertheless, it is possible to analyze and understand parts of these rules. Clearly, the rules reveal the discriminatory power of the first residual, since it was used very

often. Other rule premises are also understandable. The change of the rotor inductance is indicated by a strongly negative estimation of this change magnitude. Compare this rule to Figure 20.6a. It shows the values ΔL_{A4} for the electrical faults from
the training set. Clearly, the fault F_{LA} makes a distinct difference. Hence, it makes
sense to use ΔL_{A4} to distinguish the fault from the others. The corresponding membership functions are shown in Figure 20.6b. It must be noted that the experimental
setup allowed only a fixed deviation of the inductance by -50 % as a fault. That can be
seen in the estimation result. If, however, also positive changes are to be diagnosed,
one is able to enhance the rule manually. For instance, one could use

$$\text{IF } \bar{r}_1 \text{ is small AND } \Delta L_{A4} \text{ is not small THEN Fault } F_{LA} \qquad (20.15)$$

The corresponding membership functions for ΔL_{A4} would also have to be adapted
accordingly to allow a processing of positive values of ΔL_{A4}.

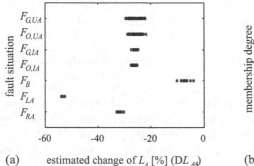 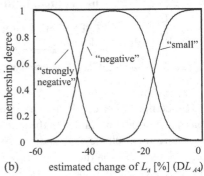

(a)   estimated change of L_A [%] (DL $_{A4}$)   (b)   estimated change of L_A [%] (DL $_{A4}$)

Fig. 20.6. Estimated rotor inductance computed from the fourth parity residual. Apparently,
most faults influence the result, however, the faulty inductance can most easily be detected due
to its strong influence: a. estimation results; b. resulting membership functions

Another interesting observation is the use of $\sigma_{est.,\psi}$ in the sixth rule of (20.14) to
distinguish offset from gain faults of the voltage sensor. This can be explained by the
fact that an offset term in the estimation equation given by an offset fault will change
the *structure* of the estimation equation, while a gain will only effect parameters.
Hence, the normal estimation equation will still be valid in case of gain faults, but
indicate a problem by a large $\sigma_{est.,\psi}$ for offset faults.

The system performed well on new experiments, showing the increased robustness through the incorporation of very simple knowledge. Additionally, the system
has a higher degree of transparency facilitating an adaptation to other motors. The
diagnostic rules can be extracted and are largely understandable.

d) Relation to fault trees

The resulting hierarchical classifier can also be interpreted as a set of fuzzy fault
trees. If one *reverses* the order of the structure and traces the decisions leading to a

particular fault back through the tree, it is possible to explicitly draw a fault tree for each individual fault. Figure 20.7 shows one fault situation (increased friction in the motor) as an example. The intermediate steps like "Mechanical Fault" from Figure 20.7 become *events* of the fault tree.

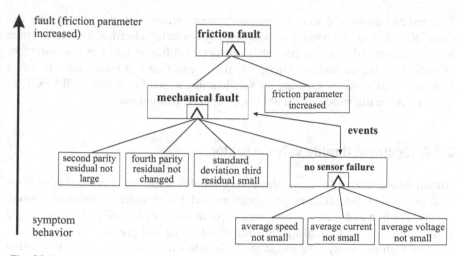

Fig. 20.7. Fault tree for one particular fault extracted from the diagnostic tree in Figure 20.5

Similar fault trees can be constructed for the other faults. This requires to analyze the rule tree and explicitly draw the trees. The resulting set of trees is a relatively redundant representation of the fault-symptom relation because the same events are used in multiple trees. They are nevertheless very intuitive and serve to understand and visualize the functionality of the diagnostic system.

e) Computational demands

The most time critical computation of the presented supervision concept is the computation of the continuous-time residuals. They require the evaluation of state-variable filters that are difficult to be implemented in fixed point arithmetic. If the computational resources are limited, also a discrete time form of the residuals is possible. This has, for instance, been implemented by [20.9].

The diagnosis on the other hand only needs to be evaluated if the fault-detection thresholds are violated. It is not time critical and can, for instance, be computed as a background job in the motor controller. Similarly, floating-point computations such as for the computation of the exponential function in the SELECT neuron can always be implemented on a lower precision fixed point controller, for instance, by using lookup tables. If the computational time is not critical, one can also implement floating point arithmetic on fixed point controllers. Since the time needed for the diagnosis compared with the time that typically is needed for personell to reach a faulty device, it is obvious that the computational demand should not be really an issue. Safety critical measures can be taken as soon as the thresholds are violated even before the diagnosis is started.

Summary

The detailed theoretical and experimental investigations with the permanently excited DC motor in idle running or with load by a rotating electrical hysteresis brake have demonstrated that it is possible to detect 14 different faults by measurement of only three signals and combining the parity equation and parameter estimation approach. Furtheron, by applying the self-learning neuro-fuzzy system SELECT all faults could be diagnosed with a 98 % correct classification rate.

20.2 Electrical throttle valve actuator

Automobile actuators have to operate very reliably under hard ambient conditions such as a wide temperature range, vibrations and disturbances in signals and power supply. Friction and time-varying process parameters, which are mainly caused by temperature influences, make it difficult to fulfil fast and precise positioning control and high reliability. The investigated throttle-valve actuator is used in ignition combustion engines to control the air mass flow through the intake manifold into the cylinders. This automotive actuator is embedded in various control systems such as engine control, traction control and velocity control, which require fast and precise operation.

20.2.1 Actuator setup

Figure 20.8 shows a schematic of the actuator. A permanently excited DC motor with a gear turns the throttle valve against the closing torque of a helical spring. The motor is driven by pulsewidth-modulated (PWM) armature voltage U_A, which is measured as well as the resulting armature current I_A. The angular position φ_K is redundantly measured in the range $[0 \ldots 90°]$ by two potentiometers.

Theoretical modelling and measurements have shown that the model structure illustrated in Figure 20.9 is a sufficient base for control design and fault detection. The gear was modelled as proportional factor with the reduction ratio v, [20.8], [20.9].

The inductance was neglected because the electrical time constant is about 1 ms and therefore much smaller than the mechanical one. Other parameters are the armature resistance R_A, the magnetic flux linkage Ψ, the inertia ratio J, the viscous friction coefficients M_{F1} and the spring constant c_F. The signal M_{ext} includes the spring pretension and external load torques. For the armature circuit then results

$$U_A(t) = R_A \, I_A(t) + \Psi \, \omega_A(t) \tag{20.16}$$

The driving torque of the DC motor is

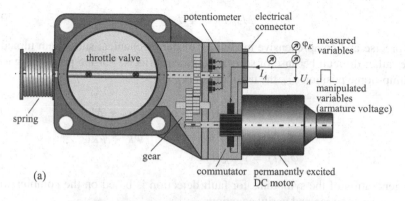

(a)

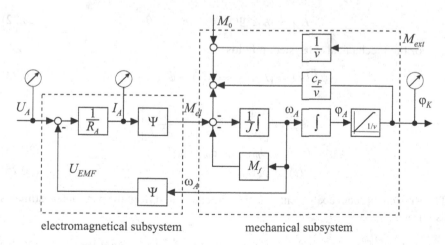

(b)

Fig. 20.8. The throttle valve actuator: (a) scheme; (b) photo

electromagnetical subsystem   mechanical subsystem

Fig. 20.9. Signal flow diagram of the throttle valve actuator

$$M_{el}(t) = \Psi \, I_A(t) \tag{20.17}$$

The precise and comprehensive modelling of the mechanical subsystem turned out to be rather difficult because of several nonlinear effects. For the positioning above the limp-home position φ_{k0} the used model is

$$v \, J \, \dot{\omega}_K(t) = \Psi \, I_A(t) - \frac{1}{v}(c_F \, \varphi_K(t) + M_0) \tag{20.18}$$
$$-M_f \, \omega_A(t) \qquad \varphi_K > \varphi_{K0}$$
$$\varphi_A(t) = v \, \varphi_K(t) \tag{20.19}$$

The generation of the symptoms for fault detection is based on the combination of parameter estimation and parity equations.

20.2.2 Parameter estimation

The parameter estimation is performed by two recursive estimators, one for the electrical and one for the mechanical subsystem. Figure 20.10 shows the overall arrangement. The parameter estimation is performed separately for the electrical and mechanical subsystem. The parameter estimation of the electrical part uses the data vector and parameter vector as follows

$$\boldsymbol{\psi}_e^T(k) = [I_A(k) \; v \; \dot{\varphi}_k(t) \; 1] \tag{20.20}$$

$$\boldsymbol{\theta}_e^T(k) = [R_A \; \Psi \; c_{oe}]^T = [\theta_{e1} \; \theta_{e2} \; \theta_{e3}]^T \tag{20.21}$$

c_{oe} is a constant for modelling additive (sensor) faults. The physical process coefficients result directly from

$$R_A = \theta_{e1}; \; \Psi = \theta_{e2}; \; c_{oe} = \theta_3 \tag{20.22}$$

For the mechanical subsystem it holds

$$\boldsymbol{\psi}_m^T(k) = [I_A(k) - \varphi_k(t) - \dot{\varphi}_k(t) - 1] \tag{20.23}$$

$$\boldsymbol{\theta}_m^T = \left[\frac{\Psi}{v \, J} \; \frac{c_F}{v^2 \, J} \; \frac{M_F}{J} \; \frac{M_0}{v^2 \, J} \right]$$
$$= [\theta_{m1} \; \theta_{m2} \; \theta_{m3} \; \theta_{m4}] \tag{20.24}$$
$$y(t) = \ddot{\varphi}_k(t) \tag{20.25}$$

The physical process coefficients can then be calculated from the parameter estimates as follows

$$J = \frac{\theta_{e2}}{v\theta_m}; \; c_F = v^2 \, J \, \theta_{m2}; \; M_F = J \, \theta_{m3}; \; M_0 = v^2 \, J \, \theta_{m4} \tag{20.26}$$

Hence, all physical process coefficients can be determined uniquely. The derivatives $\dot{\varphi}_k(t)$ and $\ddot{\varphi}_k(t)$ are generated by a state variable filter, see Figure 20.10.

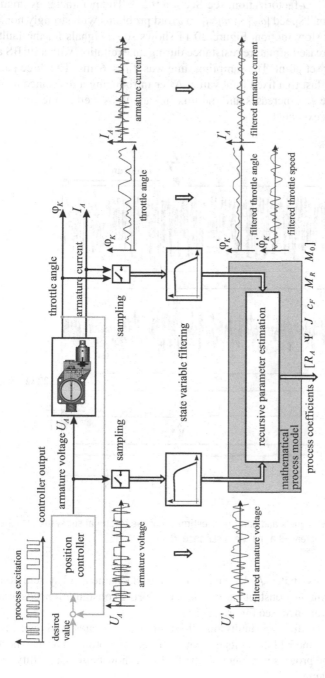

Fig. 20.10. Signal flow for continuous-time parameter estimation with state variable filters, see Chapter 23.2, in closed loop. Sampling time for state variable filters: $T_{0SVF} = 2$ ms; sampling time for parameter estimation: $T_0 = 6$ ms; corner frequency state variable filter: $f_g = 20$ Hz; critical angular velocity: $\omega_{krit} = 1.5$ rad/s

The parameter estimation was performed with recursive least squares parameter estimation and UD-factorization, see Section 9.2.3. The parameter estimation is switched off for small speed $|\omega_k| < \omega_{krit}$ to avoid problems with strongly nonlinear friction effects for slow motion. Figure 20.11 shows some signals for the fault free throttle and an increased armature resistance during an excitation with a PRBS at the position controller set point. The sampling time was $T_0 = 6$ ms. The three parameters converge very fast to a fixed final value. After introducing a resistance increase, parameter estimate $\hat{R}_A$ increases and the flux linkage $\hat{\Psi}$ as well as the constant $\hat{c}_{oe}$ remain constant as expected.

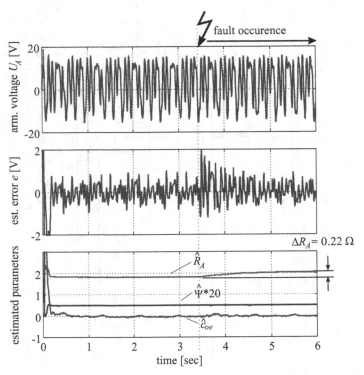

Fig. 20.11. Measured signals and parameter estimates of the electrical subsystem for the fault-free case and for an increased armature resistance R_A

The parameter estimates J, M_0 and M_F of the mechanical subsystem converge very fast to approximate constant values, only the spring constant c_F needs about 3 s and shows larger variance, see Figure 20.12.

The influence of different faults on all parameter estimates is shown in Table 20.3. Except F1, F2 and F11 all faults indicate different symptom patterns. The symptom patterns for the process parameter faults F3 to F8 show better isolability than for the sensor offset faults.

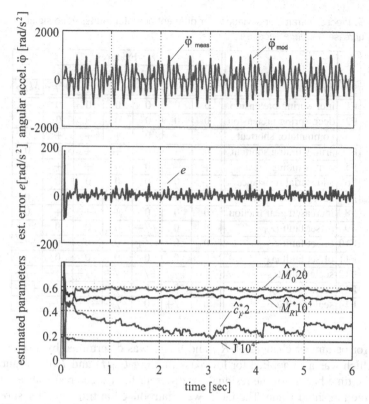

Fig. 20.12. Signals and parameter estimates of the mechanical subsystem for the fault-free case

20.2.3 Parity equations

The application of parity equations as described in Example 10.2 and in Section 20.1 requires a relatively precise process model with constant parameters. The design and practical experiences with different approaches of parity space methods for the electrical throttle has shown, however, that the residuals show too large variances because of model discrepancies and normal disturbances, [20.9]. In contrast to the DC motor example in Section 20.1, the electrical throttle includes a mechanical load with gear and a helical spring with relatively large pretension. Especially because of the friction effects at the bearings, the gear and spring adjustment, it was not possible to find an overall model of the mechanical system which can be used in parity equations. Therefore, the parity approach was finally limited to the electrical subsystem only, leading to the residual

$$r_1(t) = U_A(t) - R_A I_A(t) - \Psi v \dot{\varphi}_{1K}(t) \tag{20.27}$$

where φ_{1K} is the throttle angle of potentiometer 1 (one out of two possible positive measurements). This residual will deflect for offset faults of the sensors U_A, I_A and

Table 20.3. Process parameter deviations for different actuator faults: 0: no significant change; +: large increase; −: large decrease

	Faults	R_A	Ψ	c_{oe}	J	c_F	M_{F1}	M_{F2}
				Features parameter estimates				
F1	incr. spring pretension	0	0	0	0	0	0	+
F2	decr. spring pretension	0	0	0	0	0	0	−
F3	commutator shortcut	−	−	0	+	+	+	0
F4	arm. winding shortcut	0	−	0	+	+	+	0
F5	arm. winding break	+	−	0	0	+	+	+
F6	add. serial resistance	+	0	0	0	0	0	0
F7	add. parallel resistance	−	−	0	0	+	+	0
F8	increased gear friction	0	0	0	+	+	+	0
F9	offset fault U_A	0	0	+/−	0	0	0	0
F10	offset fault I_A	0	0	−/+	0	0	0	+/−
F11	offset fault φ_K	0	0	0	0	0	0	−/+
F12	scale fault U_A	+/−	+/−	+/−	+/−	+/−	+/−	+/−
F13	scale fault I_A	−/+	0	0	+/−	+/−	+/−	+/−
F14	scale fault φ_K	0	−/+	0	−/+	−/+	−/+	−/+

φ_1 and for parameter changes of R_A and Ψ. $\dot{\varphi}_1$ was determined by a state variable filter, which was also applied for low pass filtering $U_A(t)$ and $I_A(t)$. Figure 20.13 shows the time history of the residual for different faults. The residuals are normalized to the threshold value. The faults were introduced at time $t = 1$ s, showing an exceeding of the threshold after about 200 ms. An offset fault of the angle sensor only briefly overpasses the threshold (which is disadvantageous for fault detection). The parameter changes result in larger variances of the residual.

The experiments have further shown, that some parameters change with the temperature and load. In the electrical subsystem are these the resistance R_A and the flux linkage Ψ. They change their values after continuous operation over 30 min by +50 % and -10 %. A modification of the parity equations according to Section 10.5 can now be used to track the resistance parameter R_A and dependent on this the value Ψ. This leads to a smaller variance of residual $r_1(t)$ with small additional computational effort, [20.9].

A further residual $r_2(t)$ as difference between the two position sensors is used for a sensor fault-tolerant system, described in [20.5].

20.2.4 Diagnostic equipment for quality control

For quality control, e.g. as end-of-assembling-line testing, a special approach was developed. Figure 20.14 depicts the equipment, consisting of power electronics, an online coupled digital signal processor and a host PC, see Figure 20.14, [20.10]. As the electrical throttle is disconnected from the engine and for its control system a special test cycle can be applied.

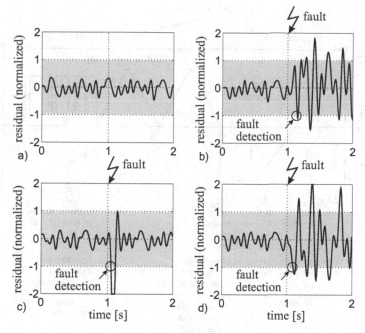

Fig. 20.13. Parity equation residual $r_1(t)$ for the electrical subsystem for PRBS excitation: a. fault free; b. increase of resistance R_A by +20 %; c. 20 % offset on throttle angle $\Delta\varphi_{1k}$; d. 20 % gain change of throttle angle φ_{1k}

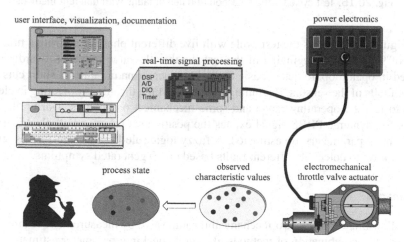

Fig. 20.14. Diagnosis equipment for electrical throttles in the frame of quality control

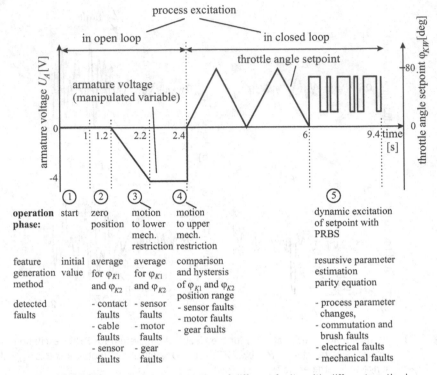

Fig. 20.15. Test cycle for the detection of different faults with different methods

Figure 20.15 shows the test cycle with five different phases requiring a time period of 9.4 s. At the beginning of the test the throttle valve is feed forwardly controlled in open loop, to detect basic faults like open armature circuit, short cuts and offset faults of the sensors. Beginning with phase 4 the throttle is operating in closed loop to test the operating range and some mechanical parameters. During the last phase 5 a dynamic PRBS signal excites the position set point and all electrical and mechanical parameters are estimated. A fuzzy logic rule-based diagnosis system finally allows to detect 38 different faults based on 30 generated symptoms.

Summary

Overall testing of the electro-mechanical throttle valve by measurement of three signals needs a combination of methods like limit checking, parameter estimation and a parity equation. Parameter estimation gave good results allowing deep fault diagnosis. However, parity equations are only suitable for the electrical part due to the difficulty to model the mechanical subsystem. Hence, the ability of parameter estimation to reduce modelling errors by the inherent regression made a comprehensive fault diagnosis possible.

Fault detection and diagnosis of a centrifugal pump-pipe-system

21.1 The pump-pipe-tank system

In order to develop an online operating fault detection and diagnosis method for a centrifugal pump-pipe-tank system over a large operating range, a plant according to Figure 21.1 and 21.2 is considered. The pump is driven by an inverter-fed, speed variable induction (squirrel cage) motor which is speed controlled by a field-oriented controller.

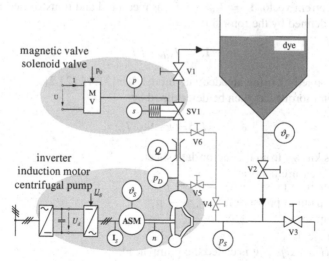

Fig. 21.1. Centrifugal pump-pipe-tank plant with measurements: AC-motor: Siemens 1 LA 5090-2AA (norm motor) $P_N = 1.5\,\text{kW}$; $n_W = 2900\,\text{rpm}$; Frequency converter: Lust MC 7404; circular pump: Hilge, $H_{max} = 130\,\text{m}$; $\dot{V}_{max} = 14\,\text{m}^3/\text{h}$; $P_{max} = 5.5\,\text{kW}$

Fig. 21.2. Photo of the investigated centrifugal pump

21.2 Mathematical models of the centrifugal pump

The stator current vector $\mathbf{I}_s = I_{s\alpha} + i\, I_{s\beta}$ is measured and transformed in the reference frame defined by the rotor flux.

$$\mathbf{I}_s = I_{sd} + i\, I_{sq} \tag{21.1}$$

which is obtained by using an adequate model, see [21.4], [21.2].

The motor torque can then be determined by

$$M_{el} = k_T\, \Psi_R\, I_{sq} \tag{21.2}$$

where k_T is known from the motor data sheet.

Further measurements are

p_1 pump inlet pressure $\Delta p = p_2 - p_1$
p_2 pump outlet pressure $\Delta p = p_2 - p_1$
ω pump speed
$\dot{V}$ volume flow

where $H = \Delta\, p/\rho\, g$ is called the pump head.

The used mathematical models of the pump have to be adapted to the pump-pipe-system. Based on the theoretically derived equations of centrifugal pumps, [21.3], following equations are used here

$$H(t) = h_{nn}\, \omega^2(t) - h_{nv}\, \omega(t)\, \dot{V}(t) - h_{vv}\, \dot{V}^2(t) \tag{21.3}$$

$$H(t) = \frac{p_2(t) - p_1(t)}{\rho\, g} = \frac{\Delta p(t)}{\rho\, g} \tag{21.4}$$

$$J \dot{\omega}(t) = M_{el} - M_{th}(t) - M_f(t) \tag{21.5}$$
$$M_{th}(t) = M_{th1}\, \omega(t)\, \dot{V}(t) - M_{th2}\, \dot{V}^2(t) \tag{21.6}$$
$$M_f(t) = M_{f0}\, \text{sign}\, \omega(t) + M_{f1}\, \omega(t) \tag{21.7}$$

A comparison of these theoretically derived equations has shown that, because of the flow $\dot{V}$ is proportional to the speed ω and neglection of the viscous friction in (21.7) following simplified relations can be used with $\Delta p = \rho g H$

$$\Delta p(t) = \tilde{h}_{nn}\, \omega^2(t) - \tilde{h}_\omega\, \omega(t) \tag{21.8}$$
$$J \dot{\omega}(t) = M_{el}(t) - M_{f0} - M_2\, \omega^2(t) \tag{21.9}$$

The dynamics of the fluid in the pipe is described by

$$H(t) = a_B \frac{d\dot{V}(t)}{dt} + h_{rr}\, \dot{V}^2(t) \tag{21.10}$$

with $a_B = l/gA$ (l: pipe length, A: cross sectional area). These models agree with [21.1] for a larger pump-pipe-system. Figure 21.3 shows the resulting signal flow diagram.

21.3 Parity equations and parameter estimation

a) Measurement of I, ω, Δp, $\dot{V}$

Nonlinear parity equations
Based on these models and after discretizing, following residuals with nonlinear equations can be obtained, compare Figure 21.4 and [21.4], [21.5]:

- Static pump model (21.8)

$$r_1(k) = \Delta p(k) - w_1\, \omega^2(k) - w_2\, \omega(k) \tag{21.11}$$

- Dynamic pipe model (21.10) and further simplifications

$$r_2(k) = \dot{V}(k) - w_3 - w_4\, \sqrt{\Delta p(k)} - w_5\, \dot{V}(k-1) \tag{21.12}$$

- Dynamic pipe-pump model (21.8), (21.10)

$$\begin{aligned} r_3(k) &= \dot{V}(k) - w_3 - w_4\, \sqrt{\Delta \hat{p}(k)} - w_5\, \dot{V}(k-1) \\ \Delta \hat{p}(k) &= w_1\, \omega^2(k) + w_2\, \omega(k) \end{aligned} \tag{21.13}$$

- Dynamic inverse pump model (21.9)

$$\begin{aligned} r_4(k) &= M_{el} - w_6 - w_7\, \omega(k) - w_8\, \omega(k-1) \\ &\quad - w_9\, \omega^2(k) - w_{10}\, M_{el}(k-1) \end{aligned} \tag{21.14}$$

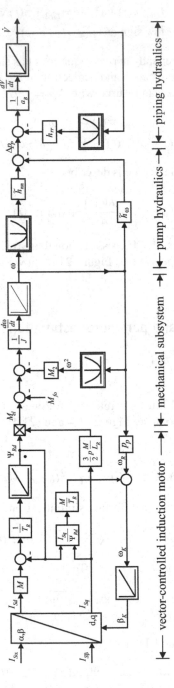

Fig. 21.3. Signal flow diagram of the simplified models for a centrifugal pump-pipe system

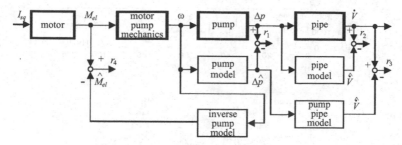

Fig. 21.4. Residual generation with parity equations for the pump-pipe system

The residuals $r_1(k)$, $r_2(k)$ and $r_3(k)$ are output residuals which follow by comparing the measured $\Delta p(k)$ and $\dot{V}(k)$ with the corresponding model outputs. However, $r_4(k)$ is an input residual, because M_{el} is compared with the output of an inverse pump model. $r_2(k)$ and $r_3(k)$ include flow sensor dynamics of first order. The sampling time is $T_0 = 10$ ms.

The parameters $w_1, \ldots, w_{10}$ for the parity equations follow directly from known physical data described in the equations above or are estimated, e.g. with methods of least squares based on measurements of $I_{sq}(t)$, $\omega(t)$, $\Delta p(t)$ and $\dot{V}(t)$. However, the parameters w_i depend, especially for low speed on the operating point. Therefore, for each residual a *multi-model approach* is used. It has turned out, that is is sufficient to consider the parameters dependent on the angular speed only.

To identify the multi-model for the parity equations, the pump system was excited by an amplitude modulated PRBS signal over the whole operating range and with the local linear model network (LOLIMOT) the parameters $w_i(\omega), i = 1, \ldots, 10$ were determined using three local models each, see Section 9.3.3. Figure 21.5 shows a comparison of measured and reconstructed values with the models. Hence, a very good agreement can be stated.

Following faults were introduced into the pump-pipe-systems:

- offset sensor faults ω, $\dot{V}$, p_1, p_2;
- increased resistance by piecewise closing of a valve after the pump;
- cavitation by piecewise closing a valve before the pump;
- increased bearing friction by removing grease and introducing iron deposits;
- defect impeller by closing one channel between two vanes with silicon;
- sealing gap losses by opening a by-pass valve;
- leakage between pump and flow measurement.

Table 21.1 shows the resulting symptoms. The residuals of the *parity equations* can be obtained without input excitation. They indicate, that the sensor offset faults, sealing gap losses and increased bearing friction are strongly isolable. However, increased flow resistance, cavitation and impeller defect are either only weakly or not isolable. This means that all the faults are detectable, but some of them cannot be differentiated. In order to avoid too large thresholds, adaptive thresholds are used. In addition to a constant value, the thresholds depend on a high-pass filtered value of the speed ω, which increases the threshold in case of a speed change, [21.5].

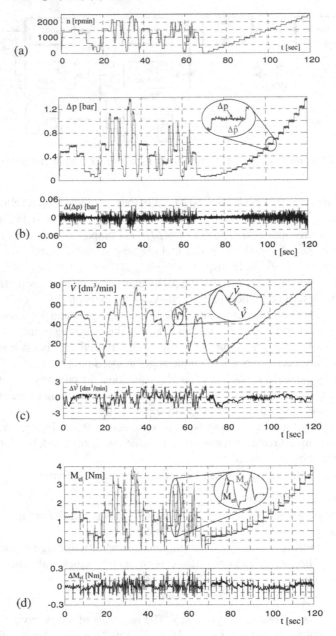

Fig. 21.5. Measured signals of the pump-pipe-system, LOLIMOT model outputs and their differences: (a) angular speed; (b) delivery pressure difference; (c) flow rate; (d) torque of AC motor

During normal operation the nonlinear parity equations with residuals r_1 and r_4 can be applied. In order to obtain a more detailed diagnosis, a parameter estimation with an amplitude modulated PRBS, an APRBS test signal excitation of the speed has to be started.

This dynamic excitation allows to estimate the parameters of the models (21.8), (21.9) and (21.10) with a recursive least squares methods as described in Section 9.2, see also [21.2], Section 6.6. The changes of these physically defined parameters are given in Table 21.1 and show that all faults are isolable and can therefore be diagnosed.

b) Measurement of I, ω

If the delivery pressure Δp and the flow rate $\dot{V}$ are not measurable, the residual r_4 can be calculated based on measured speed ω and motor current I_{sq}. This allows to detect a sensor fault in ω and some pump faults. Additional parameter estimation enables to determine parameter deviations of J, M_{f0} and M_2 with (21.8) and to isolate some more pump faults.

Similar results as described above have been obtained by [21.1] for a larger pump with $P = 3.3$ kW and $\dot{V}_{max} = 150$ m³/h and a larger pipe circulation system with two heat exchangers. Two different flow meters could be used. This allowed to generate six residuals and four parameter estimates. Together with two variances of residuals, all together 13 symptoms could be obtained, which enables to diagnose 11 different faults of sensors, pump and pipe system.

These symptoms were then used to train 20 fuzzy rules with the SELECT procedure described in Section 17.3.5, yielding a 100 % classification accuracy.

Summary

This case study has again shown that (nonlinear) parity equations are suitable for some additive faults and that parameter estimation gives much more insight and allows to detect and diagnose especially parametric (multiplicative) faults. The best fault coverage is obtained by combining parity equation and parameter estimation.

Table 21.2 enables to see which faults are only detectable and also diagnosable with combined parity equations and parameter estimation. A minimal measurement configuration with the torque $M = f(I)$ and speed ω allows to detect some few faults but not to diagnose them. By adding a sensor for p_1 and p_2, or for Δp, many more faults can be detected and diagnosed. The additional implementation of a flow rate sensor $\dot{V}$ has little influence on the number of detectable faults, but allows to diagnose many more faults. This shows that model-based detection of faults is possible with some three to four sensors, but that the fault diagnosis is improved considerably by one additional sensor (here the flow rate).

Table 21.1. Fault-symptom table for the pumpe-pipe-system: 0: small values; $+/-$: positive/negative deflection; $++/--$: strong positive/negative deflection

Fault	Symptoms										
	parity equations				parameter estimates						
	$\lvert r_1\rvert$	$\lvert r_2\rvert$	$\lvert r_3\rvert$	$\lvert r_4\rvert$	ΔJ	ΔM_{f0}	ΔM_2	Δa_B	Δh_{rr}	Δh_{nn}	Δh_ω
Sensor ω	$+/++$	0	$+/+++/++$	$+/+++/++$	0	0	0	0	0	$0/-$	$+/++$
Sensor $\dot V$	0	$+/+++/++$	$+/+++/++$	0	0	0	0	$0/-$	$+$	0	0
Sensor p_1, p_2	$+/+++/++$	$+/+++/++$	0	0	0	0	0	$0/+$	0	$0/-$	$+/++$
Gap losses	$+$	0	0	0	0	0	0	0	0	$-$	$+$
Leakage	$+$	$+$	$+$	0	0	0	0	$-$	$-$	0	0
Incr. flow resist. 20-40 %	$+$	$+$	$+$	0	0	0	0	0	$+$	0	0
Incr. flow resist. 40-60 %	$+$	$++$	$+$	$+$	0	0	0	$++$	$++$	0	0
Incr. flow resist. 60-90 %	$++$	$++$	$++$	$++$	0	0	$+$	$++$	$++$	0	0
Cavitation	$+$	$++$	$++$	$+$	0	0	$+$	$--$	$++$	$--$	$++$
Incr. bearing friction	0	0	0	$+$	0	$+$	$+$	0	0	0	0
Defect impeller	$+$	$+$	$+$	$+$	$+$	0	0	$-$	$-$	$-$	$+$

Table 21.2. Detectable and diagnosable faults in dependence on the sensors used. Assumed is a modern frequency converter that is able to reconstruct the motor torque M without additional sensors. Omitted are faults at the frequency converter itself and other electric faults in the motor. Faults in parenthesis are difficult to be identified yet not completely impossible (depends on the individual setup), [21.1].

Fault detectable	Sensor usage				
	M	M,ω	M,ω,p_2	M,ω,p_1,p_2	$M,\omega,p_1,p_2,\dot{V}$
Total breakdown	x	x	x	x	x
Defective blade wheel		(x)	x	x	x
Incr. shaft or motor friction			x	x	x
Sensor fault ω		x	x	x	x
Sensor fault $\dot{V}$					x
Sensor fault p_1,p_2			p_2	x	x
Decreased flow resistance		(x)	(x)	x	x
Increased flow resistance		(x)	(x)	x	x
Cavitation through pressure reduction		(x)	(x)	x	x
Insufficient de-ventilation of sensors p_1,p_2			p_2	x	x
Insufficient de-ventilation of sensor $\dot{V}$					x

Fault diagnosable	Sensor usage				
	M	M,ω	M,ω,p_2	M,ω,p_1,p_2	$M,\omega,p_1,p_2,\dot{V}$
Total breakdown		x	x	x	x
Defective blade wheel			x	x	x
Incr. shaft or motor friction			x	x	x
Sensor fault ω					x
Sensor fault $\dot{V}$					
Sensor fault p_1,p_2					x
Decreased flow resistance					x
Increased flow resistance					x
Cavitation through pressure reduction					x
Insufficient de-ventilation of sensors p_1,p_2			p_2	x	x
Insufficient de-ventilation of sensor $\dot{V}$					x

Fault detection and diagnosis of an automotive suspension and the tire pressures

22.1 Mathematical model of a suspension and the test rig

To perform fault diagnosis either in a service station, for example for technical inspection, or in a driving state it is important to use easy measurable variables. If the methods should be used for technical inspection the sensors must be easily applicable to the car. For on-board fault detection the existing variables for suspension control should be used. Variables which meet these requirements are the vertical accelerations of body and wheel, $\ddot{z}_B$ and $\ddot{z}_W$ and the suspension deflection $z_W - z_B$. Another important point is that the method should require only little a priori knowledge about the type of car.

A scheme for a simplified model of a car suspension system, a quarter car model, is shown in Figure 22.1. The following equations follow from force balances

$$m_B \ddot{z}_B(t) = c_B(z_W(t) - z_B(t)) + d_B(\dot{z}_W(t) - \dot{z}_B(t)) \qquad (22.1)$$

$$m_W \ddot{z}_W(t) = -c_B(z_W(t) - z_B(t)) - d_B(\dot{z}_W(t) - \dot{z}_B(t)) + c_W(r(t) - z_W(t)) \qquad (22.2)$$

In this chapter following symbols are used:

a_1, b_1	parameters of transfer functions;	F_W	wheel force;
c_B	stiffness of body spring;	m_B	body mass;
c_W	tire stiffness;	p_W	wheel pressure;
d_B	body damping coefficient;	r	road displacement;
f_r	resonance frequency;	z_B	vert. body displacement;
F_C	Coulomb friction force;	z_W	vert. wheel displacement
F_D	damper force;	Δz_{WB}	
F_S	spring and damper force;	$= z_B - z_W$	suspension deflection.

The small damping of the wheel is usually negligible. A survey of passive and semi-active suspensions and their models is given in [22.7].

The first results were obtained on a test rig, shown in Figure 22.2 which is equipped with a continuously adjustable damper. The damping is controlled by a magnetic valve, which opens or closes a bypass continuously. The test rig was constructed primarily for investigations on semi-active, parameter-adaptive suspension control, [22.2].

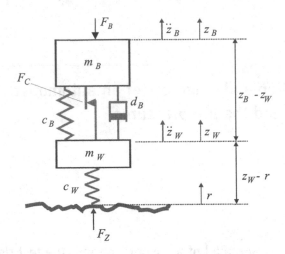

Fig. 22.1. Quarter car model

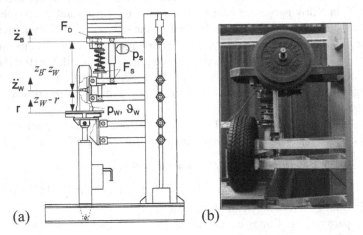

Fig. 22.2. Quarter-car test rig of IAT, TU Darmstadt: a. scheme; b. photo

22.2 Parameter estimation (test rig)

In general, the relationship between force and velocity of a shock absorber is non-linear. It is usually degressive and depends strongly on the direction of motion of the piston. In addition, the Coulomb friction of the damper should be taken into account. To approximate this behavior the characteristic damper curve can be divided into m sections as a function of the piston velocity. Considering m sections the following equation, compare (22.1), can be obtained.

$$\ddot{z}_B = \frac{d_{B,i}}{m_B}(\dot{z}_W - \dot{z}_B) + \frac{c_B}{m_B}(z_W - z_B) + \frac{1}{m_B}F_{C,i} \qquad i = 1 \ldots m \quad (22.3)$$

$F_{C,i}$ denotes the force generated by Coulomb friction, $d_{B,i}$ the damping coefficient for each section. Using (22.3) the damping curve can be estimated with a standard parameter estimation algorithm measuring the body acceleration $\ddot{z}_B$ and suspension deflection $z_W - z_B$. The velocity $\dot{z}_W - \dot{z}_B$ can be obtained by numerical differentiation. In addition either the body mass m_B or the spring stiffness c_B can be estimated. One of both variables must be a priori known. Using (22.1) and (22.2) other equations for parameter estimation can be obtained, e.g. (22.4) which can be used to estimate the tire stiffness c_W additionally

$$z_W - z_B = -\frac{d_{B,i}}{m_B}(\dot{z}_W - \dot{z}_B) - \frac{c_B}{m_B}(z_W - z_B) + \frac{c_W}{c_B}(r - z_W) - \frac{1}{c_W}F_{C,i} \quad (22.4)$$

The disadvantage of this equation is the necessity to measure the distance between road and wheel $(r - z_W)$. This variable is therefore only feasible for technical inspection on a test stand, whereas in driving cars the high sensor costs prevent the use of this estimation equation.

Figure 22.3 shows the estimated damping curve for different damper magnetic valve currents using (22.3). Because rising damper current opens the bypass the damping sinks. The damping curve was divided into four sections, two for each direction of motion. It can be seen that the damping curves at different damper currents can be distinguished. Since a worn damper results in a changed curve, a detection is possible. In addition, different other faults can be distinguished, [22.2], [22.8], [22.9], [22.11], [22.6]. Table 22.1 gives an overview. The influence of these faults on the estimated variables is obvious. Hence, there is a different parameter estimate pattern for every fault. All parametric faults are strongly isolable, except the sensor offset faults.

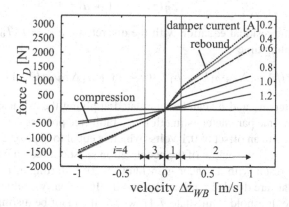

Fig. 22.3. Estimated local linear damping characteristics using (22.3)

Table 22.1. Influence of faults on the parameter estimates. + increase, − decrease, 0 no influence

	Fault	d_B	d_{B+}	c_{B-}	c_{B+}	F_{C+}	F_{C+}
Process faults	Friction +	0	0	0	0	+	−
	Damping +	+	+	0	0	0	0
	Spring stiffness +	0	0	+	+	0	0
Sensor faults	Offset $\ddot{z}_B$+	0	0	0	0	+	+
	Offset $(z_W - z_B)$+	0	0	0	0	−	−
	Gain $\ddot{z}_B$+	+	+	+	+	+	+
	Gain $(z_W - z_B)$+	−	−	−	−	0	0

22.3 Parity equations (test rig)

Parity equations do not need permanent excitation and require less computational effort than parameter estimation, but do not give the same deep insight into the process as parameter estimation. To combine the advantages of parameter estimation and parity equations it is proposed to supervise the process online with parity equations and to perform parameter estimation after a fault is detected, compare Section 14.3 and [22.10]. In the following, an example for the detection of sensor faults with parity equations is described.

With the abbreviation

$$\Delta z_{WB} = z_W - z_B \tag{22.5}$$

the z-domain transfer function of (22.1) can be calculated

$$G_B(z) = \frac{\Delta z_{WB}(z)}{\ddot{z}_B(z)} = \frac{b_1 z^{-1}}{1 + a_1 z^{-1}} \tag{22.6}$$

which leads to the residual equation with the discrete time $k = t/T_0 = 0, 1, 2, \ldots$ and T_0 the sampling time

$$r(k) = \Delta z_{WB}(k) - b_1 \ddot{z}_B(k - 1) + a_1 \Delta z_{WB}(k - 1) \tag{22.7}$$

The parameters a_1 and b_1 can be calculated applying the z-transform to (22.1) or using discrete-time parameter estimation in a fault free process state. Figure 22.4a shows the result with an offset of 0.1 Volts (which is equal to approx. 3 % of the maximum value) added to the output of the acceleration sensor at the time $t = 5$s. Figure 22.4b gives the result with a sensor gain fault of 20 % starting at $t = 5$s. In both cases the residuals are divided by their thresholds. It is clearly visible that the offset fault violates the threshold immediately. However, it cannot be distinguished which sensor, $z_W - z_B$ or $\ddot{z}_B$ leads to threshold violation. A sensor gain fault obviously only affects the variance of the residual. Hence, the results show that in both cases a detection of the sensor fault is possible in principle, if the parameters of the process remain constant. Hence, a combination of parameter estimation according to Table 22.1 and the residual equation (22.7) gives the best fault coverage. Similar results can be obtained with the wheel acceleration sensor $\ddot{z}_W$ and suspension deflection

$(z_W - z_B)$, [22.2], [22.7]. These sensors are now in series production of semi-active shock absorbers.

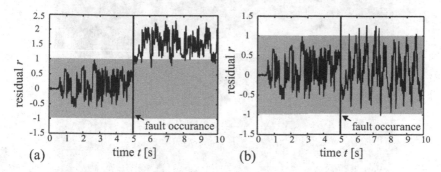

Fig. 22.4. Parity equation residual $r(k)$ for: a. acceleration sensor $\ddot{z}_B$ offset 0.1 V at 5 s; b. acceleration sensor $\ddot{z}_B$ gain offset of 20 % at 5 s

22.4 Experimental results with a driving vehicle

To test various methods in a driving car, a medium class car, an Opel Omega, Figure 22.5, was equipped with sensors to measure the vertical acceleration of body and wheel as well as the suspension deflections. To realize different damping coefficients the car is equipped with adjustable shock absorbers at the rear axle, which can be varied in three steps. In figure 22.6 the course of the estimated damping coefficients as described in Section 22.2 at different damper settings is given for driving over boards of height 2 cm, Figure 22.5.

After approximately 2.5 s the estimated values converge to their final values. The estimated damping coefficients differ approximately 10 % from the directly measured ones. In Figure 22.7 the estimated characteristic curves at the different damper settings are shown. The different settings are separable and the different damping characteristics in compression and rebound is clearly visible, although the effect is not as strong as in the directly measured characteristic curve. More results are given in [22.1].

22.5 Shock absorber fault detection during driving

All the results are carried out by considering real measurements on a highway, [22.1].

a) Recursive parameter estimation (RLS)

Figure 22.8 illustrates the suspension deflection $z_W - z_B$, the first derivative of the suspension deflection calculated with a state variable filter $\dot{z}_W - \dot{z}_B$ and the body

Fig. 22.5. Driving experiment for model validation and parameter estimation

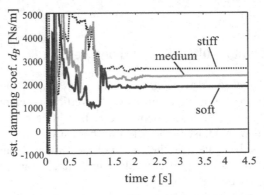

Fig. 22.6. Estimated damping coefficients for different damper settings (speed about 30 km/h)

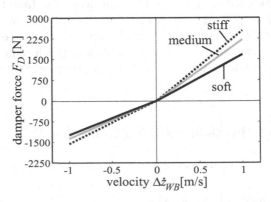

Fig. 22.7. Estimated characteristic damping characteristics for different damper settings

acceleration $\ddot{z}_B$ for the right rear wheel during a highway test drive. After 30, 60, 90, 120 seconds a change of the shock absorber damping was made.

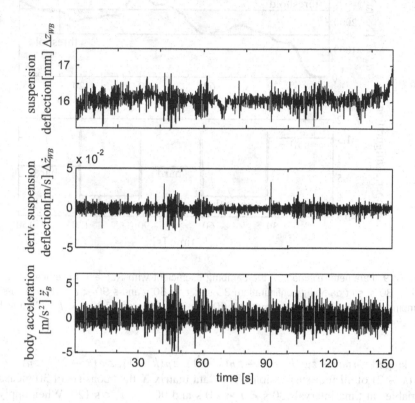

Fig. 22.8. Measured signals on a highway with variation of the damper configuration

Several estimations have shown that the recursive least squares algorithm (RLS) with exponential forgetting factor received very good results. This recursive parameter estimation is able to adapt to the different damping settings in about 10 s, see Figure 22.9.

b) Principal component analysis (PCA)

PCA, see Chapter 13, is first applied to the measured data in a normal situation, i.e. the medium damping configuration (30 s $< t <$ 60 s and 90 s $< t <$ 120 s). The measurements of all the variables that either characterize or influence the vertical dynamics of the wheel suspension system are included in the reference data matrix $\mathbf{X}$, i.e. the suspension deflection and body acceleration measured at four suspension corners, roll angular velocity, pitch angular velocity, yaw angular velocity, longitudinal body acceleration, lateral body acceleration and vertical body acceleration. Since the PCA is being applied to a dynamic system the data matrix $\mathbf{X}$ is extended to include

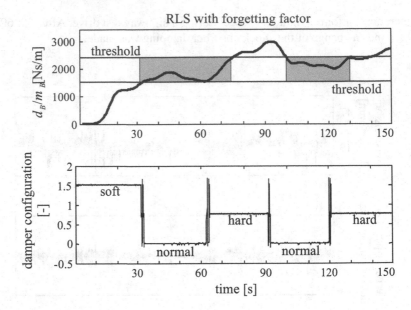

Fig. 22.9. Parameter estimate of the damping coefficient with RLS and forgetting factor. 0 s $< t <$ 30 s: soft damping configuration; 30 s $< t <$ 60 s and s 90 $< t <$ 120 s: medium damping configuration (normal situation); 60 s $< t <$ 90 s and 120 s $< t <$ 150 s: hard damping coefficient

the lagged variables $z_W(k-1) - z_B(k-1)$, $\ddot{z}_B(k-1)$, $z_W(k-2) - z_B(k-2)$, $\ddot{z}_B(k-2)$ of all four suspensions. The data matrix **X** thus consists of 30 measured variables at time intervals 30 s $< t <$ 60 s and 90 s $< t <$ s 120. When applying PCA it turns out that 24 principal components are sufficient to adequately describe the dynamic behavior of the car suspension system.

A change in the damping configuration changes the signals relationships. It is therefore expected that the measurements do not project into the same region during occurrence of this change. Figure 22.10 shows T^2 statistics

$$T^2(k) = \sum_{j=1}^{M} \frac{t_{kj}^2}{\sigma_j^2}$$

calculated for the entire data set, compare (13.24). Herewith σ_j^2 is the variance of the j−th principal component. During the periods of abnormal process operation (soft and hard damping configuration) the value T^2 statistics only occasionally exceed the threshold, thus indicating abnormal behavior (the grey areas in Figure 22.10). However, there are no false alarms (the T^2 statistics remains below the threshold when the damping configuration is normal). Hence, the parameter estimation gave better results than principal component analysis. A summary of various results for fault diagnosis of semi-active and active suspension systems is given in [22.3].

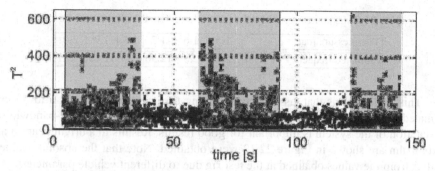

Fig. 22.10. T^2 statistical distance (dots) and the threshold (dashed line) for principal component analysis (PCA)

22.6 Tire pressure supervision with spectral analysis

Using signal spectrum analysis of the measured wheel acceleration $\ddot{z}_W$, a measurement of the spring deflection is not necessary to observe changes of the tire stiffness c_W due to tire pressure loss. Therefore only the vertical acceleration of the wheel has to be measured. Because the frequency range of the body vibrations is significantly lower than the range of the wheel vibrations the simplified model in Figure 22.11 can be applied.

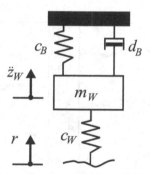

Fig. 22.11. Simplified model to determine the tire stiffness

The resonance frequency f_r of this system is

$$f_r = \frac{1}{2\pi} \sqrt{\frac{c_W + c_B}{m_W}} \sqrt{1 - \frac{d_B^2}{4m_W(c_W + c_B)}} \tag{22.8}$$

Via frequency estimation, based on an AR parameter estimation method, see Section 8.1.6, this frequency can be estimated. In Table 22.2 the results for the estimated frequency and the corresponding tire pressure with test rig experiments are given.

Table 22.2. Estimated resonance frequency for different tire pressures (test rig)

Tire pressure [bar]		1.5	1.6	1.7	1.8	1.9	2.0
estimated frequency f_r [Hz]		12.23	12.39	12.56	12.7	12.85	12.91

This table reveals that, as expected, sinking tire pressure results in a lower estimated resonance frequency. It should be noted, however, that a wide-bandwidth-excitation of the system is important for good results. Results in a driving car on an Autobahn are shown in Figure 22.12 were obtained. Note that the absolute values differ from the values obtained at the test rig due to different vehicle parameters.

The vertical wheel vibrations were measured with an accelerometer with a sampling rate of 200 Hz. The tire pressure was first set to its correct value of 2.0 bar and then reduced to 1.5 bar. Figure 22.12a shows that, although the estimated frequency varies in a range of approximately 0.5 Hz, the values for the lower pressure always stays below the values for the normal tire pressure. To reduce the influence of the road excitation, now the difference of the estimated frequency of the front wheel and the corresponding rear wheel is calculated. The results are given in Figure 22.12b. Obviously, the variation of the signals is reduced and the margin for the detection of an inflation increased. However, by calculating the difference between front and rear wheel, only relative pressure changes can be detected. If both tires loose the same amount of air, a detection with this method is infeasible. Therefore, a combination of both methods may be appropriate to improve the behavior.

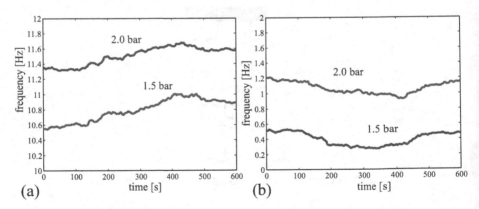

Fig. 22.12. a. Estimated resonance frequency at different tire pressure; b. Difference of the estimated frequencies between the front and rear wheel at different tire pressure

A loss of approximately 0.4 bar can be detected, which makes it possible to detect slow punctures with little computational effort. A similar method by using the body acceleration $\ddot{z}_B$ is described in [22.8], [22.5]. This method, which does not need tire pressure sensors, can be combined with observing the wheel speed difference between the wheels. With decreasing tire pressure the wheel speed increases. A survey on tire pressure monitoring with direct and indirect measurements is given in [22.4].

Summary

The applications for fault detection and diagnosis of suspensions have demonstrated, that parameter estimation is best suited. Several faults in the suspension system can be isolated because of their unique patterns of the parameter estimates. Parity equations enable to detect some faults, like sensor offsets, but not to diagnose them. Combination of both methods is therefore recommended. Tire pressure supervision is possible with vibration analysis of the wheel acceleration, especially if combined with wheel speed differences without measuring the tire pressure directly.

23

Appendix

23.1 Terminology in fault detection and diagnosis

The following definitions are the result of a coordinated action within the IFAC Technical Committee SAFEPROCESS, published in [23.8]. Some basic definitions can also be found in [23.13], [23.4] and in German standards like DIN and VDI/VDE-Richtlinien, see References at the end of this section.

1) **States and Signals**

Fault: *Unpermitted deviation* of at least one characteristic property of the system;

Failure: *Permanent* interruption of a systems ability to perform a required function under specified operating conditions;

Malfunction: *Intermittent irregularity* in fulfilment of a systems desired function;

Error: *Deviation* between a computed value (of an output variable) and the true, specified or theoretically correct value;

Disturbance: An *unknown* (and uncontrolled) *input* acting on a system;

Perturbation: An input acting on a system which results in a *temporary departure* from steady state;

Residual: *Fault indicator*, based on deviations between measurements and model equation based calculations;

Symptom: *Change* of an observable quantity from *normal behavior*.

2) **Functions**

Fault detection: Determination of faults present in a system and time of detection;

Fault isolation: Determination of kind, location and time of detection of a fault by evaluating symptoms. Follows fault detection;

Fault identification: Determination of the size and time-variant behavior of a fault. Follows fault isolation;

Fault diagnosis: Determination of kind, size, location and time of detection of a fault by evaluating symptoms. Follows fault detection. Includes fault detection, isolation and identification;

Monitoring: A continuous real-time task of determining the possible conditions of a physical system, recognizing and indicating anomalies of the behavior;

Supervision: Monitoring a physical system and taking appropriate actions to maintain the operation in the case of faults;

Protection: Means by which a potentially dangerous behavior of the system is suppressed if possible, or means by which the consequences of a dangerous behavior are avoided.

3) Models

Quantitative model: Use of static and dynamic relations among system variables and parameters in order to describe system's behavior in quantitative mathematical terms;

Qualitative model: Use of static and dynamic relations among system variables and parameters in order to describe system's behavior in qualitative terms such as causalities or if-then rules;

Diagnostic model: A set of static or dynamic relations which link specific input variables - the symptoms - to specific output variables - the faults;

Analytical redundancy: Use of two, but not necessarily identical ways to determine a quantity where one way uses a mathematical process model in analytical form

4) System properties

Reliability Ability of a system to perform a required function under stated conditions, within a given scope, during a given period of time. Measure: MTTF = Mean Time To Failure. MTTF = $1/\lambda$; λ is rate of failure [e.g. failures per hour];

Safety: Ability of a system not to cause a danger for persons or equipment or environment;

Availability: Probability that a system or equipment will operate satisfactorily and effectively at any point of time measure:

$$A = \frac{\text{MTTF}}{\text{MTTF}+\text{MTTR}}$$

MTTR Mean Time to Repair
MTTR $= 1/\mu$; μ : rate of repair

References on terminology

DIN 25424 Fehlerbaumanalyse (fault tree analysis). Beuth Verlag, Berlin, 1990.

DIN 40041 Zuverlässigkeit in der Elektrotechnik (Reliability in electrical engineering). Beuth Verlag, Berlin, 1990.

DIN 31051 Instandhaltung (Maintenance). Beuth Verlag, Berlin, 1985.

DIN 40042 Zuverlässigkeit elektrischer Geräte, Anlagen und Systeme (Reliability of electrical devices, plants and systems). Beuth Verlag, Berlin, 1989.

DIN 55350 Begriffe der Qualitätssicherung und Statistik (Terms in quality control and statistics). Beuth Verlag, Berlin, 1989.

IFIP working group 10.4. Reliable computing and fault tolerance, meeting in Como, Italy, 1983.

Laprie, J.C. (1983). On computer system dependability and un-dependability: faults, errors, and failures. IFIP WG 10.4, Como, Italy, 1983.

Lexikon Mess- und Automatisierungstechnik. (1992). VDI Verlag, Düsseldorf.

Reliability, Availability, and Maintainability Dictionary. ASQC Quality press, Milwaukee, 1988.

Robinson, A. (1982). A user-oriented perspective of fault-tolerant systems, models and terminologies. Proceedings of the 12th International Symposium on fault-tolerant computing, Los Angeles.

VDI/VDE-Richtlinie 3541. Steuerungseinrichtungen mit vereinbarter gesicherter Funktion. Beuth Verlag, Berlin, 1985.

VDI/VDE-Richtlinie 3541. Sicherheitstechnische Begriffe für Automatisierungssysteme. Beuth Verlag, Berlin, 1988.

VDI/VDE-Richtlinie 3691. Erfassung von Zuverlässigkeitswerten bei Prozessrechnereinsätzen. Beuth Verlag, Berlin, 1985.

23.2 State variable filtering of noisy signals to obtain signal derivations

Some methods as the parameter estimation for continuous models or residual equations need the derivatives $\dot{y}(t)$, $\ddot{y}(t)$, ... of the measured signal $y(t)$. If they cannot be directly measured they have to be calculated based on the mostly noisy measurement of $y(t)$. One way is the *numerical differentiation* in combination with interpolation approaches (splines, Newton's method). However, this can only be applied for very small noise. *State variable filters* (SVF) as proposed by [23.21], see Figure 23.1, with a transfer function

$$F(s) = \frac{y_f(s)}{y(s)} = \frac{1}{f_0 + f_1 s + \ldots + f_{n-1} s^{n-1} + s^n}$$

have proven to yield good results. The state variable filter is a low-pass filter that provides the derivatives as well as filters the disturbance signals. With the state filter, the input signal $u(t)$ and the output signal $y(t)$ is filtered. The choice of the filter parameters f_i is relatively free. The design of a Butterworth filter is recommended, see [23.16]. A further possibility is the application of finite impulse response filters (FIR), where the derivatives of the impulse response of a low-pass filter are convoluted with the signal, [23.14], [23.11], see also [23.9].

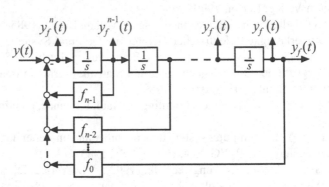

Fig. 23.1. State variable filter

23.3 Fuzzy logic – a short introduction[1]

Typical human information processing bases on rules that are not precisely formulated. They are built from both quantitative and qualitative experience and not as clearly structured as normal computer programs would require it. [23.23] has therefore combined approaches of multi-valued logics to the so-called "Fuzzy Logic" for the purpose of simulating human reasoning. This allows the processing of human experience and knowledge with digital computers. This section will present some of the basics of fuzzy logic. After an introduction into the data representation, the basic steps of fuzzy reasoning and its application for fault diagnosis will be explained.

23.3.1 Basics

The fuzzy logic processes imprecise information with the help of membership functions and if-then rules. This requires certain inference mechanisms, data representations and logic operators to enable a human-like information processing. Good overviews about these structures can be found from [23.24] or [23.10]. The current literature on the topic is immense so that a general overview cannot be given here.

The fuzzy logic approach has grown in popularity from its beginnings in the 70s and is today a standard tool for control system designers. It is used for classification problems, modelling and control systems in a variety of methods and combinations with other tools.

---

[1] compiled by Dominik Füssel, [23.3]

Fuzzy Sets

Fuzzy logics can be seen as an augmentation of the classical logic. It relies on the use of so-called fuzzy sets. While in classical logic, an element always belongs to a set or not

$$a \in A \ \oplus \ a \notin A, \tag{23.1}$$

fuzzy logic allows a gradual state in between.

An example can be seen in Figure 23.2. The temperature is assigned to the three fuzzy sets "low", "medium" and "high". These functions are usually denoted by $\mu(T)$. It should be stressed that the definition of these linguistic terms is highly application specific. In a different problem, a "low" temperature could have a completely different physical range. A temperature of 90°C has the following *membership degrees*:

$$\mu_{low}(90\,^{\circ}\mathrm{C}) = 0.5, \ \mu_{medium}(90\,^{\circ}\mathrm{C}) = 0.5, \ \mu_{high}(90\,^{\circ}\mathrm{C}) = 0, \tag{23.2}$$

The temperature can therefore not be "high" but is at the same time "low" and "medium". This meets the expectations drawn from human reasoning: Such terms as "low" or "medium" are not always precisely definable. There is rather a range in that the linguistic terms become valid, a range where they are valid and finally a range where they end to be applicable. The domain that the fuzzy sets are defined above is the complete range of possible outcomes of the variable "temperature" in this example. The range where a fuzzy set is higher than zero is also called the *area of support* of the fuzzy set. This term especially applies to multi-dimensional fuzzy sets. The membership functions that define the fuzzy sets can be described by algebraic equations:

$$\mu_{low}(T) = \begin{cases} 1 & , \ 0 \le T \langle 80\,^{\circ}\mathrm{C} \\ 1 - (T - 80)/20 & , \ 80\,^{\circ}\mathrm{C} \le T \langle 100\,^{\circ}\mathrm{C} \\ 0 & , \ 100^{o}C \le T \end{cases} \tag{23.3}$$

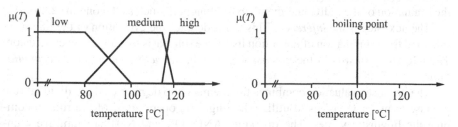

Fig. 23.2. Example of fuzzy sets (left) and the special representation of a crisp data point with fuzzy logic by a singleton (right).

The shape of the membership functions can generally be arbitrary. Typical are trapezoidals or triangular shapes. Increasingly popular are Gaussian-shaped functions, sigmoidals and B-splines for fuzzy sets. These functions are continuously differentiable which allows them to be easily used in nonlinear optimization procedures.

A special case of membership functions are the Singleton functions that represent crisp data values in the context of fuzzy logic. Figure 23.2 gives an example where the fuzzy set is defined as:

$$\mu_{boilingpoint}(T) = \begin{cases} 1 \, , \, T = 100^oC \\ 0 \, , \, T \neq 100^oC \end{cases} \qquad (23.4)$$

The membership function value is one if the temperature equals 100^oC, and otherwise equal to zero.

The membership functions are defined to have a range in the interval $0 \leq \mu \leq 1$ and are usually designed so that their sum equals unity at any value of the input domain. This allows an interpretation as probabilities. The combination of all fuzzy sets over a data range establishes a *fuzzy variable* ("temperature"). The fuzzy sets represented by membership functions ("low", ...) are named *attributes, terms* or *labels*.

Inference with If-Then Rules

The structure of a typical *fuzzy diagnosis system* is composed of a set of if-then rules. If multiple fault situations are to be distinguished, one will have a structure as can be seen in Figure 23.3. The inputs are the symptoms s_i.

The aim of the fuzzy diagnosis is now to implement *linguistic rules R* of the kind

$$R : \text{IF } \langle \mathcal{L}\{(\ldots), \ldots, (s_i \text{ is } A_{11}), \ldots, (\ldots)\} \rangle \text{ THEN } \langle f_j \text{ is } A_{F1} \rangle \qquad (23.5)$$

to draw conclusions from the symptoms s_i to the fault measures f_j. The *condition* (premise) of the rule comprises in general multiple linguistic statements that are combined by operators $\mathcal{L}$. An example is the statement "symptom s_1 increased" with the linguistic variables "symptom s_1" and the linguistic value "increased".

The first step towards the rule evaluation is the computation of the membership values $\mu_A(s_i)$ to the linguistic value A. This is called *fuzzification* and consists of the evaluation of the individual membership functions such as the one in (23.3).

The next step is the *inference*. This terms denotes the evaluation of the linguistic rules and the combination of the action list of the rule basis to a linguistic conclusion. The inference consists of the *premise evaluation*, the *activation* and the *accumulation*.

The premise evaluation combines the membership degrees of the individual rule premise terms. The task is to utilize the linguistic operators $\mathcal{L}$ of the rule to combine the linguistic values. The operators AND, OR and NOT are intuitively understandable and known from traditional logics. A minimum requirement for their implementation in fuzzy logic is that they equal the Boolean operators for the crisp membership values 0 and 1. Some of the operators used are the minimum/maximum, bounded difference/bounded sum, algebraic product/algebraic sum. In many applications, however, the choice of the operator is not vital for the function of the fuzzy system. A maximum implementation of the OR operator and a minimum of the AND is sufficient for many problems. This leads to the following rule fulfillments:

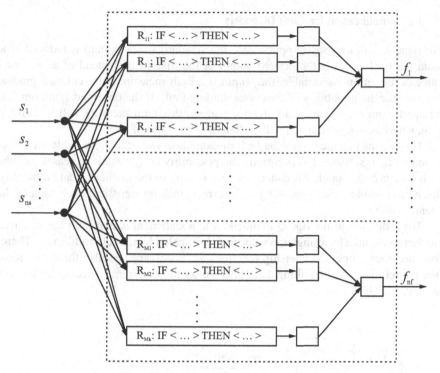

Fig. 23.3. Fuzzy logic system for fault diagnosis.

$$\text{AND} : \mu_{A \cap B} = \min(\mu_A, \mu_B)$$
$$\text{OR} : \mu_{A \cup B} = \max(\mu_A, \mu_B) \qquad (23.6)$$
$$\text{NOT} : \mu_{A'} = 1 - \mu_A$$

The *activation* is now the application of the rule fulfillments to the rule consequences. Typical methods are a limiting or a scaling (product) of the output membership functions. Possible is further a weighting with a number between 0 and 1.

The *accumulation* combines the activated output membership functions for every
linguistic output variable. This step unites the outputs of individual rules and is done
by a maximum operation. An alternative approach is a summation of the membership
functions. The result of the accumulation is then a linguistic output variable as a
fuzzy set.

The fuzzy set is transferred into a crisp output value by the opposite step to the
fuzzification: the *defuzzification*. Possible methods are the maximum, center of gravity or mean of maximum. An especially simple method is given if output singleton
sets are used. The methods will then simplify to a weighted sum of the singleton
values.

23.3.2 Simplification for Fault Diagnosis

For typical fault diagnosis applications, the standard fuzzy system is reduced. The main reason for that is the desired output of the diagnosis: Instead of an arbitrary value of a continuous variable, the output is a fault measure representing a gradual measure for the possibility of the corresponding fault. If the observed symptoms are far apart from the linguistically defined pattern, this fault measure will be close to zero, whereas a perfect match will yield a fault measure of one.

This means that the higher the fault measure becomes, the more likely the corresponding fault situation has occurred. The possibility of that event increases with the fault measure. Although this concept is very similar to the mathematical probability, this notion would strictly speaking be incorrect since no statistical evaluation of the events is done.

The reduction of the rule consequences to a statement which fault has occurred can be represented by a singleton value which is scaled by the rule fulfillment. Therefore, no other output membership functions are necessary and also the defuzzification not required. The resulting, simplified fuzzy logic system structure can be seen in Figure 23.4.

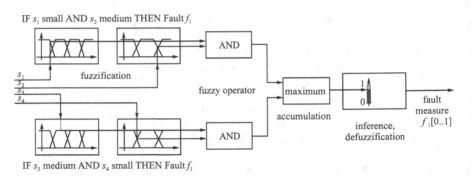

Fig. 23.4. Fuzzy logic system for fault diagnosis. Apparently, the defuzzification of a linguistic variable is not necessary. Instead, a scaling of the singleton yields a fault measure in the range $0 \dots 1$.

23.4 Estimation of physical parameters for dynamic processes

23.4.1 Introduction

Many contributions do exist for the estimation of parameters Θ of input-output models from measured input-output signals, e.g. [23.2], [23.19], [23.22], [23.6]. Some publications known for the determination of *physically defined parameters p* of the laws which govern the process dynamics are [23.20], [23.17], [23.1], see Figure 23.5. The following treatment of this problem is according to [23.5].

The physical parameters will be called *process coefficients*. For usual tasks in control engineering, e.g. the design of control systems, knowledge of the process parameters Θ is in general sufficient. However, knowledge of the process coefficients **p** is required for the following problems:

(a) determination of non-measurable coefficients in natural sciences;
(b) checking of performance data for technical systems;
(c) supervision and fault diagnosis during online operation of technical processes;
(d) quality control in manufacturing.

The determination of process parameters Θ from measured input and output signals with parameter estimation methods has obtained a mature status during the last 30 years. For linear and some nonlinear processes with stochastic disturbances, several methods are known. They result in a good convergence, if the model structure fits with the process structure, the input sufficiently excites the dynamics and the number of parameters and the number of parameters is not more than about four to six for SISO systems.

The process depends on physical process coefficients according to more or less complicated algebraic relations

$$\Theta = f(\mathbf{p}) \tag{23.7}$$

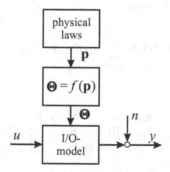

Fig. 23.5. Dynamic process model with input-output model parameters Θ and physical process coefficients **p**

These relations are known from theoretical modelling. The task now consists of determining the physical process coefficients based on measured input and output signals $u(t)$ and $y(t)$. A straightforward possibility is first to estimate the model parameters Θ and then to use the inverse relationship

$$\mathbf{p} = f^{-1}(\Theta) \tag{23.8}$$

However, the following questions arise:

(a) Can the process coefficients **p** be determined uniquely (process coefficient identifiability)?

(b) Which signals must be measured in order to determine the process coefficients **p**?

(c) What is the influence of *a priori* known process coefficients p_i on the variance of Θ and **p**?

In this context [23.17] considered the identifiability of biological models, especially compartmental models, with computer algebra. [23.1] proposed a two-step procedure to estimate physical parameters under the assumption that a considerable number of physical parameters is known. The unknown parameters are then estimated by using a gradient method with the known parameters as boundary conditions. [23.12] gave some basic relationships between physically defined process elements and the resulting model structure and show that the model parameters are the sums of products of process coefficients.

23.4.2 On the model structure for processes with lumped parameters

The basic equations of dynamic processes with lumped parameters in the form of balance equations, constitutive equations and phenomenological laws can be represented in a unified form, [23.7].

It is now assumed that the process is linear and possesses M measurable signals $\eta_j(t)$ and N non-measurable signals $\xi_j(t)$. For L elements the Laplace transformed *process element equations* become

$$\sum_{j=1}^{M} g_{ij}\, \eta_j(s) = \sum_{j=1}^{N} h_{ij}\, \xi_j(s) \quad i = 1, 2, \ldots, L \tag{23.9}$$

with

$$\left. \begin{array}{l} g_{ij} \in (0, \pm 1, \alpha_{ij}, s^\kappa) \\ h_{ij} \in (0, \pm 1, \beta_{ij}, s^\kappa) \\ \kappa \in \{-1, 0, 1\} \end{array} \right\} \tag{23.10}$$

In matrix notation this is

$$\mathbf{G}\,\eta = \mathbf{H}\,\xi \tag{23.11}$$

with

$$\eta^T = [\eta_1, \eta_2, \ldots, \eta_M] \tag{23.12}$$

$$\xi^T = [\xi_1, \xi_2, \ldots, \xi_N] \tag{23.13}$$

$$\dim(\mathbf{G}) = L \times M; \dim(\mathbf{H}) = L \times N$$

It is further assumed that the process element equations are linearly independent. For the parameter estimation a model structure with only measurable signals is required. The non-measurable signals are eliminated by transforming the matrix $\mathbf{H}$ into upper triangular form or by subsequent insertion of the single equations. One then obtains an input/output differential equation

$$a_n^* y^{(n)}(t) + \cdots + a_1^* y^{(1)}(t) + a_0^* y(t) = b_0^* u(t) + b_1^* u^{(1)}(t) + \cdots + b_m^* u^{(m)}(t)$$

(23.14)

However, for continuous-time parameter estimation the following form is required

$$a_n y^{(n)}(t) + \cdots + a_1 y^{(1)}(t) + y(t) = b_0 u(t) + b_1 u^{(1)}(t) + \cdots + b_m u^{(m)}(t) \quad (23.15)$$

such that the regression model

$$y(t) = \boldsymbol{\psi}^T(t)\, \boldsymbol{\theta} \tag{23.16}$$

with

$$\boldsymbol{\psi}^T(t) = [-y^{(1)}(t) \quad \cdots \quad -y^{(n)}(t) \;\vdots\; u(t)\, u^{(1)}(t) \quad \cdots \quad u^{(m)}(t)]$$

$$\boldsymbol{\theta}^T(t) = [\theta_1, \theta_2, \quad \cdots \quad \theta_r] = [a_1 \quad \cdots \quad a_n \;\vdots\; b_0 \quad \cdots \quad b_m]$$

can be obtained. Hence, all parameters in (23.14) have to be multiplied by $1/a_0^*$ and the number of parameters θ_i reduces by one.

In the system of basic equations (23.11) the process coefficients

$$\mathbf{p}^T = [p_1, p_2, \ldots p_l] \tag{23.17}$$

appear separately in the original form. The elements of $\mathbf{G}$ and $\mathbf{H}$ are

$$g_{ij} = p_{gij}\, s^\kappa; \quad h_{ij} = p_{hij}\, s^\kappa \quad \kappa \in (-1, 0, 1) \tag{23.18}$$

After transformation into upper triangular form the input-output model appears in the last row, [23.12]. The model parameters then take the form

$$\theta_i = \sum_{\mu=1}^{q} c_{i\mu} \prod_{v=1}^{l} p_v^{\varepsilon 11}\, p_2^{\varepsilon 12} \cdots p_l^{\varepsilon 1l} + c_{i2}\, p_1^{\varepsilon 21}\, p_2^{\varepsilon 22} \cdots p_l^{\varepsilon 2l} + \cdots$$

Hereby it is assumed that a_0^* can be represented by

$$a_0^* = \prod_{v=1}^{l} p_v^{\varepsilon \mu v}$$

With the abbreviation

$$z_\mu = \prod_{v=1}^{l} p_v^{\varepsilon \mu v} \tag{23.19}$$

the model parameters can be written as

$$\Theta_i = \sum_{\mu=1}^{q} c_{i\mu}\, z_\mu \tag{23.20}$$

Therefore the parameters are algebraic functions of the process coefficients p_v. For the second order electrical circuit in the Appendix these relations are

$$p_1 = R_1; \quad p_2 = C_1; \quad p_3 = R_2; \quad p_4 = C_2$$
$$\theta_1 = a_1 = z_1 + z_2 + z_3 = p_1 p_2 + p_3 p_4 + p_1 p_4$$
$$\theta_2 = a_2 = z_4 = p_1 p_2 p_3 p_4 \tag{23.21}$$
$$\theta_3 = b_1 = z_5 = p_2 p_3 p_4$$
$$\theta_4 = b_2 = z_6 + z_7 = p_2 + p_4$$

Hence the z_μ are abbreviations for the existing products and single values of the process coefficients $p_1, \ldots, p_l$.

(23.20) can now be turned into matrix form

$$
\underbrace{\begin{bmatrix} \theta_1 \\ \theta_2 \\ \vdots \\ \theta_r \end{bmatrix}}_{\boldsymbol{\theta}}
=
\underbrace{\begin{bmatrix} c_{11} \cdots c_{1q} \\ c_{21} \cdots c_{2q} \\ \vdots \quad \vdots \\ c_{r1} \cdots c_{rq} \end{bmatrix}}_{\mathbf{C}}
\underbrace{\begin{bmatrix} z_1 \\ z_2 \\ \vdots \\ z_q \end{bmatrix}}_{\mathbf{z}}
\tag{23.22}
$$

Generally, first the process model parameters θ_i are determined by parameter estimation via the measured signals. This was treated in Section 9.2.5 using (23.16), compare (9.116).

If some parameters $\boldsymbol{\theta}''$ are known, the parameter vector is split

$$\boldsymbol{\theta} = [\boldsymbol{\theta}' \ \boldsymbol{\theta}'']^T \tag{23.23}$$

where $\boldsymbol{\theta}'$ are the unknown parameters to be estimated and the signal derivatives with known parameters are separated

$$y''(t) = y(t) - \boldsymbol{\psi}''^T(t)\,\boldsymbol{\theta}'' = \boldsymbol{\psi}'^T(t)\,\boldsymbol{\theta}' + e(t) \tag{23.24}$$

The LS-estimate then becomes

$$\hat{\boldsymbol{\theta}} = [\boldsymbol{\Psi}'^T \boldsymbol{\Psi}']^{-1} \boldsymbol{\Psi}'^T \, \mathbf{y}'' \tag{23.25}$$

Simulations with a second order process have shown that a considerable improvement of the convergence is obtained by:

(i) one parameter is known: a_2 or b_1 is known, i.e., parameters with largest variances are known;

(ii) more parameters are known: a_2 or b_1 must be contained in the set of known parameters. If other parameters are known, they must be known very precisely.

23.4.3 Calculation of the physical process coefficients

As the process parameters $\boldsymbol{\theta}$ are nonlinear algebraic functions of the process coefficients $\mathbf{p}$ no general applicable solution for the unknown process coefficients p_v in the form of (23.8) can be given. For models of first or second order, in most cases a direct solution can be found. For higher order systems, successive resolution for the unknown p_v can be tried or the use of computer algebra, [23.20], [23.18].

However, an *identifiability condition* for the process coefficients can be given independently of the solution method. The basic relation (23.22) is written in implicit form as

$$q = \theta - Cz = 0 \qquad (23.26)$$

where

$$z = g(\mathbf{p})$$

The implicit function theorem, [23.15], now states that a necessary condition for a solution for $\mathbf{p}$ in the neighborhood of the solution $\mathbf{p}_0$ is that the functional determinant

$$\det \mathbf{Q}_p \neq 0 \qquad (23.27)$$

where $\mathbf{Q}_p$ is the functional matrix

$$\mathbf{Q}_p = \frac{\partial \mathbf{p}^T}{\partial \mathbf{p}} = \begin{bmatrix} \frac{\partial q_1}{\partial p_1} & \frac{\partial q_2}{\partial p_1} & \cdots & \frac{\partial q_r}{\partial p_1} \\ \frac{\partial q_1}{\partial p_2} & \frac{\partial q_2}{\partial p_2} & \cdots & \frac{\partial q_r}{\partial p_2} \\ \vdots & & & \vdots \\ \frac{\partial q_1}{\partial p_l} & \frac{\partial q_2}{\partial p_l} & \cdots & \frac{\partial q_r}{\partial p_l} \end{bmatrix} \qquad (23.28)$$

This implies $r = l$, which means that the number l of process coefficients must be equal the number r of model parameters. In the case of the example in Section 23.4.4, this gives

$$\det \mathbf{Q}_p = C_1\, C_2^3\, R_2^2 \neq 0$$

The process coefficients are identifiable if

$$C_1 \neq 0 \cap C_2 \neq 0 \cap R_2 \neq 0$$

23.4.4 Example: Second order electrical circuit

The basic equations are, Figure 23.6,

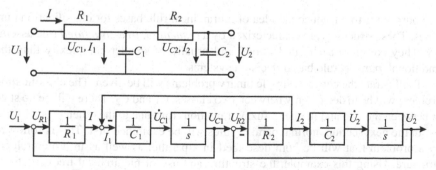

Fig. 23.6. Electrical second order network with block diagram

$$U_{R1}(t) = R_1 I(t)$$
$$I_1(t) = C_1 \dot{U}_{C1}(t)$$
$$I(t) = I_1(t) + I_2(t)$$
$$U_1(t) + U_2(t) - U_{C1}(t) = 0$$
$$R_{R2}(t) = R_2 I_2(t)$$
$$I_2(t) = C_2 \dot{U}_{C2}(t)$$
$$U_{C1}(t) + U_{R2}(t) - U_{C2}(t) = 0$$
$$U_2(t) = U_{C2}(t)$$

For input $U_1(t)$ and output $U_2(t)$ one obtains:

$$a_2 \ddot{I}(t) + a_1 \dot{I}(t) + I(t) = b_1 \dot{U}_1(t) + b_2 \ddot{U}_1(t)$$
$$a_2 = R_1 C_1 R_2 C_2$$
$$a_1 = R_1 C_1 + R_2 C_2 + R_1 C_2$$
$$b_1 = C_1 + C_2$$
$$b_2 = R_2 C_1 C_2$$

The process coefficients follow from

$$R_1 = a_2 / b_2$$
$$R_2 = \frac{(a_1^2 b_2^2 - 2a_1 b_1 a_2 b_2 + b_1^2 a_2^2)}{b_2 (b_2^2 - a_1 b_1 b_2 + b_1^2 a_2)}$$
$$C_1 = b_2^2 / (a_1 b_2 - a_1 b_2)$$
$$C_2 = \frac{-(b_2^2 - a_1 b_1 b_2 + b_1^2 a_2)}{(a_1 b_2 - b_1 a_2)}$$

An example for the axis of an industrial robot is shown in [23.5].

23.5 From Parallel to Hierarchical Rule Structures[2]

The purpose is to introduce the idea of hierarchical rule bases for classification purposes. These structures are characterized by a *sequential, tree-structured* rule assembly. They are more easily understood and usually of smaller complexity than the traditional, *parallel* rule based-decision systems.

To illustrate the concept, simple binary problems will be given. The classification problem will be to decide only between two classes, C_1 and C_2. There will be no state in between the two and neither a fuzzy transition from one to the other.

In the next section, normal parallel rules will be examined first. This is followed by a problem that will be both described in a parallel as well as in a hierarchical structure. Using this example, the structure and use of hierarchical tree-structured rule bases, including the place of the OR-operator, will be examined.

---

[2] compiled by Dominik Füssel, [23.3]

23.5.1 Parallel Rule Bases

Figure 23.7 gives a first example. The classification problem consists of distinguishing two classes, C_1 and C_2, based on the given values s_1 and s_2.

Visible is a problem with a grid-like partition of the space spanned by s_1 and s_2. This partition is *crisp*. However, to keep a consistent notation, the wording is taken from fuzzy logics: The intervals along the axes are named with $A_{11}, A_{12}, \ldots$ where this stands for *attribute A_{11}, attribute A_{12}* and so on. For a better understanding, one can also keep the fuzzy descriptions like *small, medium, large* etc. in mind. But it is again emphasized that there is no fuzzy set involved in this example.

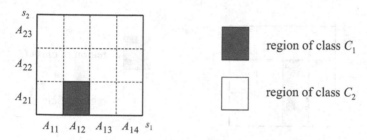

Fig. 23.7. Elementary binary classification problem

The crisp rule that describes the problem from Figure 23.7 is simple. The rule describes only the region of class C_1. If C_1 is not detected, the result is automatically C_2. One can for instance imagine C_1 being a fault situation, whereas C_2 is the fault-free case. The rule is:

$$R_1 : \quad \text{IF } s_1 \text{ is } A_{12} \text{ AND } s_2 \text{ is } A_{21} \text{ THEN class } C_1 \qquad (23.29)$$

This rule R_1 describes the situation completely. A rule like (23.29) is similar to the prototype Mamdani fuzzy rule. Indeed, normal fuzzy classifiers are composed of similar rules (see Chapter 23.3).

Typically, classifiers will be built from a *set* of rules. A second example problem is given in Figure 23.8. The corresponding rule base is now:

$$\begin{aligned} R_1 : \quad &\text{IF } s_1 \text{ is } A_{12} \text{ AND } s_2 \text{ is } A_{21} \text{ THEN class } C_1 \\ R_2 : \quad &\text{IF } s_1 \text{ is } A_{12} \text{ AND } s_2 \text{ is } A_{22} \text{ THEN class } C_1 \end{aligned} \qquad (23.30)$$

Each of the two rules describes one of the shaded rectangular regions. To get the result of the rule base for a new data point, one has to evaluate the two rules and finally combine their result. This step is called *accumulation*.

For classification problems, one finds that the accumulation is simply a *union* of the results of all rules with the same conclusion. As such, it is simply an OR-operation. The simplest implementation is the *maximum*-operator.

This accumulation combines the degrees of fulfillment of R_1 and R_2. For this simple binary example where a data point can only be inside or outside the shaded region from Figure 23.8, it is identical with the binary (boolean) OR-operator.

The interesting result for classification rule bases is that the *accumulation* (i.e., the combination of the degrees of fulfillment of different rules) is identical to an *OR-operation in the rule premise*. Indeed, one can write the two rules from (23.30) as only one:

$$R_1 : \text{ IF } (s_1 \text{ is } A_{12} \text{ AND } s_2 \text{ is } A_{21}) \text{ OR} \\ (s_1 \text{ is } A_{12} \text{ AND } s_2 \text{ is } A_{22}) \text{ THEN class } C_1 \qquad (23.31)$$

That way, the meanings of the two operators become clear: An AND is used to *shrink the selected region*, whereas the OR *combines regions* in the input (symptom) space to describe the complete region belonging to C_1.

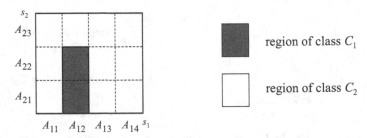

Fig. 23.8. First example: Binary classification requiring more than one rule.

23.5.2 Hierarchical Rule Bases

With the same methodology one can now address the second example. It is pictured in Figure 23.9. This time, the class regions have different shapes. In particular, one can decompose the shaded region into axis-orthogonal elements. This means that for example the upper region of the symptom space depends only on one of the two inputs, namely s_2 in this example. Such a situation occurs often: If higher-dimensional spaces are considered, a decision might not always depend on all inputs at the same time. The true meaning in fault diagnosis applications is that some symptoms are irrelevant for certain faults in certain regions of the symptom space.

The question arises how this can be explained. It essentially reflects the fact that the problem is intrinsically of a lower complexity than potentially possible based on the given input dimensionality.

Coming back to the example from Figure 23.9, one easily determines the rules similar to (23.30):

$$\begin{array}{ll} R_1 : & \text{IF } s_1 \text{ is } A_{11} \text{ AND } s_2 \text{ is } A_{23} \text{ THEN class } C_1 \\ R_2 : & \text{IF } s_1 \text{ is } A_{12} \text{ AND } s_2 \text{ is } A_{23} \text{ THEN class } C_1 \\ R_3 : & \text{IF } s_1 \text{ is } A_{13} \text{ AND } s_2 \text{ is } A_{23} \text{ THEN class } C_1 \\ R_4 : & \text{IF } s_1 \text{ is } A_{14} \text{ AND } s_2 \text{ is } A_{23} \text{ THEN class } C_1 \\ R_5 : & \text{IF } s_1 \text{ is } A_{12} \text{ AND } s_2 \text{ is } A_{21} \text{ THEN class } C_1 \\ R_6 : & \text{IF } s_1 \text{ is } A_{12} \text{ AND } s_2 \text{ is } A_{22} \text{ THEN class } C_1 \end{array} \qquad (23.32)$$

Again, the individual rule fulfillments must be combined by the accumulation.

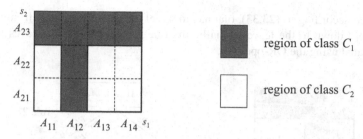

Fig. 23.9. Second example: Binary classification problem with more difficult shape.

The example has risen in complexity: Now, six rules are necessary. The problem is still understandable because it is 2-dimensional and can be visualized easily. Typical, however, are problems of more than 2 or 3 dimensions. They can not be visualized and are difficult to comprehend. A promising approach to enable an understanding is then to divide the complex problem into smaller and simpler subproblems. This creates a *hierarchy* that is typically easier to comprehend.

In this example, one can write the rules in the following form:

$$\begin{aligned} R_1^* &: \quad \text{IF } s_2 \text{ is } A_{23} \text{ THEN class } C_1 \\ R_2^* &: \quad \text{ELSE IF } s_1 \text{ is } A_{12} \text{ THEN class } C_1 \end{aligned}$$

(23.33)

They describe the exact same picture from Figure 23.9.

Figure 23.10 shows the hierarchy of the two rules. The first describes the upper shaded part, the second applies only on the lower region. It must be computed only if the first rule does not apply.

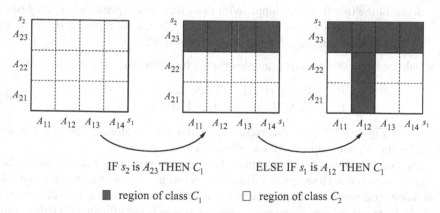

IF s_2 is A_{23} THEN C_1 ELSE IF s_1 is A_{12} THEN C_1

■ region of class C_1 □ region of class C_2

Fig. 23.10. Sequential decomposition of rules for classification of second example problem.

This hierarchy of the two rules can be pictured as a tree structure. This is visualized in Figure 23.11. The tree is composed of rule premises at the nodes and the conclusions (here class C_1 or C_2) at the leaves of the tree. To compute the output of the tree according to (23.33), one has to travel the tree downwards from the root (at the top) down to the leaves. Finally, the rule fulfillments at the leaves must be accumulated using the OR-operator.

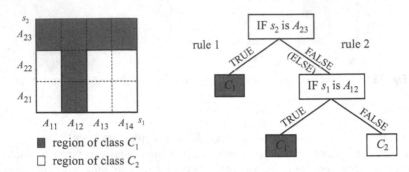

Fig. 23.11. Tree representation of sequential rule base.

But how can the large complexity reduction of (23.33) compared to (23.32) be explained? There are essentially two reasons for the lower complexity:

1) The use of the hierarchy expressed by the "ELSE" in the second rule of (23.33) essentially *hides complexity*. It comprises the opposite of all rules above. In parallel notation, one would have to write:

$$
\begin{aligned}
R_1' &: \text{ IF } s_2 \text{ is } A_{23} \text{ THEN class } C_1 \\
R_2' &: \text{ IF NOT } (s_2 \text{ is } A_{23}) \text{ AND } s_1 \text{ is } A_{12} \text{ THEN class } C_1
\end{aligned} \tag{23.34}
$$

2) Rules with a premise that does not contain a certain input reflect the independence of the rule from this input. With these *incomplete premises*, a *set of rules is combined*. In the example, $R_1 \ldots R_4$ from (23.32) are replaced by R_1^* from (23.33). In the same way, R_5 and R_6 collapse to R_2^*.

It should be noted that for this example also the following rule tree can be used:

$$
\begin{aligned}
R_1 &: \text{ IF } s_2 \text{ is } A_{12} \text{ THEN class } C_1 \\
R_2 &: \text{ ELSE IF } s_2 \text{ is } A_{23} \text{ THEN class } C_1
\end{aligned} \tag{23.35}
$$

Analog to Figure 23.10 one can picture a similar rule sequence. The result is the same but the decomposition with the rules different.

The example shows which potential of complexity reduction can be found in hierarchical structures. However, not in all cases a complexity reduction is guaranteed. The worst case would be a pattern that resembles a chess board in two dimensions with the two class regions alternating. Such a problem is not efficiently handled with tree structures. A parallel rule base would be equally appropriate for this problem.

With the SELECT method that is described in Section 17.3.5, a neuro-fuzzy system is given that works with the hierarchical rule structure presented. However, different from the binary example in this section the SELECT method will use the following features:

- There will be fuzzy transitions between the individual class regions.
- The AND-operator will be a continuous fuzzy-AND neuron.
- The resulting decision boundaries will not always be strictly axis-orthogonal.
- The rule fulfillments will have a continuous value (output membership) between 0 and 1. Therefore, the complete rule structure will always have to be computed.

Concluding remarks

This book treats basic methods for fault detection and diagnosis based on measured process signals. Various approaches of signal model and process-based model methods are described, and use of examples is made. Furtheron, basic structures of fault-tolerant systems are given. Some application examples show experimental results with different methods.

In order to meet process specific properties, like static or dynamic behavior, input excitation, precision of modelling and practical requirements like computational expense, sampling time, diagnosis depth and costs appropriate fault detection and diagnosis methods have to be selected. To reach a certain fault coverage in most cases different methods have to be combined properly. Therefore, the book offers a selection of FDD-methods which can be used to meet the practical needs of special processes.

As FDD-methods can only be judged by applying them to concrete processes, a broad class of applications is treated in a second book:

Fault diagnosis of technical processes
– Applications –

published also with Springer-Verlag to appear around 2006. The treated examples are as follows

A COMPONENTS
  - Fault detection of electrical drives;
  - Fault detection of electrical actuators;
  - Fault detection of fluidic actuators.

B MACHINES AND PLANTS
- Fault detection of pumps;
- Leak detection of pipelines;
- Fault detection of industrial robots;
- Fault detection of machine tools;
- Fault detection of heat exchangers;
- Fault detection for medical engineering devices.

C AUTOMOTIVE SYSTEMS
- Fault detection of combustion engines;
- Fault detection of automobiles.

D FAULT-TOLERANT SYSTEMS
- Fault-tolerant systems;
- Fly-by-wire and drive-by-wire systems.

Then applications show which methods are applicable and demonstrate with experiments of real processes how successful results can be obtained but also shows which limits do exist.

References

Chapter 1

1.1 Bakiotis, C., Raymond, J., and Rault, A. Parameter and discriminiant analysis for jet engine mechanical state diagnosis. In *Proc. of The 1979 IEEE Conf. on Decision & Control*, Fort Lauderdale, USA, 1979.

1.2 Beard, R. Failure accommodation in linear systems through self-reorganization. Technical Report MVT-71-1, Man Vehicle Laboratory, Cambridge, Mass, 1971.

1.3 Blanke, M., Izadi-Zamenabadi, R., Bogh, S., and Lunan, C. Fault-tolerant control systems. *Control Engineering Practice – CEP*, 5(5):693–702, 1997.

1.4 Brunet, J., Jaume, D., Labarrère, M., Rault, A., and Vergé, M. *Détection et diagnostic de panne - approche par modélisation*. Hermes Press, Paris, 1990.

1.5 Chen, J. and Patton, R. *Robust model-based fault diagnosis for dynamic systems*. Kluwer, Boston, 1999.

1.6 Clark, R. A simplified instrument detection scheme. *IEEE Trans. Aerospace Electron. Systems*, 14(3):558–563, 1990.

1.7 Filbert, D. Fault diagnosis in nonlinear electromechanical systems by continuous-time parameter estimation. *ISA Trans.*, 24(3):23–27, 1985.

1.8 Filbert, D. and Metzger, L. Quality test of systems by parameter estimation. In *Proc. 9th IMEKO-Congress*, Berlin, Germany, May 1982.

1.9 Frank, P. Advanced fault detection and isolation schemes using nonlinear and robust observers. In *10th IFAC Congress*, volume 3, pages 63–68, München, Germany, 1987.

1.10 Frank, P. Fault diagnosis in dynamic systems using analytical and knowledge-based redundancy. *Automatica*, 26(3):459–474, 1990.

1.11 Frank, P. and Wünnenberg, J. Sensor fault detection via robust observers. In Tzafestas, S., Singh, M., and Schmidt, G., editors, *System fault diagnostics, reliability & related knowledge-based approaches*, volume 1, pages 147–160. D. Reidel Press, Dordrecht, 1987.

1.12 Gertler, J. A survey of model-based failure detection and isolation in complex plants. *IEEE Control Systems Magazine*, 8(6):3–11, 1988.

1.13 Gertler, J. *Fault detection and diagnosis in engineering systems.* Marcel Dekker, New York, 1998.

1.14 Gertler, J. and Singer, D. Augmented models for statistical fault isolation in complex dynamic systems. In *Proc. American Control Conference (ACC)*, volume 1, pages 317–322, Boston, MA,, 1985.

1.15 Himmelblau, D. *Fault detection and diagnosis in chemical and petrochemical processes.* Elsevier, New York, 1978.

1.16 Hohmann, H. *Automatische Überwachung und Fehlerdiagnose an Werkzeugmaschinen.* PhD thesis, Technische Hochschule, Darmstadt, 1987.

1.17 IFAC. *Symposium on Fault Detection, Supervision and Safety of Technical Processes (SAFEPROCESS). Baden-Baden (1991), Helsinki (1994), Hull (1997), Budapest (2000), Washington (2003), Beijing (2006).* Pergamon, London, 1991–2006.

1.18 IFAC-Workshop. *On-line fault detection and supervision in the chemical process industries. Kyoto (1986), Newark (1992), Newcastle (1995), Folaize (1998), Cheju (2001).* Pergamon, London, 1986–2001.

1.19 Isermann, R. Methoden zur Fehlererkennung für die Überwachung technischer Prozesse. *Regelungstechnische Praxis*, (9 & 10):321–325 & 363–368, 1980.

1.20 Isermann, R. Parameter-adaptive control algorithms - a tutorial. *Automatica*, 18(5):513–528, 1982.

1.21 Isermann, R. Process fault detection on modeling and estimation methods - a survey. *Automatica*, 20(4):387–404, 1984.

1.22 Isermann, R. Fault diagnosis of machines via parameter estimation and knowledge processing. In *Proc. IFAC Symposium on Fault Detection, Supervision and Safety for Technical Processes (SAFEPROCESS)*, volume 1, pages 121–133, Baden-Baden, Germany, September 1993.

1.23 Isermann, R. Integration of fault-detection and diagnosis methods. In *Proc. IFAC Symposium on Fault Detection, Supervision and Safety for Technical Processes (SAFEPROCESS)*, pages 597–609, Espoo, Finland, June 1994.

1.24 Isermann, R., editor. *Überwachung und Fehlerdiagnose - Moderne Methoden und ihre Anwendungen bei technischen Systemen.* VDI-Verlag, Düsseldorf, 1994.

1.25 Isermann, R. *Fault diagnosis of technical processes – applications.* Springer, Heidelberg, 2006.

1.26 Isermann, R. and Ballé, P. Trends in the application of model-based fault detection and diagnosis in technical processes. *Control Engineering Practice – CEP*, 5(5):638–652, 1997.

1.27 Jones, H., editor. *Failure detection in linear systems.* Dept. of Aeronautics, M.I.T., Cambridge, 1973.

1.28 Mehra, R. and Peschon, J. An innovations approach to fault detection and diagnosis in dynamic systems. *Automatica*, 7:637–640, 1971.

1.29 Patton, R. Robust fault detection using eigenstructure assignment. In *Proc. 12th IMACS World Congress on Scientific Computation*, volume 2, pages 431–434, Paris, France, 1988.

1.30 Patton, R. Robust model-based fault diagnosis: the state of the art. In *Proc. IFAC Symposium on Fault Detection, Supervision and Safety for Technical Processes (SAFEPROCESS)*, pages 1–24, Helsinki, Finland, 1994.

1.31 Patton, R., Frank, P., and Clark, P., editors. *Fault diagnosis in dynamic systems, theory and application*. Prentice Hall, London, 1989.

1.32 Patton, R., Frank, P., and Clark, P., editors. *Issues of fault diagnosis for dynamic systems*. Springer, New York, 2000.

1.33 Pau, L., editor. *Failure diagnosis and performance monitoring*, volume 11 of *Contr. & Syst. Theory Series*. Marcel Dekker, New York, 1975.

1.34 Poulizeos, A. and Stravlakakis, G. *Real time fault monitoring of industrial processes*. Kluwer Academic Publishers, Dordrecht, 1994.

1.35 Siebert, H. and Isermann, R. Leckerkennung und -lokalisierung bei Pipelines durch Online-Korrelation mit einem Prozeßrechner. *Regelungstechnik*, 25(3):69–74, 1977.

1.36 Watanabe, U. and Himmelblau, D. Instrument fault detection in systems with uncertainties. *Int. Journal of Systems Science*, 13:137–158, 1982.

1.37 Willsky, A. A survey of design methods for failure detection systems. *Automatica*, 12:601–611, 1976.

Chapter 2

2.1 Blanke, M., Kinnaert, M., Lunze, J., and Staroswiecki, M. *Diagnosis and fault-tolerant control*. Springer, Berlin, 2003.

2.2 Bonnett, A. Understanding motor shaft failures. *IEEE Industry Application Magazine*, (September–October):25–41, 1999.

2.3 Chen, J. and Patton, R. *Robust model-based fault diagnosis for dynamic systems*. Kluwer, Boston, 1999.

2.4 Dalton, T., Patton, R., and Chen, J. An application of eigenstructure assignement to robust residual design for fdi. In *Proc. UKACC Int. Conf. on Control (CONTROL '96)*, pages 78–83, Exter, UK, 1996.

2.5 Ericsson, S., Grip, N., Johannson, E., Persson, L., Sjöberg, R., and Strömberg, J. Towards automatic detection of local bearing defects in rotating machines. *Mechanical Systems and Signal Processing*, 9:509–535, 2005.

2.6 Filbert, D. Fault diagnosis in nonlinear electromechanical systems by continuous-time parameter estimation. *ISA Trans.*, 24(3):23–27, 1985.

2.7 Geiger, G. *Technische Fehlerdiagnose mittels Parameterschätzung und Fehlerklassifikation am Beispiel einer elektrisch angetriebenen Kreiselpumpe*, volume Fortschr.-Ber. VDI Reihe 8, 91. VDI Verlag, Düsseldorf, 1985.

2.8 Gertler, J. *Fault detection and diagnosis in engineering systems*. Marcel Dekker, New York, 1998.

2.9 Grimmelius, H., Meiler, P., Maas, H., Bonnier, B., Grevink, J., and Kuilenburg, R. van. Three state-of-the-art method for condition monitoring. *IEEE Trans. on Industrial Electronics*, 46(2):401–416, 1999.

438   References

2.10 Hermann, O. and Milek, J. Modellbasierte Prozessüberwachung am Beispiel eines Gasverdichters. *Technisches Messen*, 66(7–8):293–300, 1995.

2.11 Higham, E. and Perovic, S. Predictive maintenance of pumps based on signal analysis of pressure and differential pressure (flow) measurements. *Trans. of the Institute of Measurement and Control*, 23(4):226–248, 2001.

2.12 Himmelblau, D. Fault detection and diagnosis - today and tomorrow. In *Proc. IFAC Workshop on Fault Detection and Safety in Chemical Plants*, pages 95–105, Kyoto, Japan, 1986.

2.13 IEC 61508. *Functional safety of electrical/electronic/programmable electronic systems*. International Electrotechnical Commission, Switzerland, 1997.

2.14 IFIP. *Proc. of the IFIP 9th World Computer Congress, Paris, France, September 19-23*. Elsevier, 1983.

2.15 Isermann, R. Integration of fault-detection and diagnosis methods. In *Proc. IFAC Symposium on Fault Detection, Supervision and Safety for Technical Processes (SAFEPROCESS)*, pages 597–609, Espoo, Finland, June 1994.

2.16 Isermann, R., editor. *Überwachung und Fehlerdiagnose - Moderne Methoden und ihre Anwendungen bei technischen Systemen*. VDI-Verlag, Düsseldorf, 1994.

2.17 Isermann, R. Supervision, fault-detection and fault-diagnosis methods – an introduction. *Control Engineering Practice – CEP*, 5(5):639–652, 1997.

2.18 Isermann, R. Fehlertolerante Komponenten für Drive-by-wire Systeme. *Automobiltechnische Zeitschrift – ATZ*, 104(4):382–392, 2002.

2.19 Isermann, R. and Ballé, P. Trends in the application of model-based fault detection and diagnosis in technical processes. *Control Engineering Practice – CEP*, 5(5):638–652, 1997.

2.20 Kiencke, U. Diagnosis of automotive systems. In *Proc. IFAC Symposium on Fault Detection, Supervision and Safety for Technical Processes (SAFEPROCESS)*, Hull, United Kingdom, August 1997. Pergamon Press.

2.21 Kolerus, J. *Zustandsüberwachung von Maschinen*. expert Verlag, Renningen-Malmsheim, 2000.

2.22 Leveson, N. *Safeware. System safety and computer*. Reading, Wesely Publishing Company, 1995.

2.23 Melody, J., Basar, T., Perkins, W., and Voulgaris, P. Parameter estimation for inflight detection of aircraft icing. In *Proc. 14th IFAC World Congress*, pages 295–300, Beijing, P.R. China, 1991.

2.24 Musgrave, J., Guo, T.-H., Wong, E., and Duyar, A. Real-time accommodation of actuator faults on a reusable rocket engine. *IEEE Trans. on Control Systems Technology*, 5(1):100–109, 1997.

2.25 Nold, S. *Wissensbasierte Fehlererkennung und Diagnose mit den Fallbeispielen Kreiselpumpe und Drehstrommotor*, volume Fortschr.-Ber. VDI Reihe 8, 273. VDI Verlag, Düsseldorf, 1991.

2.26 Omdahl, T., editor. *Reliability, availability and maintainability (RAM) dictionary*. ASQC Quality Press, Milwaukee, 1988.

2.27 Patton, R. Fault detection and diagnosis in aerospace systems using analytical redundancy. *IEE Computing & Control Eng. J.*, 2(3):127–136, 1991.

2.28 Patton, R., Frank, P., and Clark, P., editors. *Fault diagnosis in dynamic systems, theory and application.* Prentice Hall, London, 1989.

2.29 Patton, R., Frank, P., and Clark, P., editors. *Issues of fault diagnosis for dynamic systems.* Springer, New York, 2000.

2.30 Rasmussen, J. Diagnostic reasoning in action. *IEEE Trans. on System, Man and Cybernetics,* 23(4):981–991, 1993.

2.31 Rizzoni, G., Soliman, A., and Passino, K. A survey of automotive diagnostic equipment and procedures. SAE 930769. In *Proc. International Congress and Exposition,* Detroit, MI, USA, 1993. SAE.

2.32 Russell, E., Chiang, L., and Baatz, R. *Data-driven techniques for fault detection and diagnosis in chemical processes.* Springer, London, 2000.

2.33 Schneider-Fresenius, W. *Technische Fehlerfrühdiagnose-Einrichtungen: Stand der Technik und neuartige Einsatzmöglichkeiten in der Maschinenbauindustrie.* Oldenbourg, München, 1985.

2.34 Sill, U. and Zörner, W. *Steam turbine generators process control and diagnostics – modern instrumentation for the greatest economy of power plants.* Wiley-VCH, Weinheim, 1996.

2.35 Simani, S., Fantuzzi, C., and Patton, R. *Model-based fault diagnosis in dynamic systems using identification techniques.* Springer, London, 2003.

2.36 STARTS Guide. *The STARTS purchases Handbook: software tools for application to large real-time systems.* National Computing Centre Publications, Manchester, 2nd edition, 1989.

2.37 Storey, N. *Safety-critical computer systems.* Addison Wesely Longman Ltd., Essex, 1996.

2.38 Struss, P., Malik, A., and Sachenbacher, M. Qualitative modeling is the key to automated diagnosis. In *13th IFAC World Congress,* San Francisco, CA, USA, 1996.

2.39 Sturm, A. and Förster, B. *Maschinen und Anlagendiagnostik.* B.G. Teubner, Stuttgart, 1986.

2.40 Sturm, A., Förster, B., Hippmann, N., and Kinsky, D. *Wälzlaufdiagnose an Maschinen und Anlagen.* Verlag TÜV Rheinland, Köln, 1986.

2.41 VDI 2206. *Design methodology for mechatronic systems.* Beuth Verlag, Berlin, 2003.

2.42 VDMA   Fachgemeinschaft   Pumpen.   *Betreiberumfrage   zur Störungsfrüherkennung bei Pumpen.* VDMA, Frankfurt, 1995.

2.43 Wowk, V. *Machinery vibrations.* McGraw Hill, New York, 1991.

Chapter 3

3.1 ADAC. *Pannenstatistik 1999.* ADAC e.V., München, 2002.

3.2 Aggarwal, K. *Reliability engineering.* Kluwer Academics Publishers, Dordrecht, 1993.

3.3 Allianz. *Handbuch der Schadensverhütung.* VDI, Düsseldorf, 3rd edition, 1984.

3.4   Birolini, A. *Qualität und Zuverlässigkeit technischer Systeme: Theorie, Praxis, Management*. Springer, Berlin, 3rd edition, 1991.

3.5   DEKRA Automobil GmbH. Technische Sicherheit und Mobilität. Bericht über Untersuchungen von Fahrzeugen auf unfallrelevenate technische Mängel im Zeitraum 1996–2000. *DEKRA Fachzeitschrift*, (55/01), 2001.

3.6   DIN 40041. *Dependability concepts (Zuverlässigkeitsbegriffe)*. Beuth Verlag, Berlin, 1990.

3.7   Dressler, T. and Trilling, U. Instandhaltung in der prozessnahen technik - cost of ownership. In *Proc. GMA-Kongreß '96 Meß- und Automatisierungstechnik*, volume VDI Ber. 1282, pages 551–561, Düsseldorf, 1996. VDI Verlag.

3.8   Electric Power Research Institute (EPRI). *Improved motors for utility applications*, volume EL-2678, 1. General Electric Company, 1982.

3.9   Feick, S. and Pandit, M. Steer-by-wire as a mechatronic implementation. In *Proc. SAE World Congress 2000*, Detroit, MI, USA, 2000.

3.10   IEC 50-191. *International electrotechnical dictionary (Internationales Elektrotechnische Wörterbuch), Teil 191: Zuverlässigkeit und Dienstgüte*.

3.11   Kafka, T., editor. *Aufbau eines Störungsfrüherkennungssystems für Pumpen der Verfahrenstechnik mit Hilfe maschinellen Lernens*. Univ. Kaiserlautern, 1999.

3.12   Kroll, A. Prioritizing asset management activities for chemical plants: learning from event statistics (in german). *at - Automatisierungstechnik*, pages 228–238, 2004.

3.13   Meyna, A. *Zuverlässigkeitsbewertung zukunftsorientierter Technologien*. Vieweg, Wiesbaden, 1994.

3.14   MIL-HDBK-217. *Design analysis procedure for failure modes, effects and criticality analysis (FMECA)*, volume 217-D. National Technical Information Service, Springfield, VA, 1982.

3.15   Motor Reliability Working Group. Report of large motor reliability survey of industrial and commercial installations. *IEEE Trans. on Industry Applications*, IA-21(4):853–872, 1985.

3.16   Nold, S. *Wissensbasierte Fehlererkennung und Diagnose mit den Fallbeispielen Kreiselpumpe und Drehstrommotor*, volume Fortschr.-Ber. VDI Reihe 8, 273. VDI Verlag, Düsseldorf, 1991.

3.17   Omdahl, T., editor. *Reliability, availability and maintainability (RAM) dictionary*. ASQC Quality Press, Milwaukee, 1988.

3.18   Reichel, R. and Boos, F. *Redundantes Rechnersystem für Fly-by-wire Steuerungen*. Bodensee-Gerätewerk, Überlingen, 1986.

3.19   Reliability toolkit. *Reliability toolkit. Commercial Practices Edition. A practical guide for commericial products and military systems under acquisition reform*. Rome Laboratory & RAC, Rome, NY, 1995.

3.20   Schrüfer, E. *Zuverlässigkeit von Mess- und Automatisierungsanlagen*. Hanser Verlag, München, 1984.

3.21   Schrüfer, E. *VDI-Lexikon Mess- und Automatisierungstechnik*. VDI Verlag, Düsseldorf, 1992.

3.22 Storey, N. *Safety-critical computer systems*. Addison Wesely Longman Ltd., Essex, 1996.

3.23 Straky, H. *Modellgestützter Funktionsentwurf für Kfz-Stellglieder; Regelung der elektromechanischen Ventiltriebaktorik und Fehlerdiagnose der Bremssystemhydraulik*, volume Fortschr.-Ber. VDI Reihe 12, 546. VDI Verlag, Düsseldorf, 1996.

3.24 Thorsen, O. and Dalva, M. A survey of the reliability with an analysis of faults on variable frequency drives in industry. In *Proc. European Conference on Power Electronics and Applications EPE '95*, pages 1033–1038, 1995.

3.25 VDMA Fachgemeinschaft Pumpen. *Betreiberumfrage zur Störungsfrüherkennung bei Pumpen*. VDMA, Frankfurt, 1995.

3.26 Wolfram, A. *Komponentenbasierte Fehlerdiagnose industrieller Anlagen am Beispiel frequenzumrichtergespeister Asynchronmaschinen und Kreiselpumpen*, volume Fortschr.-Ber. VDI Reihe 8, 967. VDI Verlag, Düsseldorf, 2002.

3.27 Wondrak, W., Boos, A., and Constapel, R. Design for reliability in automotive electronics. In *Proc. Microtec 2000*, Berlin, Germany, 2000. VDE, Frankfurt.

Chapter 4

4.1 IEC 60812. *Analysis techniques for system reliability procedure for failure mode and effects analysis (FMEA)*. International Electrotechnical Commission, Switzerland, 1985.

4.2 IEC 61508. *Functional safety of electrical/electronic/programmable electronic systems*. International Electrotechnical Commission, Switzerland, 1997.

4.3 IQ FMEA. *Handbuch für das Softwarepaket IQ-FMEA zur Erstellung einer System-FMEA*. APIS Informationstechnologien, Fa. Peter Rosenbeck - The Knowledge Base, Version 3.1, 1996.

4.4 Isermann, R. *Fault diagnosis of technical processes – applications*. Springer, Heidelberg, 2006.

4.5 Isermann, R., Schwarz, R., and Stölzl, S. Fault-tolerant drive-by-wire systems – concepts and realizations. In *Proc. IFAC Symposium on Fault Detection, Supervision and Safety for Technical Processes (SAFEPROCESS)*, Budapest, Hungary, 2000.

4.6 Leveson, N. *Safeware. System safety and computer*. Reading, Wesely Publishing Company, 1995.

4.7 Onodera, K. Effective techniques of FMEA at each life-cycle stage. In *1997 Proc. Annual Reliability and Maintainability Symposium*, pages 50–56. IEEE, 1997.

4.8 Prometheus WG5. *A recommended practice of safety and reliability engineering of future automotive systems*. Internal report. BMW, München, 1998.

4.9 Reichart, G. Sichere Elektronik im Kraftfahrzeug. *Automatisierungstechnik – at*, 46(2):78–83, 1998.

4.10 Reichow, D. Fault management in a modern airliner. In *Prepr. IFAC Symposium on Fault Detection, Supervision and Safety for Technical Processes (SAFEPROCESS'91)*, volume 1, pages 1–8, Baden-Baden, Germany, September 1991. Pergamon Press.

4.11 SAE. Design analysis procedure for failure modes, effects and criticality analysis (FMECA). *Aerospace Recommended Practice*, SAE ARP 926, 1967.

4.12 Stölzl, S. *Fehlertolerante Pedaleinheit für ein elektromechanisches Bremssystem (Brake-by-Wire)*, volume Fortschr.-Ber. VDI Reihe 12, 462. VDI Verlag, Düsseldorf, 2000.

4.13 Storey, N. *Safety-critical computer systems*. Addison Wesely Longman Ltd., Essex, 1996.

4.14 Verband der Automobilindustrie (VDA). *Sicherung der Qualität vor Serieneinsatz*, volume 4. TÜV-Verlag, Köln, 1996.

Chapter 5

5.1   Åström, K. *Introduction to stachastic control theory*. New York, Academic Press, 1970.

5.2   Cellier, F. *Continuous system modeling*. Springer, New York, 1991.

5.3   Eykhoff, P. *System identification*. John Wiley, London, 1974.

5.4   Gawthrop, P. and Smith, L. *Metamodelling: bond graphs and dynamic systems*. Prentice Hall, Hemel Hempstead, 1996.

5.5   Haber, R. and Unbehauen, H. Structure identification of nonlinear dynamic systems - a survey on input/output approaches. *Automatica*, 26(4):651–677, 1990.

5.6   Höfling, T. and Isermann, R. Adaptive parity equations and advanced parameter estimation for fault detection and diagnosis. In *Proc. 13th IFAC World Congress*, San Francisco, CA, USA, 1996.

5.7   Isermann, R. Beispiele für die Fehlerdiagnose mittels Parameterschätzung. *Automatisierungstechnik – at*, 37:342–343 & 445–447, 1980.

5.8   Isermann, R. Methoden zur Fehlererkennung für die Überwachung technischer Prozesse. *Regelungstechnische Praxis*, (9 & 10):321–325 & 363–368, 1980.

5.9   Isermann, R. Process fault detection on modeling and estimation methods - a survey. *Automatica*, 20(4):387–404, 1984.

5.10 Isermann, R. Estimation of physical parameters for dynamic processes with application to an industrial robot. *International Journal of Control*, 55(6):1287–1298, 1992.

5.11 Isermann, R. *Mechatronic systems – fundamentals*. Springer, London, 2003.

5.12 Isermann, R. and Freyermuth, B. Process fault diagnosis based on process model knowledge. *Journal A*, 31(4), 1991.

5.13 Isermann, R., Lachmann, K.-H., and Matko, D. *Adaptive control systems*. Prentice Hall International UK, London, 1992.

5.14 Karnopp, D., Margolis, D., and Rosenberg, R. *System dynamics: a unified approach*. Wiley, New York, 1990.

5.15 Karnopp, D. and Rosenberg, R. *System dynamics: a unified approach*. Wiley, New York, 1975.

5.16 Ljung, L. *System identification – theory for the user*. Prentice Hall, Englewood Cliffs, 1987.

5.17 Mohler, R. *Bilinear control processes – with application to engineering, ecology, and medicine*. Academic Press, New York, 1973.

5.18 Nold, S. *Wissensbasierte Fehlererkennung und Diagnose mit den Fallbeispielen Kreiselpumpe und Drehstrommotor*, volume Fortschr.-Ber. VDI Reihe 8, 273. VDI Verlag, Düsseldorf, 1991.

5.19 Profos, P. and Pfeifer, T. *Handbuch der industriellen Messtechnik*. Oldenbourg, München, 1992.

5.20 Siebert, H. and Isermann, R. Leckerkennung durch online Korrelationsanalyse mit Prozeßrechnern. In *Proc. VDI-VDE Aussprachetage Korrelationsverfahren in der Mess- und Regelungstechnik*, 1976.

5.21 Thoma, J. *Simulation by bond graphs*. Springer, Berlin, 1990.

5.22 Wellstead, P. *Introduction to physical system modelling*. Addison-Wesley, Reading, MA, 1947.

5.23 Yu, D., Shields, D., and Daley, S. A hybrid fault diagnosis approach using neural networks. *Neural Computing and Applications*, 4:21–26, 1996.

Chapter 6

6.1 Åström, K. *Introduction to stachostic control theory*. New York, Academic Press, 1970.

6.2 Bendal, J. and Piersol, A. *Random data: analysis measurement procedures*. John Wiley & Sons, Inc., New York, 1971.

6.3 Box, G. and Jenkins, G. *Time series analysis: forecasting and control*. Holden-Day, San Francisco, 1970.

6.4 Hänsler, E. *Statistische Signale – Grundlagen und Anwendungen*. Springer, Berlin, 3rd edition, 2001.

6.5 Isermann, R. *Identifikation dynamischer Systeme*. Springer, Berlin, 1992.

6.6 Isermann, R. *Mechatronic systems – fundamentals*. Springer, London, 2003.

6.7 Papoulis, A., editor. *Probability, random variables, and stochastic processes*. McGraw-Hill, New York, 3rd edition, 1991.

Chapter 7

7.1 Basseville, M. Detecting changes in signals and systems – a survey. *Automatica*, 24(3):309–326, 1988.

7.2 Basseville, M. On-board component fault detection and isolation using the statistical local appraoch. *Automatica*, 34(11):1391–1416, 1998.

7.3   Basseville, M. Model-based statistical signal processing and decision theoretic approaches to monitoring. In *Prepr. 5th Symposium on Fault Detection, Supervision and Safety for Technical Processes (SAFEPROCESS)*, pages 1–12, Washington, DC, 2003.

7.4   Bronstein, I. and Semendjajew, K. *Handbook of mathematics*. Springer, Berlin, 2004.

7.5   Clark, R. State estimation schemes for instrument fault detection. In Patton, R., Frank, P., and Clark, R., editors, *Fault diagnosis in dynamic systems*, chapter 2, pages 21–45. Prentice Hall, New York, 1989.

7.6   DIN 13303. *Stochastik*. Beuth Verlag, Berlin, 1980.

7.7   Frank, P. Enhancement of robustness in observer-based fault detection. In *Prepr. IFAC Symposium on Fault Detection, Supervision and Safety for Technical Processes (SAFEPROCESS)*, volume 1, pages 275–287, Baden-Baden, Germany, September 1991. Pergamon Press.

7.8   Füssel, D. *Fault diagnosis with tree-structured neuro-fuzzy systems*, volume Fortschr.-Ber. VDI Reihe 8, 957. VDI Verlag, Düsseldorf, 2002.

7.9   Grant, E. *Statistical quality control*. McGraw Hill, New York, 1964.

7.10 Hancock, J. and Wintz, P. *Signal detection theory*. McGraw-Hill, New York, 1966.

7.11 Hartung, J. *Statistik*. Oldenbourg, München, 13th edition, 2002.

7.12 Himmelblau, D. *Process analysis by statistical methods*. J. Wiley, New York, 1970.

7.13 Himmelblau, D. *Fault detection and diagnosis in chemical and petrochemical processes*. Elsevier, New York, 1978.

7.14 Himmelblau, D. and Bischoff, K. *Process analysis and simulation. Deterministic systems*. J. Wiley, New York, 1968.

7.15 Höfling, T. *Methoden zur Fehlererkennung mit Parameterschätzung und Paritätsgleichungen*, volume Fortschr.-Ber. VDI Reihe 8, 546. VDI Verlag, Düsseldorf, 1996.

7.16 Höfling, T. and Isermann, R. Fault detection based on adaptive parity equations and single-parameter tracking. *Control Engineering Practice – CEP*, 4(10):1361–1369, 1996.

7.17 Isermann, R. Methoden zur Fehlererkennung für die Überwachung technischer Prozesse. *Regelungstechnische Praxis*, (9 & 10):321–325 & 363–368, 1980.

7.18 Isermann, R. *Digital control systems*, volume 2. Springer, Berlin, 1991.

7.19 Isermann, R., editor. *Überwachung und Fehlerdiagnose - Moderne Methoden und ihre Anwendungen bei technischen Systemen*. VDI-Verlag, Düsseldorf, 1994.

7.20 Isermann, R. On fuzzy logic applications for automatic control, supervision and fault diagnosis. In *Proc. 3rd European Congress on Fuzzy and Intelligent Technologies (EUFIT)*, volume 2, pages 738–753, Aachen, Germany, 1995.

7.21 Page, E. Continuous inspection schemes. *Biometrika*, 41:100–115, 1954.

7.22 Roberts, S. A comparison of some control chart procedures. *Technometrics*, 8:411–430, 1966.

7.23 Scharf, L. *Statistical signal processing*. Addison-Wesley, Reading, 1991.

7.24 Stark, H. and Woods, J. *Probability, random processes and estimation theory for engineers*. Prentice Hall, 1994.
7.25 Wald, A. *Sequential analysis*. J. Wiley, New York, 1947.
7.26 Zurmühl, R. *Praktische Mathematik*. Springer, Berlin, 5th edition, 1965.

Chapter 8

8.1   Akaike, H. A new look at the statistical model identification. *IEEE Trans. on Automatic Control*, 19(6):716–723, 1974.
8.2   Best, R. Wavelets: Eine praxisorientierte Einführung mit Beispielen. Teile 2 & 8. *Technisches Messen*, 67(4 & 11):182–187, 491–505, 2000.
8.3   Bogert, B., Healy, M., and Tukey, J. The quefrency analysis of time series for echoes. In Rosenblatt, M., editor, *Proc. Symp. Time Series Analysis*, pages 209–243. Wiley, 1968.
8.4   Box, G. and Jenkins, G. *Time series analysis: forecasting and control*. Holden-Day, San Francisco, 1970.
8.5   Brigham, E. *The fast Fourier transform*. Prentice Hall, Englewood Cliffs, 2nd edition, 1974.
8.6   Burg, J. A new analysis technique for time series data. In *NATO Advanced Study Institute on Signal Processing with Emphasis on Underwater Acoustics*, August 1968.
8.7   Cooley, J. and Tukey, J. An algorithm for the machine calculation of complex fourier series. *Math. of Computation*, 19:297–301, 1965.
8.8   Edward, J. and Fitelson, M. Notes on maximum entropy processing. *IEEE Trans. Inform. Theory*, IT-19:232–234, 1973.
8.9   Ericsson, S., Grip, N., Johannson, E., Persson, L., Sjöberg, R., and Strömberg, J. Towards automatic detection of local bearing defects in rotating machines. *Mechanical Systems and Signal Processing*, 9:509–535, 2005.
8.10 Friedmann, A. An introduction to linear and nonlinear systems and their relation to machinery faults. *www.DLIengineering.com*, 2001.
8.11 Führer, J., Sinsel, S., and Isermann, R. Erkennung von Zündaussetzern aus Drehzahlsignalen mit Hilfe eines Frequenzbereichsverfahrens. *Proc. 13. Tagung Elektronik im Kraftfahrzeug, Essen*, 1993.
8.12 Hänsler, E. *Statistische Signale – Grundlagen und Anwendungen*. Springer, Berlin, 3rd edition, 2001.
8.13 Harris, T. *Rolling bearing analysis*. J. Wiley & Sons, New York, 4th edition, 2001.
8.14 Hess, W. *Digitale Filter*. Teubner Studienbücher, Stuttgart, 1989.
8.15 Hippenstiel, R. *Detection theory*. CRC Press, Boca Raton, 2002.
8.16 Isermann, R. *Digitale Regelsysteme*, volume 1 &, 2. Springer, Berlin, 1988.
8.17 Isermann, R. *Identifikation dynamischer Systeme*. Springer, Berlin, 1992.
8.18 Isermann, R. *Fault diagnosis of technical processes – applications*. Springer, Heidelberg, 2006.

8.19 Isermann, R., Lachmann, K.-H., and Matko, D. *Adaptive control systems*. Prentice Hall International UK, London, 1992.

8.20 Janik, W. *Fehlerdiagnose des Außenrund-Einstechschleifens mit Prozeß- und Signalmodellen*, volume Fortschr.-Ber. VDI Reihe. VDI Verlag, Düsseldorf, 1992.

8.21 Janik, W. and Fuchs. Process- and signal-model based fault detection of the grinding process. In *Prepr. IFAC Symposium on Fault Detection, Supervision and Safety for Technical Processes (SAFEPROCESS)*, volume 2, pages 299–304, Baden-Baden, Germany, September 1991.

8.22 Janik, W. and Isermann, R. Signal model-based diagnosis system for the supervision of periodically and intermittant working machines tools. In *Proc. 11th IFAC World Congress*, volume 1, pages 130–134, Tallinn, USSR, 1990.

8.23 Kammeyer, K. and Kroschel, K. *Digitale Signalverarbeitung: Filterung und Spektralanalyse*. Teubner, Stuttgart, 3rd edition, 1996.

8.24 Kay, S. *Modern spectral estimation - theory and applications*. Prentice Hall, Englewood Cliffs, 1987.

8.25 Kolerus, J. *Zustandsüberwachung von Maschinen*. expert Verlag, Renningen-Malmsheim, 2000.

8.26 Makhoul, J. Linear prediction: a tutorial review. *Proc. of IEEE*, 63:561–580, 1975.

8.27 Marple, S. *Digital spectral analysis with applications*. Prentice Hall, Englewood Cliffs, 1987.

8.28 Meyer-Bäse, U. *Digital signal processing with field programmable gate arrays*. Springer, Berlin, 2004.

8.29 Mitra, S. *Digital signal processing: a computer-based approach*. McGraw Hill & Irwin, Boston, 2nd edition, 2001.

8.30 Muffert, K.-H. Ursachen und Beispiele von Schäden an Verbrennungsmotoren. *Der Maschinenschaden*, 53:95–102, 1980.

8.31 Neumann, D. Fault diagnosis of machine-tools by estimation of signal spectra. In *Preprints IFAC Symposium on Fault Detection, Supervision and Safety for Technical Processes (SAFEPROCESS)*, volume 1, pages 73–78, Baden-Baden, Germany, September 1991.

8.32 Neumann, D. Analyse periodischer Signale zur Fehlererkennung. In Isermann, R., editor, *Überwachung und Fehlerdiagnose*, pages 43–71. VDI, Düsseldorf, 1994.

8.33 Neumann, D. and Janik, W. Fehlerdiagnose an spanenden Werkzeugmaschinen mit parametrischen Signalmodellen von Spektren. In *VDI-Schwingungstagung*, Mannheim, Germany, 1990.

8.34 Nussbaumer, H. *Fast Fourier transform and convolution algorithms*. Springer, Heidelberg, 1981.

8.35 Oppenheim, A., Schafer, R., and Buck, J. *Discrete-time signal processing*. Prentice Hall, Englewood Cliffs, 2nd edition, 1999.

8.36 Pandit, S. and Wu, S.-M., editors. *Time series and system analysis with applications*. Wiley, New York, 1983.

8.37 Papoulis, A. *Probability, random variables, and stochastic processes.* McGraw-Hill, New York, 2nd edition, 1994.

8.38 Platz, R. *Untersuchungen zur modellgestützten Diagnose von Unwuchten und Wellenrissen in Rotorsystemen*, volume Fortschr.-Ber. VDI Reihe 11, 325. VDI Verlag, Düsseldorf, 2004.

8.39 Platz, R., Markert, R., and Seidler, M. Validation of online diagnosis of malfunctions in rotor systems. In *Trans. 7th ImechE-Conf. on Vibrations in Rotating Machines*, pages 581–590, University of Nottingham, 2000.

8.40 Porat, B. *A course on digital signal processing.* J. Wiley & Sons Inc., New York, 1997.

8.41 Press, W., Flannery, B., Teukolsky, W., and Vetterling, S. *Numerical recipes in C.* Cambrigde University Press, Cambridge, 1988.

8.42 Qian, S. and Chen, D. *Joint time-frequency analysis: methods and applications.* Prentice Hall, Upper Saddle River, 1996.

8.43 Randall, R. *Frequency analysis.* Bruel & Kjaer, Naerum, 3rd edition, 1987.

8.44 Ribbens, W. and Rizzoni, G. Onboard diagnosis of engine misfires. In *Proc. SAE 90*, number SAE 901768, Warrendale, USA, 1990.

8.45 Schüßler, H. *Digitale Signalverarbeitung 1 - Analyse diskreter Signale und Systeme.* Springer, Berlin, 4th edition, 1994.

8.46 Stearns, S. *Digital signal analysis.* Hayden Book Company, Rochelle Park, 1975.

8.47 Stearns, S. and Hush, D. *Digital signal analysis.* Prentice Hall, Englewood Cliffs, 1990.

8.48 Ulrych, T. and Bishop, T. Maximum entropy spectral analysis and autoregressive decomposition. *Reviews of Geophysics and Space Physics*, 13(February):183–200, 1975.

8.49 Williams, A. and Taylor, F. *Electronic filter design handbook.* McGraw Hill, 3rd edition, 1995.

8.50 Willimowski, M. *Verbrennungsdiagnose von Ottomotoren mittels Abgasdruck und Ionenstrom.* Shaker Verlag, Aachen, 2003.

8.51 Willimowski, M., Füssel, D., and Isermann, R. Diagnose von Verbrennungsaussetzern in Ottomotoren durch Messung des Abgasdrucks. *Motortechnische Zeitschrift – MTZ*, 10:654–663, 1999.

8.52 Willimowski, M. and Isermann, R. A time domain based diagnostic system for misfire detection in spark-ignition engines by exhaust-gas pressure analysis. In *SAE 2000 World Congress*, volume SP-1501, pages 33–43, Detroit, MI, USA, March 2000.

8.53 Wirth, R. Maschinendiagnose an Industriegetrieben – Grundlagen. *Antriebstechnik*, 37(10 & 11):75–80 & 77–81, 1998.

8.54 Wowk, V. *Machinery vibrations.* McGraw Hill, New York, 1991.

8.55 Zoubir, A. and Iskander, D. R. *Bootstrap techniques for signal processing.* University Press, Cambridge, 2004.

Chapter 9

9.1   Armstrong-Hélouvry, B. *Control of machines with friction.* Kluwer, Boston, 1991.

9.2   Ayoubi, M. *Nonlinear system identification based on neural networks with locally distributed dynamics and application to technical processes*, volume Fortschr.-Ber. VDI Reihe 8, 591. VDI Verlag, Düsseldorf, 1996.

9.3   Babuska, R. and Verbruggen, H. An overview of fuzzy modeling for control. *Control Engineering Practice – CEP*, 4(11):1593–1606, 1996.

9.4   Bakiotis, C., Raymond, J., and Rault, A. Parameter and discriminiant analysis for jet engine mechanical state diagnosis. In *Proc. of The 1979 IEEE Conf. on Decision & Control*, Fort Lauderdale, USA, 1979.

9.5   Ballé, P. Fuzzy-model-based parity equations for fault isolation. *Control Engineering Practice – CEP*, 7:261–270, 1998.

9.6   Bierman, G. *Factorization methods for discrete sequential estimation*, volume 128 of *Mathematics in Science and Engineering*. Academic Press, New York, 1977.

9.7   Bishop, C. *Neural networks for pattern recognition.* Oxford University Press, Oxford, 1995.

9.8   Bosch. *Automotive brake systems.* SAE, Warrendale, 1995.

9.9   Canudas de Wit, C. *Adaptive control of partially known system.* Elsevier, Boston, 1988.

9.10   Drewelow, W. Parameterschätzung nach der ausgangsfehlermethods. *Messen, Steuern, Regeln – MSR*, 33(1):15–23, 1990.

9.11   (Ernst), S. T. Hinging hyperplane trees for approximation and identification. In *37th IEEE Conference on Decision and Control*, Tampa, FL, USA, December 1998.

9.12   Eykhoff, P. *System identification.* John Wiley, London, 1974.

9.13   Filbert, D.   Fault diagnosis in nonlinear electromechanical systems by continuous-time parameter estimation. *ISA Trans.*, 24(3):23–27, 1985.

9.14   Filbert, D. and Metzger, L. Quality test of systems by parameter estimation. In *Proc. 9th IMEKO-Congress*, Berlin, Germany, May 1982.

9.15   Golub, G. H. and Loan, C. F. V. *Matrix computations.* Johns Hopkins Studies in the Mathematical Sciences. Academic Press, Baltimore and London, 1996.

9.16   Goodwin, G. and Payne, R. *Dynamic system identification: experiment design and data analysis*, volume 136 of *Mathematics in science and engineering*. Academic Press, New York, 1977.

9.17   Gustavsson, I., Ljung, L., and Söderström, T. Identification of processes in closed loop identifiability and accuracy aspects. *Automatica*, 13(1):59–75, 1977.

9.18   Haber, R. and Unbehauen, H. Structure identification of nonlinear dynamic systems - a survey on input/output approaches. *Automatica*, 26(4):651–677, 1990.

9.19 Hafner, S., Geiger, H., and Kreßel, U. Anwendung künstlicher neuronaler Netze in der Automatisierungstechnik. Teil 1: Eine Einführung. *Automatisierungstechnische Praxis – atp*, 34(10):592–645, 1992.

9.20 Haykin, S. *Neural networks*. Macmillan College Publishing Company, Inc., New York, 1994.

9.21 Hecht-Nielson, R. *Neurocomputing*. Addison-Wesley, Reading, 1990.

9.22 Held, V. *Parameterschätzung und Reglersynthese für Industrieroboter*, volume Fortschr.-Ber. VDI Reihe 8, 275. VDI Verlag, Düsseldorf, 1991.

9.23 Held, V. and Maron, C. Estimation of friction characteristics, inertial and coupling coefficients in robotic joins based on current and speed measurements. In *Proc. IFAC Symposium on Robot Control*, Karlsruhe, Germany, 1988. Pergamon, Oxford.

9.24 Hohmann, H. *Automatische Überwachung und Fehlerdiagnose an Werkzeugmaschinen*. PhD thesis, Technische Hochschule, Darmstadt, 1987.

9.25 Holzmann, H., Halfmann, C., Germann, S., Würtenberger, M., and Isermann, R. Longitudinal and lateral control and supervision of autonomous intelligent vehicles. *Control Engineering Practice – CEP*, 5(11):1599–1605, 1997.

9.26 Isermann, R. Methoden zur Fehlererkennung für die Überwachung technischer Prozesse. *Regelungstechnische Praxis*, (9 & 10):321–325 & 363–368, 1980.

9.27 Isermann, R. Parameter-adaptive control algorithms - a tutorial. *Automatica*, 18(5):513–528, 1982.

9.28 Isermann, R. Process fault detection on modeling and estimation methods - a survey. *Automatica*, 20(4):387–404, 1984.

9.29 Isermann, R. *Digitale Regelsysteme*, volume 1 &, 2. Springer, Berlin, 1988.

9.30 Isermann, R. *Identifikation dynamischer Systeme*. Springer, Berlin, 1992.

9.31 Isermann, R., Lachmann, K.-H., and Matko, D. *Adaptive control systems*. Prentice Hall International UK, London, 1992.

9.32 Isermann, R., (Töpfer), S. E., and Nelles, O. Identification with dynamic neural networks - architecture, comparisons, applications -. In *Proc. IFAC Symposium on System Identification*, Fukuoka, Japan, July 1997.

9.33 Kaminski, P., Bryson, A., and Schmidt, S. Discrete square root filtering: A survey of current technologies. *IEEE Trans. Auto. Control*, 16:727–736, 1971.

9.34 Kofahl, R. Self-tuning of PID controllers based on process parameter estimation. *Journal A*, 27(3), 1986.

9.35 Lachmann, K.-H. *Parameteradaptive Regelalgorithmen für bestimmte Klassen nichtlinearer Prozesse mit eindeutigen Nichtlinearitäten*, volume Fortschr.-Ber. VDI Reihe 8, 66. VDI Verlag, Düsseldorf, 1983.

9.36 Ljung, L., Morf, M., and Falconer, D. Fast calculation of gain matrices for recursive estimate schemes. *Int. J. Control*, 27:1–19, 1978.

9.37 Ljung, L. and Söderström, T. *Theory and practive of recursive identification*. MIT Press, Cambridge, MA, 1983.

9.38 Maron, C. *Methoden zur Identifikation und Lageregelung mechanischer Prozesse mit Reibung*, volume Fortschr.-Ber. VDI Reihe 8, 246. VDI Verlag, Düsseldorf, 1996.

9.39 McCulloch, W. and Pitts, W. A logical calculus of the ideas immanent in nervous activity. *Bull. Math. Biophys.*, 5:115–133, 1943.

9.40 Müller, N. *Adaptive Motorregelung beim Ottomotor unter Verwendung von Brennraumdruck-Sensoren*, volume Fortschr.-Ber. VDI Reihe 12, 545. VDI Verlag, Düsseldorf, 2003.

9.41 Murray-Smith, R. and Johansen, T. *Multiple model approaches to modelling and control.* Taylor & Francis, London, 1997.

9.42 Nelles, O. LOLIMOT – Lokale, lineare Modelle zur Identifikation nichtlinearer, dynamischer Systeme. *Automatisierungstechnik – at*, 45(4):163–174, 1997.

9.43 Nelles, O. *Nonlinear system identification.* Springer, Heidelberg, 2001.

9.44 Nelles, O., Hecker, O., and Isermann, R. Automatic model selection in local linear model trees (LOLIMOT) for nonlinear system identification of a transport delay process. In *Proc. 11th IFAC Symposium on System Identification (SYSID)*, Kitakyushu, Fukuoka, Japan, 1991.

9.45 Nelles, O. and Isermann, R. Identification of nonlinear dynamic systems – classical methods versus radial basis function networks. In *Proc. American Control Conference (ACC)*, Seattle, USA, 1995.

9.46 Oppenheim, A., Schafer, R., and Buck, J. *Discrete-time signal processing.* Prentice Hall, Englewood Cliffs, 2nd edition, 1999.

9.47 Panuska, V. An adaptive recursive least square identification algorithm. In *Proc. IEEE Symp. Adaptive Processes, Decision and Control*, 1969.

9.48 Peter, K.-H. *Parameteradaptive Regelalgorithmne auf der Basis zeitkontinuierlicher Prozessmodelle*, volume Fortschr.-Ber. VDI Reihe 8, 348. VDI Verlag, Düsseldorf, 1993.

9.49 Peterka, V. A square root filter for real time multivariate regression. *Kybernetika*, 11:53–67, 1975.

9.50 Pfeufer, T. *Modellgestützte Fehlererkennung und Diagnose am Beispiel eines Fahrzeugaktors*, volume Fortschr.-Ber. VDI Reihe 8, 749. VDI Verlag, Düsseldorf, 1999.

9.51 Preuß, H.-P. and Tresp, V. Neuro-Fuzzy. *Automatisierungstechnische Praxis - at*, 36(5):10–24, 1994.

9.52 Raab, U. *Modellgestützte digitale Regelung und Überwachung von Kraftfahrzeugaktoren*, volume Fortschr.-Ber. VDI Reihe 8, 313. VDI Verlag, Düsseldorf, 1993.

9.53 Rosenblatt, F. The perceptron: a probabilistic model for information storage & organization in the brain. *Psychological Review*, 65:386–408, 1958.

9.54 Schmitt, M. *Untersuchungen zur Realisierung mehrdimensionaler lernfähiger Kennfeler in Großserien- Steuergeräten*, volume Fortschr.-Ber. VDI Reihe 12, 246. VDI Verlag, Düsseldorf, 1995.

9.55 Strejc, V. Least square parameter estimation. *Automatica*, 16:535–550, 1980.

9.56 Töpfer, S. *Hierarchische neuronale Modelle für die Identifikation nichtlinearer Systeme*, volume Fortschr.-Ber. VDI Reihe 10, 705. VDI Verlag, Düsseldorf, 2002.

9.57 Töpfer, S., Wolfram, A., and Isermann, R. Semi-physical modelling of nonlinear processes by means of local model approaches. In *Proc. 15th IFAC World Congress*, Barcelona, Spain, July 2002.

9.58 Widrow, B. and Hoff, M. Adaptive switching circuits. *IRE WESCON Conv. Rec.*, pages 96–104, 1960.

9.59 Wolfram, A. and Moseler, O. Design and application of digital FIR differentiators using modulating functions. In *Proc. 12th IFAC Symposium on System Identification (SYSID)*, Santa Barbara, CA, USA, June 2001.

9.60 Wolfram, A. and Vogt, M. Zeitdiskrete Filteralgorithmen zur Erzeugung zeitlicher Ableitungen. *Automatisierungstechnik – at*, 7:346–353, 2002.

9.61 Young, P. The use of linear regression and related procedures for the identification of dynamic processes. In *Proc. 7th IEEE Symposium on Adaptive Processes*, Los Angeles, CA, USA, 1968.

9.62 Young, P. *Recursive estimation and time-series analysis.* Springer, Berlin, 1984.

Chapter 10

10.1 Babuska, R. and Verbruggen, H. An overview of fuzzy modeling for control. *Control Engineering Practice – CEP*, 4(11):1593–1606, 1996.

10.2 Ballé, P. Fuzzy-model-based parity equations for fault isolation. *Control Engineering Practice – CEP*, 7:261–270, 1998.

10.3 Ballé, P. *Modellbasierte Fehlererkennung für nichtlineare Prozesse mit linear-parameterveränderlichen Modellen*, volume Fortschr.-Ber. VDI Reihe 8, 960. VDI Verlag, Düsseldorf, 2002.

10.4 Chen, J. and Patton, R. *Robust model-based fault diagnosis for dynamic systems.* Kluwer, Boston, 1999.

10.5 Chow, E. and Willsky, A. Analytical redundancy and the design of robust failure detection systems. *IEEE Trans. on Automatic Control*, 29(7):603–614, 1984.

10.6 Fischer, M. and Nelles, O. Fuzzy model-based predictive control of nonlinear processes with fast dynamics. In *2nd International ICSC Symposium on Fuzzy Logic and Applications - ISFL'97*, Zürich, Switzerland, February 1997.

10.7 Gertler, J. Analytical redundancy methods in failure detection and isolation - survey and synthesis. In *Prepr. IFAC Symposium on Fault Detection, Supervision and Safety for Technical Processes (SAFEPROCESS)*, volume 1, pages 9–21, Baden-Baden, Germany, September 1991.

10.8 Gertler, J. *Fault detection and diagnosis in engineering systems.* Marcel Dekker, New York, 1998.

10.9 Gertler, J. and Kunwer, M. Optimal residual decoupling for robust fault diagnosis. In *Proc. Int. Conf. on Fault Diagnosis: TOOLDIAG 1993*, pages 436–452, Toulouse, France, Apirl 1993.

10.10 Gertler, J. and Singer, D. Augmented models for statistical fault isolation in complex dynamic systems. In *Proc. American Control Conference (ACC)*, volume 1, pages 317–322, Boston, MA,, 1985.

10.11 Gertler, J. and Singer, D. A new structural framework for parity equation-based failure detection and isolation. *Automatica*, 26(2):381–388, 1990.

10.12 Höfling, T. *Methoden zur Fehlererkennung mit Parameterschätzung und Paritätsgleichungen*, volume Fortschr.-Ber. VDI Reihe 8, 546. VDI Verlag, Düsseldorf, 1996.

10.13 Höfling, T. and Deibert, R. Estimation of parity equations in nonlinear systems. In *Proc. IEEE Conference on Control Applications*, Glasgow, UK, 1994.

10.14 Höfling, T. and Isermann, R. Adaptive parity equations and advanced parameter estimation for fault detection and diagnosis. In *Proc. 13th IFAC World Congress*, San Francisco, CA, USA, 1996.

10.15 Höfling, T. and Isermann, R. Fault detection based on adaptive parity equations and single-parameter tracking. *Control Engineering Practice – CEP*, 4(10):1361–1369, 1996.

10.16 Isermann, R., editor. *Überwachung und Fehlerdiagnose - Moderne Methoden und ihre Anwendungen bei technischen Systemen*. VDI-Verlag, Düsseldorf, 1994.

10.17 Isermann, R. *Mechatronic systems – fundamentals*. Springer, London, 2003.

10.18 Isermann, R. *Fault diagnosis of technical processes – applications*. Springer, Heidelberg, 2006.

10.19 Münchhof, M. and Unger, I. Evaluation of parity equations with respect to process parameters and additive input/output faults. internal report. Technical report, Darmstadt, 2003.

10.20 Nelles, O. LOLIMOT – Lokale, lineare Modelle zur Identifikation nichtlinearer, dynamischer Systeme. *Automatisierungstechnik – at*, 45(4):163–174, 1997.

10.21 Nelles, O. *Nonlinear system identification with local linear neuro-fuzzy models*. Shaker, Aachen, 1999.

10.22 Nelles, O. and Isermann, R. Basis function networks for interpolation of local linear models. In *IEEE Conference on Decision and Control (CDC)*, pages 470–475, Kobe, Japan, 1996.

10.23 Patton, R. and Chen, J. A review of parity space approaches to fault diagnosis. In *Prepr. IFAC Symposium on Fault Detection, Supervision and Safety for Technical Processes (SAFEPROCESS)*, volume 1, pages 239–255, Baden-Baden, Germany, September 1991.

10.24 Sugeno, M. and Kang, G. Structure identification of fuzzy model. *Fuzzy Sets and Systems*, 26(1):15–33, 1985.

10.25 Takagi, T. and Sugeno, M. Fuzzy identification of systems and its applications to modelling and control. *IEEE Trans. on Syst. Man and Cyb.*, 15(1):116–132, 1985.

10.26 Wolfram, A. *Komponentenbasierte Fehlerdiagnose industrieller Anlagen am Beispiel frequenzumrichtergespeister Asynchronmaschinen und Kreiselpumpen*, volume Fortschr.-Ber. VDI Reihe 8, 967. VDI Verlag, Düsseldorf, 2002.

10.27 Wolfram, A., Füssel, D., Brune, T., and Isermann, R. Component-based multi-model approach for fault detection and diagnosis of a centrifugal pump. In *Proc. American Control Conference (ACC)*, Arlington, VA, USA, 2001.

Chapter 11

11.1   AGARD. *Theory and applications of Kalman filtering*, volume AGAR-DOgraph 139. Zentralstelle für Luftfahrtdokumentation & ESRO/ELDO, München & Neuilly-Seine, 1970.

11.2   Åström, K. and Wittenmark, B. *Computer controlled systems – theory and design*. Prentice-Hall, Englewood Cliffs, 1984.

11.3   Beard, R. Failure accommodation in linear systems through self-reorganization. Technical Report MVT-71-1, Man Vehicle Laboratory, Cambridge, Mass, 1971.

11.4   Brown, R. and Hwang, Y. *Introduction to random signals and applied Kalman filtering*. John Wiley & Sons, 2nd edition, 1992.

11.5   Chen, J. and Patton, R. *Robust model-based fault diagnosis for dynamic systems*. Kluwer, Boston, 1999.

11.6   Chen, J. and Zhang, H. Robust detection of faulty actuators via unknown input observers. *Int. J. Sys. Sci.*, 22(10):1829–1839, 1991.

11.7   Clark, R. A simplified instrument detection scheme. *IEEE Trans. Aerospace Electron. Systems*, 14(3):558–563, 1990.

11.8   Frank, P. and Wünnenberg, J. Sensor fault detection via robust observers. In Tzafestas, S., Singh, M., and Schmidt, G., editors, *System fault diagnostics, reliability & related knowledge-based approaches*, volume 1, pages 147–160. D. Reidel Press, Dordrecht, 1987.

11.9   Frank, P. and Wünnenberg, J. Robust fault diagnosis using unknown input observers schemes. In Patton, R., Frank, P., and Clark, R., editors, *Fault diagnosis in dynamic systems*, pages 47–98. Prentice Hall, New York, 1989.

11.10 Gelb, A., editor. *Applied optimal estimation*. MIT press, Cambridge, 1974.

11.11 Gertler, J. *Fault detection and diagnosis in engineering systems*. Marcel Dekker, New York, 1998.

11.12 Höfling, T. Zustandsgrößenschätzung zur Fehlererkennung. In Isermann, R., editor, *Überwachung und Fehlerdiagnose*, pages 89–108. VDI, Düsseldorf, 1994.

11.13 Höfling, T. *Methoden zur Fehlererkennung mit Parameterschätzung und Paritätsgleichungen*, volume Fortschr.-Ber. VDI Reihe 8, 546. VDI Verlag, Düsseldorf, 1996.

11.14 Isermann, R. Fault diagnosis of machines via parameter estimation and knowledge processing. In *Proc. IFAC Symposium on Fault Detection, Supervision and Safety for Technical Processes (SAFEPROCESS)*, volume 1, pages 121–133, Baden-Baden, Germany, September 1993.

11.15 Jacobs, O. *Introduction to control theory*. Oxford University Press, Oxford, 2nd edition, 1993.

11.16 Jones, H., editor. *Failure detection in linear systems*. Dept. of Aeronautics, M.I.T., Cambridge, 1973.

11.17 Kalman, R. A new approach to linear filtering and prediction problems. *Trans. ASME, Series D, J. Basic Eng.*, 82(March):35–45, 1960.

11.18 Kinnaert, M. Robust fault detection based on observers for bilinear systems. *Automatica*, 35:1829–1824, 1999.

11.19 Lewis, R. *Optimal estimation with an introduction to stochastic control theory*. John Wiley & Sons, 1986.

11.20 Magni, J. and Mayon, P. On residual generation by observer and parity space approaches. *IEEE Trans. on Automatic Control*, 39(2):441–447, 1994.

11.21 Massoumnia, M. A geometric approach to the synthesis of failure detection filters. *IEEE Trans. on Automatic Control*, 31:839–846, 1986.

11.22 Mehra, R. and Peschon, J. An innovations approach to fault detection and diagnosis in dynamic systems. *Automatica*, 7:637–640, 1971.

11.23 Tomizuka, M. *Advanced control systems II, Class Notes for ME233*. Dept of Mechanical Engineering, University of California at Berkeley, 1998.

11.24 Tsui, C.-C. A general failure detection, isolation and accomodation system with model uncertainty and measurement noise. In *Preprints of the 12th IFAC World Congress*, Sydney, Australia, 1993.

11.25 Unger, I. Vergleich von modellgestützten Fehlererkennungsmethoden durch Simulation. Master's thesis, TU Darmstadt, Institute of Automatic Control, Darmstadt, 2003.

11.26 Viswanadham, N. and Srichander, R. Fault detection using unknown input observers. *Control - Theory and Advanced Technology*, 3:91–101, 1987.

11.27 Watanabe, U. and Himmelblau, D. Instrument fault detection in systems with uncertainties. *Int. Journal of Systems Science*, 13:137–158, 1982.

11.28 Welch, G. and Bishop, G. *Introduction to Kalman filter*. Department of Computer Science, University of North Carolina, Chapel Hill, update monday, april 5 edition, 2004.

11.29 Wenzel, L. Kalman-Filter - Ein mathematische Modell zur Auswertung von Messdaten für die Regelungstechnik. *Elektronik*, (6 & 8 & 11):64–75 & 50–55 & 52–58, 2000.

11.30 White, J. and Speyer, J. Detection filter design: spectral theory and algorithms. *IEEE Trans. on Automatic Control*, 32:593–603, 1987.

11.31 Willsky, A. A survey of design methods for failure detection systems. *Automatica*, 12:601–611, 1976.

Chapter 12

12.1   Åström, K. and Wittenmark, B. *Adaptive control*. Addison Wesely, Reading, MA, 1995.

12.2   Deibert, R. *Methoden zur Fehlererkennung an Komponenten im geschlosse-nen Regelkreis*, volume Fortschr.-Ber. VDI Reihe 8, 650. VDI Verlag, Düsseldorf, 1997.

12.3   Dittmar, R., Bebar, M., and Reinig, C. Control loop performance monitoring. *Automatisierungstechnische Praxis – atp*, 45:94–103, 2003.

12.4   Ender, D. Process control performance: not as good as you think. *Control Engineering Practice – CEP*, 40(10):180–190, 1993.

12.5   Hägglund, T. A control-loop performance monitor. *Control Engineering Practice – CEP*, 3:1543–1551, 1995.

12.6   Hägglund, T. and Åström, K. Supervision of adaptive control algorithms. *Automatica*, 36(8):1171–1180, 2000.

12.7   Harris, T. Assessment of control loop performance. *Can. J. Chem. Eng*, 67:856–861, 1968.

12.8   Horch, A. *Condition monitoring of control loops*. PhD thesis, Royal Institute of Technology, Stockholm, 2000.

12.9   Horch, A. A simple method for detection of stiction in control valves. *Control Engineering Practice – CEP*, 7:1221–1231, 2004.

12.10   Horch, A. and Dumont, G. Special issue on control performance monitor-ing. *International Journal of Adaptive Control and Signal Processing*, 17(7-9):523–727, 2003.

12.11   Isermann, R. Beispiele für die Fehlerdiagnose mittels Parameterschätzung. *Automatisierungstechnik – at*, 37:342–343 & 445–447, 1980.

12.12   Isermann, R. Fault diagnosis of machines via parameter estimation and knowledge processing. In *Proc. IFAC Symposium on Fault Detection, Super-vision and Safety for Technical Processes (SAFEPROCESS)*, volume 1, pages 121–133, Baden-Baden, Germany, September 1993.

12.13   Isermann, R. *Mechatronic systems – fundamentals*. Springer, London, 2003.

12.14   Isermann, R. and Lachmann, K.-H. Parameter-adaptive control with configu-ration aids and supervision functions. *Automatica*, 21(6):625–638, 1985.

12.15   Isermann, R., Lachmann, K.-H., and Matko, D. *Adaptive control systems*. Prentice Hall International UK, London, 1992.

12.16   Ko, B.-S. and Edgar, T. Assessment of achievable PI control performance for linear processes with deadtime. In *Proc. American Control Conference*, pages 1548–1552, Philadelphia, PA, USA, 1998.

12.17   Lynch, C. and Dumont, G. Control loop performance monitoring. *IEEE Trans. Control Syst. Techn.*, 4:185–192, 1996.

12.18   Morari, M. and Morari, M. Performance monitoring of control systems using likelihood methods. *Automatica*, 32:1145–1162, 1996.

12.19   Swanda, A. and Seborg, D. Evaluating the performance of PID-type feedback control loops using normalized settling time. In *Proc. of the IFAC ADCHEM Symposium*, pages 283–288, Banff, Canada, 1997.

Chapter 13

13.1   Dunia, R. and Qin, S. Joint diagnosis of process and sensor faults using pca. *Control Engineering Practice – CEP*, 6:457–469, 1998.

13.2   Gertler, J. and Cao, J. PCA-based process diagnosis in the presence of control. *Proc. 5th IFAC Symposium on Fault Detection, Supervision and Safety for Technical Processes (SAFEPROCESS)*, pages 849–854, 2003.

13.3   Gertler, J. and McAvoy, T. Principal component analysis and parity relations - a strong duality. In *Proc. 3rd IFAC-Symposium IFAC Symposium on Fault Detection, Supervision and Safety for Technical Processes (SAFEPROCESS)*, volume 2, page 837, Hull, GB, 1997.

13.4   Haus, F. Examples for principal component analysis. internal report. Technical report, Darmstadt, 2003.

13.5   Jackson, J. *A user's guide to principal components.* J. Wiley, New York, 1991.

13.6   Joliffe, I. *Principle component analysis.* Springer-Verlag, New York, 2002.

13.7   Kresta, J., MacGregor, J., and Marlin, T. Multivariate statistical monitoring of process operation performance. *Canadian Journal of Chemical Engineering*, 69:35–47, 1991.

13.8   MacGregor, J. Data-based methods for process analysis, monitoring and control. In *Proc. 13th IFAC Symposium System Identification*, Rotterdam, NL, 2003.

13.9   MacGregor, J. and Kourti, F. Statistical process control of multivariate processes. *Control Engineering Practice – CEP*, (3):403–414, 1995.

13.10 Piovoso, M., Kosanovich, K., and Pearson, R. Monitoring process performance in real time. In *Proc. of the American Control Conference*, pages 2359–2363, 1992.

13.11 Seborg, D. A perspective on advanced strategies for process control (revisited). In Frank, P., editor, *Advances in Control. Highlights of ECC´99*, chapter 4, pages 104–134. Springer, London, 1999.

13.12 Wise, B. and Gallagher, N. The process chemometrics approach to process monitoring and fault detection. *Journal of Process Control*, 6(6):329–348, 1996.

13.13 Wise, B., Ricker, N., Veltkamp, D., and Kowalski, B. A theoretical basis for the use of principal component models for monitoring multivariate processes. *Process Control and Quality*, 1:41–51, 1990.

Chapter 14

14.1   Chen, J. and Patton, R. *Robust model-based fault diagnosis for dynamic systems.* Kluwer, Boston, 1999.

14.2   Isermann, R. On the applicability of model-based fault detection for technical processes. In *Preprints of the 12th IFAC World Congress*, volume 9, pages 195–200, Sydney, Australia, 1993.

14.3 Isermann, R. Integration of fault-detection and diagnosis methods. In *Proc. IFAC Symposium on Fault Detection, Supervision and Safety for Technical Processes (SAFEPROCESS)*, pages 597–609, Espoo, Finland, June 1994.

Chapter 15

15.1 Freyermuth, B. Knowledge-based incipient fault diagnosis of industrial robots. In *Prepr. IFAC Symposium on Fault Detection, Supervision and Safety for Technical Processes (SAFEPROCESS)*, volume 2, pages 31–37, Baden-Baden, Germany, September 1991. Pergamon Press.

15.2 Füssel, D. *Fault diagnosis with tree-structured neuro-fuzzy systems*, volume Fortschr.-Ber. VDI Reihe 8, 957. VDI Verlag, Düsseldorf, 2002.

15.3 Isermann, R. On fuzzy logic applications for automatic control, supervision and fault diagnosis. In *Proc. 3rd European Congress on Fuzzy and Intelligent Technologies (EUFIT)*, volume 2, pages 738–753, Aachen, Germany, 1995.

15.4 Isermann, R. and Ulieru, M. Integrated fault detection and diagnosis. In *Proc. IEEE/SMC Conference Systems Engineering in the Service of Humans*, Le Touquet, France, 1993.

15.5 Späth, H. *Cluster-Analyse-Algorithmen zur Objektklassifizierung und Datenreduktion*. Oldenbourg, München, Wien, 1975.

Chapter 16

16.1 Barschdorff, D., Gerhardt, D., and Kronmüller, M. Vergleich verschiedener Klassifikationsverfahren für den Bereich der technischen Fehlerdiagnose und der Biomedizin. In *Proc. of the FNC'97*, pages 510–517, Soest, 1997.

16.2 Bauer, B., George, D., Geropp, B., and Poschmann, M. Lathe tool wear monitoring with neural networks. In *Proc. European Congress on Fuzzy and Intelligent Techniques (EUFIT)*, pages 245–248, Aachen, Germany, 1996.

16.3 Bishop, C. *Neural networks for pattern recognition*. Oxford University Press, Oxford, 1995.

16.4 Dalmi, I., Kovacs, L., Lorant, I., and Terstyansky, G. Diagnosing priori unknown faults by radial basis function neural networks. In *Proc. of the 4th IFAC Symposium on Fault Detection, Supervision and Safety for Technical Processes (SAFEPROCESS)*, pages 405–409, Budapest, Hungaria, 2000.

16.5 Fochem, M., Wischnewski, P., and Hofmeier, R. Quality control systems on the production line of tape deck chassis using self organizing feature maps. In *Proc. European Congress on Fuzzy and Intelligent Techniques (EUFIT)*, Aachen, Germany, 1997.

16.6 Fritzke, B. Incremental neuro-fuzzy systems. In *Proc. SPIE International Symposium on Optical Science, Engineering and Instrumentation*, San Diego, USA, 1997.

16.7   Füssel, D. *Fault diagnosis with tree-structured neuro-fuzzy systems*, volume Fortschr.-Ber. VDI Reihe 8, 957. VDI Verlag, Düsseldorf, 2002.

16.8   Hänsler, E. *Statistische Signale – Grundlagen und Anwendungen*. Springer, Berlin, 3rd edition, 2001.

16.9   Hart, P. The condensed nearest neighbor rule. *IEEE Trans. on Information Theory*, IT-14:515–516, 1968.

16.10   Isermann, R. and Ballé, P. Trends in the application of model-based fault detection and diagnosis in technical processes. In *13th IFAC World Congress*, volume N, pages 1–12, San Francisco, CA, USA, 1996.

16.11   Leonard, J. and Kramer, M. Radial basis function networks for classifying process faults. *Control System Magazine*, April:31–38, 1991.

16.12   Leonhardt, S. *Modellgestützte Fehlererkennung mit neuronalen Netzen - Überwachung von Radaufhängungen und Diesel-Einspritzanlagen*, volume Fortschr.-Ber. VDI Reihe 12, 295. VDI Verlag, Düsseldorf, 1996.

16.13   Leonhardt, S., Bußhardt, J., Rajamani, R., Hedrick, J., and Isermann, R. Parameter estimation of shock absorbers with artificial neural network. In *Proc. American Control Conference (ACC)*, San Francisco, CA, USA, 1993.

16.14   Nelles, O. *Nonlinear system identification with local linear neuro-fuzzy models*. Shaker, Aachen, 1999.

16.15   Press, W., Flannery, B., Teukolsky, W., and Vetterling, S. *Numerical recipes in C*. Cambrigde University Press, Cambridge, 1988.

16.16   Schürmann, J. *Statistical pattern classification*. DATAKONTEXT-Verlag, Köln, 1998.

16.17   Simani, S. Fault diagnosis of an industrial plant at different operating points using neural networks. In *Proc. 4th IFAC Symposium on Fault Detection, Supervision and Safety for Technical Processes (SAFEPROCESS)*, pages 192–196, Budapest, Hungaria, 2000.

16.18   Sorsa, T. and Koivo, H. Application of artificial neural networks in process fault diagnosis. *Automatica*, 29(4):843–849, 1993.

16.19   Sorsa, T. and Koivo, H. Artificial neural networks in fault diagnosis: Practical considerations. In *Proc. of the First Asian Control Conference*, Tokyo, Japan, 1994.

16.20   Sorsa, T., Suontausta, J., and Koivo, H. Dynamic fault diagnosis using radial basis function networks. In *Proc. of the TOOLDIAG 1993*, pages 160–169, Toulouse, France, 1993.

16.21   Terra, M.-H. and Tinos, R. Fault detection and isolation in robotic system via artificial neural networks. In *Proc. 4th IFAC Symposium on Fault Detection, Supervision and Safety for Technical Processes (SAFEPROCESS)*, pages 180–185, Budapest, Hungaria, 2000.

16.22   Tzafestas, S. and Dalianis, P. Fault diagnosis in complex systems using artificial neural networks. In *Proc. of the 3rd IEEE CCA*, pages 877–882, 1994.

16.23   Weispfenning, T. and Isermann, R. Fehlererkennung an semiaktiven und konventionellen Radaufhängungen. In *Proc. Tagung Aktive Fahrwerkstechnik, Haus der Technik*, Essen, Germany, 1995.

16.24 Willimowski, M., Füssel, D., and Isermann, R. Misfire detetion for spark-ignition engines by exhaust gas pressure analysis. *MTZ worldwide*, pages 8–12, 1999.

16.25 Yu, D., Shields, D., and Daley, S. A hybrid fault diagnosis approach using neural networks. *Neural Computing and Applications*, 4:21–26, 1996.

16.26 Zhang, J. and Morris, A. Process fault diagnosis using fuzzy neural networks. In *Proc. American Control Conference (ACC)*, pages 971–975, Baltimore, ML, 1994.

Chapter 17

17.1 Abe, S. A method for fuzzy rule extraction directly from numerical data and its application to pattern classification. *IEEE Trans. on Fuzzy Systems*, pages 18–28, 1995.

17.2 Altug, S. and Chow, M.-Y. Comparative analysis of fuzzy inference systems implemented on neural structures. In *Proc. Int. Conf. on Neural Networks*, pages 426–431, Houston, TX, USA, 1997.

17.3 Ayoubi, M. *Nonlinear system identification based on neural networks with locally distributed dynamics and application to technical processes*, volume Fortschr.-Ber. VDI Reihe 8, 591. VDI Verlag, Düsseldorf, 1996.

17.4 Ballé, P. *Modellbasierte Fehlererkennung für nichtlineare Prozesse mit linear-parameterveränderlichen Modellen*, volume Fortschr.-Ber. VDI Reihe 8, 960. VDI Verlag, Düsseldorf, 2002.

17.5 Barlow, R. and Proschan, F. *Statistical theory of reliability and life testing*. Holt, Rinehart & Winston, Inc., 1975.

17.6 Bezdek, J. *Fuzzy Mathematics in Pattern Classification. PhD Theses*. Applied Math Center, Cornell University, Ithaca, 1973.

17.7 Bitterlich, N. Maschinenüberwachung mittels Fuzzy Pattern Classification. In *Beiträge des Innovations- und Bildungszentrum Hohen Luckow e.V.* Hohen Luckow, 1995.

17.8 Bothe, H.-H. *Neuro-Fuzzy-Methoden*. Springer, Heidelberg, 1997.

17.9 Cayrac, D., Dubois, D., and Prade, H. Handling uncertainty with possiblity theory and fuzzy sets in a satellite fault-diagnosis application. *IEEE Trans. on Fuzzy Systems*, 4(3):251–269, 1996.

17.10 DIN 24519. *Störfallablaufanalyse*. Beuth Verlag, Berlin, 1979.

17.11 DIN 25424. *Fehlerbaumanalyse, Methode und Bildzeichen*. Beuth Verlag, Berlin, 1981.

17.12 DIN 40042. *Zuverlässigkeit elektrischer Geräte, Anlagen und Systeme – Begriffe*. Beuth Verlag, Berlin, 1970.

17.13 Driankov, D., Hellendoorn, H., and Reinfrank, M. *An introduction to fuzzy control*. Springer, Berlin, 1993.

17.14 Fink, P. and Lusth, J. Expert systems and diagnostic expertise in the mechanical and electrical domains. *IEEE Trans. on Systems, Man, and Cybernetics*, 17(3):340–349, 1987.

17.15 Freyermuth, B. Knowledge-based incipient fault diagnosis of industrial robots. In *Prepr. IFAC Symposium on Fault Detection, Supervision and Safety for Technical Processes (SAFEPROCESS)*, volume 2, pages 31–37, Baden-Baden, Germany, September 1991. Pergamon Press.

17.16 Freyermuth, B. *Wissensbasierte Fehlerdiagnose am Beispiel eines Industrieroboters*, volume Fortschr.-Ber. VDI Reihe 8, 315. VDI Verlag, Düsseldorf, 1993.

17.17 Fritzke, B. Fast learning with incremental rbf networks. *Neural Processing Letters*, 1(1):1–5, 1994.

17.18 Frost, R. *Introduction to knowledge base systems*. Collins, London, 1986.

17.19 Füssel, D. *Fault diagnosis with tree-structured neuro-fuzzy systems*, volume Fortschr.-Ber. VDI Reihe 8, 957. VDI Verlag, Düsseldorf, 2002.

17.20 Füssel, D. and Isermann, R. Hierarchical motor diagnosis utilizing structural knowledge and a self-learning neuro-fuzzy scheme. In *Proc. 24th Annual Conference of the IEEE Industrial Electronics Society, IECON'98*, Aachen, Germany, 1998.

17.21 Gmytrasiewicz, P., Durfee, E., and Wehe, D. A decision theoretic approach to co-ordinating multi-agent interactions. In *Proc. of International Joint Conference on Artificial Intelligence*, 1991.

17.22 Harmon, P. and King, D. *Expert systems – artifical intelligence in business*. J. Wiley, New York, 1985.

17.23 Himmelblau, D. *Fault detection and diagnosis in chemical and petrochemical processes*. Elsevier, New York, 1978.

17.24 IEC 271. *List of basic terms, definitions and related mathematics*. IEC-Publications.

17.25 Isermann, R. On fuzzy logic applications for automatic control, supervision and fault diagnosis. In *Proc. 3rd European Congress on Fuzzy and Intelligent Technologies (EUFIT)*, volume 2, pages 738–753, Aachen, Germany, 1995.

17.26 Isermann, R. and Füssel, D. Supervision, fault detection and fault-diagnosis methods - advanced methods and applications. In Zimmermann, H.-J., editor, *Practical applications of fuzzy technologies*, pages 119–159. Kluwer Academic, Boston, 1999.

17.27 Isermann, R. and Ulieru, M. Integrated fault detection and diagnosis. In *Proc. IEEE/SMC Conference Systems Engineering in the Service of Humans*, Le Touquet, France, 1993.

17.28 Jang, J.-S. R., Sun, C.-T., and Mizutani, E. *Neuro-fuzzy and soft computing*. Prentice Hall, Upper Saddle River, 1997.

17.29 Kitowski, J. and Bargiel, M. Diagnostics of faulty states in complex physical systems using fuzzy relational equations. In Sanchez, E. and Zadeh, L., editors, *Approximate reasoning in intelligent systems, decision and control*. Pergamon, 1987.

17.30 Kleer, J. de. An assumption-based TMS. *Artificial Intelligence*, 28:127–162, 1986.

17.31 Lee, W., Grosh, D., Tillmann, F., and Lie, C. Fault tree analysis, methods and applications - a review. *IEEE Trans. on Reliability*, 34(3):194–202, 1991.

17.32 Lin, C.-T. and Lee, C. Neural-network-based fuzzy logic control and decision system. *IEEE Trans. on Computers*, 40(12):1320–1336, 1991.

17.33 Lopez, C., Patton, R., and Daley, S. Fault tolerant traction system control using fuzzy inference modelling. In *Proc. of the IFAC Conference on On-Line Fault Detection and Supervision in the Chemical Process Industries*, pages 137–142, Lyon, France, 1998.

17.34 Lunze, J. Diagnosis of quantised systems. In *Proc. of the IFAC Symposium on Fault Detection, Supervision and Safety for Technical Processes (SAFE-PROCESS)*, pages 28–39, 2000.

17.35 Mamdani, E. and Assilian, S. An experiment in linguistic synthesis with a fuzzy logic controller. *Int. Journal of Man-Machine Studies*, 7(1):1–13, 1975.

17.36 Milne, R. Strategies for diagnosis. *IEEE Trans. on Sys., Men & Cybernetics*, 17(3):333–339, 1987.

17.37 Mitra, S. and Hayashi, Y. Neuro-fuzzy rule generation: survey in soft computing framework. *IEEE Trans. on Neural Networks*, 11(3):748–768, 2000.

17.38 Nauck, D. Neuro-fuzzy systems: review and prospects. In *Proc. European Congress on Fuzzy and Intelligent Techniques (EUFIT)*, pages 1044–1053, Aachen, Germany, 1997.

17.39 Nauck, D. and Kruse, R. Nefclass - a neuro-fuzzy approach for the classification of data. In *Proc. of the 1995 ACM Symposium on Applied Computing*, pages 461–465, Nashville, TN, USA, 1995.

17.40 Nelles, O. *Nonlinear system identification with local linear neuro-fuzzy models*. Shaker, Aachen, 1999.

17.41 Nold, S. *Wissensbasierte Fehlererkennung und Diagnose mit den Fallbeispielen Kreiselpumpe und Drehstrommotor*, volume Fortschr.-Ber. VDI Reihe 8, 273. VDI Verlag, Düsseldorf, 1991.

17.42 Oezyurt, B. and Kandel, A. A hybrid hierarchical neural network-fuzzy expert system approach to chemical process fault diagnosis. *Fuzzy Sets and Systems*, 83:11–25, 1996.

17.43 Pearl, J. *Probabilistic reasoning in intelligent systems: networks of plausible inference*. Morgan Kaufmann Publishers, 1988.

17.44 Peng, Y. and Reggia, J. A probabilistic causal model for diagnostic problem solving – part 2, diagnostic strategy. *IEEE Trans. on Systems, Men and Cybernetics*, 17:395–406, 1987.

17.45 Peng, Y. and Reggia, J. *Abductive inference models for diagnostic problem-solving*. Springer, New York, 1990.

17.46 Pistauer, M. and Steger, C. Multistep parameterlearning in a neural network based fuzzy diagnosis module. In *Proc. IEE Int. Conf. On System Man and Cybernetics*, pages 3249–3254, 1995.

17.47 Reiter, R. A theory if diagnosis from first principles. *Artificial Intelligence*, 32:57–95, 1987.

17.48 Sanchez, E. Medical diagnostics applications in a linguistic approach using fuzzy logic. In *Proc. Int. Workshop on Fuzzy System Applications*, pages 38–50, Iizukam Japan, 1977.

17.49 Sanchez, E. Solutions in composite fuzzy relation equation – application to medical diagnosis in browerian logic. In Gupta, M., Saridis, G., and Gaines, B., editors, *Fuzzy automata and decision processes*, pages 221–234. North-Holland, Amsterdam, 1977.

17.50 Schneeweiss, W. *Zuverlässigkeitstechnik - von den Komponenten zum System.* DATAKONTEXT-Verlag, Köln, 1992.

17.51 Schram, G. and Verbruggen, H. A fuzzy logic approach to fault-tolerant control. *Journal A*, 39(3):11–21, 1998.

17.52 Schullerus, G., Supavatanakul, P., Krebs, V., and Lunze, J. Relations of timed event graphs and timed automata in fault diagnosis. In *Proc. 5th IFAC Symposium on Fault Detection, Supervision and Safety of Technical Processes (SAFEPROCESS)*, pages 819–824, Washington, DC, USA, 2003.

17.53 Schullerus, G., Supavatanakul, P., Krebs, V., and Lunze, J. Diagnose zeitbewerteter ereignisdiskreter Systeme. *Automatisierungstechnik – at*, 52(4):164–173, 2004.

17.54 Shortliffe, E. *Computer-based medical consulations, MYCIN*, volume 2 of *Artificial Intelligence Series*. Elsevier, Amsterdam, 1976.

17.55 Sun, G., Lee, Y., and Chen, H. A novel net that learns sequential decision process. In *Proc. of the Neural Information Procesing Conference*, pages 761–766, Dencer, CO, USA, 1987.

17.56 Supavatanakul, P., Falkenberg, C., and Lunze, J. Identification of timed discrete-event model for diagnosis. In *Proc. 14th International Workshop on Principles of Diagnosis*, Washington, DC, USA, 2003.

17.57 Teneketzis, D., Sinnamohideen, K., Sampath, M., Sengupta, R., and Lafortune, S. Failure diagnosis using discrete event models. *IEEE Trans. on Control Systems Technology*, 2(2):105–124, 1996.

17.58 Torasso, P. and Console, L. *Diagnostic problem solving*. North Oxford Academic, Oxford, 1989.

17.59 Tzafestas, S. and Dalianis, P. Fault diagnosis in complex systems using artificial neural networks. In *Proc. of the 3rd IEEE CCA*, pages 877–882, 1994.

17.60 Ulieru, M. *Fuzzy Logik für die Diagnose: Possibilistische Netze*. PhD thesis, Institute of Automatic Control, University of Technology.

17.61 Ulieru, M. From fault trees to fuzzy relations in managing heuristics for technical diagnosis. In *Proc. European Congress on Fuzzy and Intelligent Techniques (EUFIT)*, Aachen, Germany, 1993.

17.62 Willimowski, M. *Verbrennungsdiagnose von Ottomotoren mittels Abgasdruck und Ionenstrom*. Shaker Verlag, Aachen, 2003.

17.63 Zadeh, L. Outline of a new approach to the analysis of complex systems and decision processes. *IEEE Trans. on System, Man and Cybernetics*, 3:28–44, 1973.

17.64 Zimmermann, H.-J. *Fuzzy set theory - and its applications*. Kluwer, Boston, 2nd edition, 1991.

Chapter 18

18.1 Favre, C. Fly-by-wire for commercial aircraft: the airbus experience. *Int. Journal of Control*, 59(1):139–157, 1994.

18.2 IEC 61508. *Functional safety of electrical/electronic/programmable electronic systems*. International Electrotechnical Commission, Switzerland, 1997.

18.3 Isermann, R., Schwarz, R., and Stölzl, S. Fault-tolerant drive-by-wire systems – concepts and realizations. In *Proc. IFAC Symposium on Fault Detection, Supervision and Safety for Technical Processes (SAFEPROCESS)*, Budapest, Hungary, 2000.

18.4 Lauber, R. and Göhner, P. *Prozessautomatisierung*. Springer, Berlin, 3rd edition, 1999.

18.5 Leveson, N. *Safeware. System safety and computer*. Reading, Wesely Publishing Company, 1995.

18.6 Reichel, R. Modulares Rechnersystem für das Electronic Flight Control System (EFCS). In *DGLR-Jahrestagung, Deutsche Luft- und Raumfahrtkongress*, Berlin, Germany, 1999.

18.7 Reichel, R. and Boos, F. *Redundantes Rechnersystem für Fly-by-wire Steuerungen*. Bodensee-Gerätewerk, Überlingen, 1986.

18.8 Storey, N. *Safety-critical computer systems*. Addison Wesely Longman Ltd., Essex, 1996.

18.9 Tosunoglu, S. Fault-tolerant control of mechanical systems. In *Proc. IEEE Industrial Electronics Society, IECON '95*, volume 1, pages 127–132, Orlando, Fl, USA, 1995.

Chapter 19

19.1 Banda, S. Special issue on reconfigurable flight control. *Int. J. of Robust and Nonlinear Control*, 9(14):997–1115, 1999.

19.2 Blanke, M., Izadi-Zamenabadi, R., Bogh, S., and Lunan, C. Fault-tolerant control systems. *Control Engineering Practice – CEP*, 5(5):693–702, 1997.

19.3 Caglayan, A., Allen, S., and Wehmuller, K. Evaluation of a second generation reconfiguration strategy for aircraft flight control systems subjected to actuator failure surface damage. In *Proc. Nat. Aero. & Electron Conf.*, pages 520–529, Dayton, USA, May 1988.

19.4 Chandler, P. Reconfigurable flight control at wright laboratory. *AGARD Advisory Report*, (Report 360, Aersorpace 2020), 1997.

19.5 Clark, R. State estimation schemes for instrument fault detection. In Patton, R., Frank, P., and Clark, R., editors, *Fault diagnosis in dynamic systems*, chapter 2, pages 21–45. Prentice Hall, New York, 1989.

19.6 Clarke, D. Sensor, actuator, and loop validation. *IEE Control Systems*, 15(August):39–45, 1995.

19.7   Gelderloos, H. and Young, D. Redundancy management of shuttle flight control rate gyroscopes and accelerometers. In *Proc. American Control Conference (ACC)*, pages 808–811, 1982.

19.8   Hajiyev, C. and Caliskan, F. *Fault diagnosis and reconfiguration in flight control systems*. Kluwer Academic Publishers, 1966.

19.9   Heiner, G. and Thurner, T. Time-triggered architecture for safety-related distributed real-time systems in transportation systems. In *Proc. Fault-tolerant computing (FTCD 28)*, München, Germany, 1998.

19.10   Henry, M. and Clarke, D. The self-validating sensor: rationale, definitions, and examples. *Control Engineering Practice – CEP*, 1(2):585–610, 1993.

19.11   Isermann, R. *Fault diagnosis of technical processes – applications*. Springer, Heidelberg, 2006.

19.12   Isermann, R. and Börner, M. Characteristic velocity stability indicator for passengers cars. In *Proc. IFAC Symposium on Advances in Automotive Control*, Salerno, Italy, 2004.

19.13   Isermann, R., Lachmann, K.-H., and Matko, D. *Adaptive control systems*. Prentice Hall International UK, London, 1992.

19.14   Isermann, R. and Raab, U. Intelligent actuators – ways to autonomous actuating systems. *Automatica*, 29(5):1315–1331, 1993.

19.15   Isermann, R., Schwarz, R., and Stölzl, S. Fault-tolerant drive-by-wire systems. *IEEE Control Systems Magazine*, (October):64–81, 2002.

19.16   Kopetz, K. *Real-time systems*. Kluwer, Boston, 1997.

19.17   Krautstrunk, A. and Mutschler, P. Remedial strategy for a permanent magnet synchronous motor drive. In *8th European Conference on Power Electronics and Applications, EPE'99*, Lausanne, Switzerland, Sept 1999.

19.18   M. Aslam, D. M. abd. Reconfigurable flight control systems. In *Proc. CONTROL'91, IEE Pub. 332*, volume 1, pages 234–242, 1991.

19.19   Mesch, F. Strukturen zur Selbstüberwachung von Messsystemen. *Automatisierungstechnische Praxis – atp*, 43(8):62–67, 2001.

19.20   Moseler, O. *Mikrocontrollerbasierte Fehlererkennung für mechatronische Komponenten am Beispiel eines elektromechanischen Stellantriebs*, volume Fortschr.-Ber. VDI Reihe 8, 980. VDI Verlag, Düsseldorf, 2001.

19.21   Moseler, O., Heller, T., and Isermann, R. Model-based fault detection for an actuator driven by a brushless DC motor. In *14th IFAC-World Congress*, volume P, pages 193–198, Beijing, China, 1999.

19.22   Oehler, R., Schoenhoff, A., and Schreiber, M. Online model-based fault detection and diagnosis for a smart aircraft actuator. In *Prepr. IFAC Symposium on Fault Detection, Supervision and Safety for Technical Processes (SAFEPROCESS)*, volume 2, pages 591–596, Hull, United Kingdom, August 1997. Pergamon Press.

19.23   Patton, R. Fault-tolerant control: the 1997 situation. In *Prepr. IFAC Symposium on Fault Detection, Supervision and Safety for Technical Processes (SAFEPROCESS)*, volume 2, pages 1033–1055, Hull, United Kingdom, August 1997. Pergamon Press.

19.24 Pfeufer, T. *Modellgestützte Fehlererkennung und Diagnose am Beispiel eines Fahrzeugaktors*, volume Fortschr.-Ber. VDI Reihe 8, 749. VDI Verlag, Düsseldorf, 1999.

19.25 Poledna, S. and Kroiss, G. TTD: Drive-by-wire in greifbarer Nähe. *Elektronik*, 14:36–43, 1999.

19.26 Rauch, H. Intelligent fault diagnosis and control reconfiguration. *IEEE Control Systems*, (June):13–20, 1994.

19.27 Rauch, H. Autonomous control reconfiguration. *IEEE Control Systems Magazine*, 15(6):37–48, 1995.

19.28 Spitzer, C., editor. *The avionics handbook*. Contributions on fly-by-wire flight control. CRC Press, Boca Raton, 2001.

19.29 Stengel, R. Intelligent failure-tolerant control. *IEEE Control Systems Magazine*, pages 14–23, 1991.

19.30 Suryanaryanan, S. and Tomizuka, M. Fault-tolerant lateral control of automated vehicles based on simultaneous stabilization. In *Proc. IFAC Conference om Mechatronic Systems*, Darmstadt, Germany, 2000.

19.31 X-By-Wire. *Safety-related fault-tolerant systems in vehicles. Final Report of EU Project Brite-EuRam III, 3 B.5, 3 B. 6, Prog. No BE 95/1329. Contract No. BRPR-CT95-0032*. 1998.

19.32 Zhang, Y. and Jiang, J. Bibliographical review on reconfigurable fault-tolerant control systems. In *Proc. IFAC-Symposium on Fault Detection, Supervision and Safety for Technical Processes (SAFEPROCESS)*, pages 329–348, Washington, DC, USA, June 2003.

Chapter 20

20.1 Füssel, D. *Fault diagnosis with tree-structured neuro-fuzzy systems*, volume Fortschr.-Ber. VDI Reihe 8, 957. VDI Verlag, Düsseldorf, 2002.

20.2 Höfling, T. Zustandsgrößenschätzung zur Fehlererkennung. In Isermann, R., editor, *Überwachung und Fehlerdiagnose*, pages 89–108. VDI, Düsseldorf, 1994.

20.3 Höfling, T. *Methoden zur Fehlererkennung mit Parameterschätzung und Paritätsgleichungen*, volume Fortschr.-Ber. VDI Reihe 8, 546. VDI Verlag, Düsseldorf, 1996.

20.4 Höfling, T. and Isermann, R. Fault detection based on adaptive parity equations and single-parameter tracking. *Control Engineering Practice – CEP*, 4(10):1361–1369, 1996.

20.5 Isermann, R. *Fault diagnosis of technical processes – applications*. Springer, Heidelberg, 2006.

20.6 Isermann, R., Lachmann, K.-H., and Matko, D. *Adaptive control systems*. Prentice Hall International UK, London, 1992.

20.7 Pfeuer, T. Improvement of flexibility and reliability of automobiles actuators by model-based algorithms. In *IFAC SICICA*, Budapest, Hungary, 1994.

20.8   Pfeufer, T. Application of model-based fault detection and diagnosis to the quality assurance of an automotive actuator. *Control Engineering Practice – CEP*, 5(5):703–708, 1997.

20.9   Pfeufer, T. *Modellgestützte Fehlererkennung und Diagnose am Beispiel eines Fahrzeugaktors*, volume Fortschr.-Ber. VDI Reihe 8, 749. VDI Verlag, Düsseldorf, 1999.

20.10   Pfeufer, T., Isermann, R., and Rehm, L. Quality assurance of mechanical-electronical automobile actuator using an integrated model-based diagnosis control (in german). volume VDI-Bericht Nr. 1287, pages 145–159, Baden-Baden, Germany, September 1996.

Chapter 21

21.1   Füssel, D. *Fault diagnosis with tree-structured neuro-fuzzy systems*, volume Fortschr.-Ber. VDI Reihe 8, 957. VDI Verlag, Düsseldorf, 2002.

21.2   Isermann, R. *Mechatronic systems – fundamentals*. Springer, London, 2003.

21.3   Pfleiderer, C. and Petermann, H. *Strömungsmaschinen*. Springer, Berlin, 7th edition, 2005.

21.4   Wolfram, A. *Komponentenbasierte Fehlerdiagnose industrieller Anlagen am Beispiel frequenzumrichtergespeister Asynchronmaschinen und Kreiselpumpen*, volume Fortschr.-Ber. VDI Reihe 8, 967. VDI Verlag, Düsseldorf, 2002.

21.5   Wolfram, A., Füssel, D., Brune, T., and Isermann, R. Component-based multi-model approach for fault detection and diagnosis of a centrifugal pump. In *Proc. American Control Conference (ACC)*, Arlington, VA, USA, 2001.

Chapter 22

22.1   Börner, M., Zele, M., and Isermann, R. Comparison of different fault-detection algorithms for active body control components: automotive suspension system. In *American Control Conference*, Arlington, VA, USA, June 2001.

22.2   Bußhardt, J. *Selbsteinstellende Feder-Dämpfer-Last-Systeme für Kraftfahrzeuge*, volume Fortschr.-Ber. VDI Reihe 12, 240. VDI Verlag, Düsseldorf, 1995.

22.3   Fischer, D. and Isermann, R. Mechatronic semi-active and active vehicle suspensions. *Control Engineering Practice – CEP*, 12(11):1353–1367, 2004.

22.4   Fischer, M. *Tire pressure monitoring (BT 243)*. Verlag moderne industrie, Landsberg, 2003.

22.5   Halfmann, C., Ayoubi, M., and Holzmann, H. Supervision of vehicles' tyre pressures by measurement of body accelerations. *Control Engineering Practice – CEP*, 5(8):1151–1159, 1997.

22.6   Isermann, R. Fault-tolerant components for drive-by-wire systems. In *VDA Technischer Kongress*, Bad Homburg, Germany, May 2001.

22.7   Isermann, R. *Mechatronic systems – fundamentals*. Springer, London, 2003.

22.8   Isermann, R., Germann, S., Würtenberger, M., Halfmann, C., and Holzmann, H. Model-based control and supervision of vehicle dynamics. In *FISITA Congress*, Prague, Czech Republic, June 1996.

22.9   Leonhardt, S., Bußhardt, J., Rajamani, R., Hedrick, J., and Isermann, R. Parameter estimation of shock absorbers with artificial neural network. In *Proc. American Control Conference (ACC)*, San Francisco, CA, USA, 1993.

22.10   Weispfenning, T. Fault detection and diagnosis of components of the vehicle vertical dynamics. In *1st Int. Conference on Control and Diagnostics in Automotive Applications*, Genua, Italy, October 1996.

22.11   Weispfenning, T. and Leonhardt, S. Model-based identification of a vehicle suspension using parameter estimation and neural networks. In *Proc. 13th IFAC World Congress*, San Francisco, CA, USA, 1996.

Chapter 23

23.1   Dasgupta, S., Anderson, B., and Kaye, R. Identification of physical parameters in structured systems. *Automatica*, 24(2):217–225, 1988.

23.2   Eykhoff, P. *System identification*. John Wiley, London, 1974.

23.3   Füssel, D. *Fault diagnosis with tree-structured neuro-fuzzy systems*, volume Fortschr.-Ber. VDI Reihe 8, 957. VDI Verlag, Düsseldorf, 2002.

23.4   IFIP. *Proc. of the IFIP 9th World Computer Congress, Paris, France, September 19-23*. Elsevier, 1983.

23.5   Isermann, R. Estimation of physical parameters for dynamic processes with application to an industrial robot. *International Journal of Control*, 55(6):1287–1298, 1992.

23.6   Isermann, R. *Identifikation dynamischer Systeme*. Springer, Berlin, 1992.

23.7   Isermann, R. *Mechatronic systems – fundamentals*. Springer, London, 2003.

23.8   Isermann, R. and Ballé, P. Trends in the application of model-based fault detection and diagnosis in technical processes. *Control Engineering Practice – CEP*, 5(5):638–652, 1997.

23.9   Isermann, R., Lachmann, K.-H., and Matko, D. *Adaptive control systems*. Prentice Hall International UK, London, 1992.

23.10   Kruse, R., Gebhardt, J., and Klawonn, F. *Foundations of fuzzy systems*. Wiley, Chichester, 1994.

23.11   Moseler, O. and Vogt, M. FIT- filtering and identification. In *Proc. 12th IFAC Symposium on System Identification (SYSID)*, Santa Barbara, CA, USA, 2000.

23.12   Nold, S. and Isermann, R. Identifiability of process coefficients for technical failure diagnosis. In *25th IEEE Conference on Decision and Control*, Athens, Greece, 1986.

23.13   Omdahl, T., editor. *Reliability, availability and maintainability (RAM) dictionary*. ASQC Quality Press, Milwaukee, 1988.

23.14 Oppenheim, A., Schafer, R., and Buck, J. *Discrete-time signal processing*. Prentice Hall, Englewood Cliffs, 2nd edition, 1999.

23.15 Ortega, J. and Rheinboldt, W. *Iterative solution of nonlinear equations in several variables*. Academic Press, New York, 1970.

23.16 Peter, K.-H. *Parameteradaptive Regelalgorithmne auf der Basis zeitkontinuierlicher Prozessmodelle*, volume Fortschr.-Ber. VDI Reihe 8, 348. VDI Verlag, Düsseldorf, 1993.

23.17 Raksanyi, A., Lecourtier, Y., Walter, E., and Venot, A. Identifibility and distinguishability testing via computer algebra. *Mathematical Biosciences*, 77:245–266, 1985.

23.18 Schumann, A. Computer algebra - a key technique for computer-aided theoretical modelling. In *9th IASTED International Conference on Modelling, Identification and Control*, Innsbruck, Austria, 1990.

23.19 Sorenson, H. *Parameter estimation: principles and problems*. Dekker, New York, 1980.

23.20 Walter, E. *Identifiability of state space models*, volume 46 of *Springer Lecture Notes in Biomathematics*. Springer, Berlin, 1982.

23.21 Young, P. An instrumental variable method for real-time identification of a noise process. *Automatica*, 31:134–144, 1970.

23.22 Young, P. *Recursive estimation and time-series analysis*. Springer, Berlin, 1984.

23.23 Zadeh, L. Fuzzy sets. *Information and Control*, 8:338–353, 1965.

23.24 Zimmermann, H.-J. *Fuzzy set theory - and its applications*. Kluwer, Boston, 2nd edition, 1991.

Index